Alan F. Hegarty • Natalia Kopteva
Eugene O'Riordan • Martin Stynes
Editors

BAIL 2008 - Boundary and Interior Layers

Proceedings of the International Conference on Boundary and Interior Layers - Computational and Asymptotic Methods, Limerick, July 2008

Alan F. Hegarty
Department of Mathematics and Statistics
University of Limerick
Limerick
Ireland
alan.hegarty@ul.ie

Natalia Kopteva
Department of Mathematics and Statistics
University of Limerick
Limerick
Ireland
natalia.kopteva@ul.ie

Eugene O'Riordan
School of Mathematical Sciences
Dublin City University
Glasnevin
Dublin 9
Ireland
eugene.oriordan@dcu.ie

Martin Stynes
Department of Mathematics
National University of Ireland
Cork
Ireland
m.stynes@ucc.ie

ISSN 1439-7358
ISBN 978-3-642-00604-3 e-ISBN 978-3-642-00605-0
DOI: 10.1007/978-3-642-00605-0
Springer Dordrecht Heidelberg London New York

Library of Congress Control Number: 2009926245

Mathematics Subject Classification Numbers (2000): 34, 35, 65, 76

Cover design: deblik, Heidelberg

The front cover photograph of the Poulnabrone dolmen (a megalithic monument located in the west of Ireland) was taken by Kathy Coyle.

Printed on acid-free paper

Springer is part of Springer Science + Business Media (www.springer.com)

Lecture Notes in Computational Science and Engineering

69

Editors

Preface

These Proceedings contain a selection of the lectures given at the conference *BAIL 2008: Boundary and Interior Layers – Computational and Asymptotic Methods*, which was held from 28th July to 1st August 2008 at the University of Limerick, Ireland. The first three BAIL conferences (1980, 1982, 1984) were organised by Professor John Miller in Trinity College Dublin, Ireland. The next seven were held in Novosibirsk (1986), Shanghai (1988), Colorado (1992), Beijing (1994), Perth (2002), Toulouse (2004), and Göttingen (2006). With BAIL 2008 the series returned to Ireland. BAIL 2010 is planned for Zaragoza.

The BAIL conferences strive to bring together mathematicians and engineers whose research involves layer phenomena, as these two groups often pursue largely independent paths. BAIL 2008, at which both communities were well represented, succeeded in this regard. The lectures given were evenly divided between applications and theory, exposing all conference participants to a broad spectrum of research into problems exhibiting solutions with layers.

The Proceedings give a good overview of current research into the theory, application and solution (by both numerical and asymptotic methods) of problems that involve boundary and interior layers. In addition to invited and contributed lectures, the conference included four mini-symposia devoted to stabilized finite element methods, asymptotic scaling of wall-bounded flows, systems of singularly perturbed differential equations, and problems with industrial applications (supported by MACSI, the Mathematics Applications Consortium for Science and Industry). These titles exemplify the mix of interests among the participants.

All papers in the Proceedings were subject to a standard refereeing process. We are grateful to the authors and to the unnamed referees for their valuable contributions, without which this volume would not exist.

Thanks are also due to the organizers of the mini-symposia at BAIL 2008, to the judges of the Hemker prize, and to all the attendees for their enthusiastic participation in the conference.

January 2009

Alan F. Hegarty (Limerick)
Natalia Kopteva (Limerick)
Eugene O'Riordan (Dublin)
Martin Stynes (Cork)

Contents

List of Contributors

L. Beneš
Department of Technical Mathematics, Faculty of Mechanical Engineering
Czech Technical University in Prague, Karlovo náměstí 13, CZ-12135 Praha 2
Czech Republic, benes@marian.fsik.cvut.cz

H.C. Boisson
Université de Toulouse, Institut de Mécanique des Fluides de Toulouse, UMR
CNRS/INPT/UPS, Allée Camille Soula, 31400 Toulouse, France

Philippe Caillol
Department of Applied Mathematics, Sheffield University, Sheffield
S3 7RH, UK, P.L.Caillol@shef.ac.uk

P. Cathalifaud
Institut de Mécanique des Fluides de Toulouse UMR-CNRS
Alleé Camille Soula, 31400
and
Université Paul Sabatier, 118 route de Narbonne, 31062 Toulouse Cedex
France, catalifo@cict.fr

C. Clavero
Department of Applied Mathematics, University of Zaragoza, Zaragoza, Spain
clavero@unizar.es

J. Cousteix
Département Modèles pour l'Aérodynamique et l'énergétique, ONERA
2 avenue Édouard Belin, B.P. 4025, 31055 Toulouse Cedex 4, France
and
École Nationale Supérieure de l'Aéronautique et de l'Espace
10, avenue Édouard Belin, 31055 Toulouse Cedex, France
Jean.Cousteix@onera.fr

P.A. Farrell
Department of Computer Science, Kent State University, OH 44242, USA
farrell@cs.kent.edu

C. de Falco
School of Mathematical Sciences, Dublin City University, Dublin 9, Ireland
carlo.defalco@imati.cnr.it

Ph. Fraunié
LSEET/CNRS Université de Toulon et du Var, France
Philippe.Fraunie@lseet.univ-tln.fr

J. Fürst
Department of Technical Mathematics, Faculty of Mechanical Engineering
Czech Technical University in Prague, Karlovo náměstí 13, CZ-12135 Praha 2
Czech Republic, furst@marian.fsik.cvut.cz

A. Giovannini
Université de Toulouse, Institut de Mécanique des Fluides de Toulouse, UMR CNRS/INPT/UPS, Allée Camille Soula, 31400 Toulouse, France

M. Gonzalez
University of Limerick, Limerick, Ireland, maria.gonzalez@ul.ie

J.L. Gracia
Department of Applied Mathematics. University of Zaragoza, Zaragoza
Spain, jlgracia@unizar.es

A.F. Hegarty
University of Limerick, Limerick, Ireland, alan.hegarty@ul.ie

N. Ipek
Avesta Research Centre, Koppardalsvägen 65, P. O. Box 74, SE-774 22 Avesta
Sweden, nulifer.ipek@outokumpu.com

Volker John
FR 6.1 – Mathematik, Universität des Saarlandes, Postfach 15 11 50, 66041
Saarbrücken, Germany, john@math.uni-sb.de

Alfred Kluwick
Institute of Fluid Mechanics and Heat Transfer
Vienna University of Technology, Resselgasse 3/E322, A-1040 Vienna, Austria
alfred.kluwick@tuwien.ac.at

David J. Knezevic
OUCL, University of Oxford, Parks Road, Oxford, OX1 3QD, UK
davek@comlab.ox.ac.uk

Petr Knobloch
Charles University, Faculty of Mathematics and Physics
Sokolovská 83, 186 75 Praha 8, Czech Republic
knobloch@karlin.mff.cuni.cz

Bernhard Kotesovec
Vienna University of Technology, Institute of Fluid Mechanics and Heat Transfer
Resselgasse 3, 1040 Vienna, Austria

F.J. Lisbona
Departamento de Matemática Aplicada, Universidad de Zaragoza
Spain, lisbona@unizar.es

J. Löwe
Institute for Numerical and Applied Mathematics
Georg-August-University of Göttingen, D-37083 Göttingen, Germany
loewe@math.uni-goettingen.de

G. Lube
Mathematics Department, NAM, Georg-August-University Göttingen
Lotzestrasse 16-18, D-37083 Göttingen, Germany
lube@math.uni-goettingen.de

M. Madaune-Tort
Laboratoire de Mathématiques Appliquées, Université de Pau et des Pays de l'Adour, France, monique.madaune-tort@univ-pau.fr

Niall Madden
Department of Mathematics, National University of Ireland, Galway
Republic of Ireland, Niall.Madden@NUIGalway.ie

Sherwin A. Maslowe
Department of Mathematics and Statistics, McGill University, Montréal
Canada, maslowe@math.mcgill.ca

J. Mauss
Institut de Mécanique des Fluides de Toulouse UMR-CNRS
Allée Camille Soula, 31400 Toulouse, France
and
Université Paul Sabatier, 118 route de Narbonne, 31062 Toulouse Cedex
France, mauss@cict.fr

J.J.H. Miller
Institute for Numerical Computation and Analysis, Dublin, Ireland
jm@incaireland.org

C. Moldoveanu
Université de Toulouse, Institut de Mécanique des Fluides de Toulouse
UMR CNRS/INPT/UPS, Allée Camille Soula, 31400 Toulouse, France
and
Military Technical Academy, Department of Mechanical Engineering
81 George Cosbuc, Bucharest, Romania

V. Ya. Neyland
TsAGI, Zhukovsky, Moscow Region, 140180, Russia
Neyland@tsagi.ru

E. O'Riordan
School of Mathematical Sciences, Dublin City University, Ireland
eugene.oriordan@dcu.ie

Bernhard Scheichl
Institute of Fluid Mechanics and Heat Transfer
Vienna University of Technology, Resselgasse 3/E322, A-1040 Vienna
Austria, bernhard.scheichl@tuwien.ac.at

Ellen Schmeyer
FR 6.1 – Mathematik, Universität des Saarlandes, Postfach 15 11 50, 66041
Saarbrücken, Germany, schmeyer@math.uni-sb.de

V.V. Shvedchenko
TsAGI, Zhukovsky, Moscow Region, 140180, Russia

L.A. Sokolov
TsAGI, Zhukovsky, Moscow Region, 140180, Russia

Herbert Steinrück
Vienna University of Technology, Institute of Fluid Mechanics and Heat Transfer
Resselgasse 3, 1040 Vienna, Austria herbert.steinrueck@tuwien.ac.at

Meghan Stephens
Department of Mathematics, National University of Ireland, Galway
Republic of Ireland, Meghan.Stephens@NUIGalway.ie

Endre Süli
OUCL, University of Oxford, Parks Road, Oxford, OX1 3QD, UK
endre.suli@comlab.ox.ac.uk

P. Sváček
Czech Technical University Prague, Faculty of Mechanical Engineering
Karlovo nám. 13, 121 35 Praha 2, Czech Republic, Petr.Svacek@fs.cvut.cz

B. Tews
Mathematics Department, NAM, Georg-August-University Göttingen
Lotzestrasse 16-18, D-37083 Göttingen, Germany

S.R. Thomas
IBISC, University of Evry, Boulevard François Mitterand
91 025 Evry Cedex, France, randy.thomas@ibisc.univ-evry.fr

Lutz Tobiska
Institute for Analysis and Computational Mathematics
Otto von Guericke University Magdeburg, PF 4120, 39016 Magdeburg
Germany, tobiska@ovgu.de

S. Valarmathi
Department of Mathematics, Bishop Heber College(Autonomous)
Tiruchirappalli-620 017, Tamil Nadu, India, valarmathi07@gmail.com

Relja Vulanović
Department of Mathematical Sciences
Kent State University Stark Campus, 6000 Frank Ave NW, North Canton
OH 44720, USA, rvulanov@kent.edu

M. Vynnycky
MACSI, Department of Mathematics and Statistics
University of Limerick, Limerick, Ireland
michael.vynnycky@ul.ie

Paolo Zunino
MOX, Department of Mathematics, Politecnico di Milano
Italy, paolo.zunino@polimi.it

Part I
Invited Papers

High-Reynolds-Number Asymptotics of Turbulent Boundary Layers: From Fully Attached to Marginally Separated Flows

Dedicated to Professor Klaus Gersten on the occasion of his 80th birthday

Alfred Kluwick and Bernhard Scheichl

Abstract This paper reports on recent efforts with the ultimate goal to obtain a fully self-consistent picture of turbulent boundary layer separation. To this end, it is shown first how the classical theory of turbulent small-defect boundary layers can be generalised rigorously to boundary layers with a slightly larger, i.e. moderately large, velocity defect and, finally, to situations where the velocity defect is of $O(1)$. In the latter case, the formation of short recirculation zones describing marginally separated flows is found possible, as described in a rational manner.

1 Introduction

Despite the rapid increase of computer power in the recent past, the calculation of turbulent wall-bounded flows still represents an extremely challenging and sometimes insolvable task. Direct-Numerical-Simulation computations based on the full Navier–Stokes equations are feasible for moderately large Reynolds numbers only. Flows characterised by much higher Reynolds numbers can be investigated if one resorts to turbulence models for the small scales, as accomplished by the method of Large Eddy Simulation, or for all scales, as in computer codes designed to solve the Reynolds-averaged Navier–Stokes equations. Even then, however, the numerical efforts rapidly increase with increasing Reynolds number. This strongly contrasts the use of asymptotic theories, the performance of which improves as the values of the Reynolds number become larger and, therefore, may be considered to complement purely numerically based work.

With a few exceptions (e.g. [7, 21]), studies dealing with the high-Reynolds-number properties of turbulent boundary layers start from the time- or, equivalently, Reynolds-averaged equations. By defining non-dimensional variables

B. Scheichl (✉)
Institute of Fluid Mechanics and Heat Transfer, Vienna University of Technology, Resselgasse 3/E322, A-1040 Vienna, Austria, E-mail: bernhard.scheichl@tuwien.ac.at

A.F. Hegarty et al. (eds.), *BAIL 2008 - Boundary and Interior Layers.*
Lecture Notes in Computational Science and Engineering,
DOI: 10.1007/978-3-642-00605-0,

in terms of a representative length $\tilde{L}$ and flow speed $\tilde{U}$ and assuming incompressible nominally steady two-dimensional flow they take on the form

$$\frac{\partial u}{\partial x} + \frac{\partial v}{\partial y} = 0, \tag{1a}$$

$$u\frac{\partial u}{\partial x} + v\frac{\partial u}{\partial y} = -\frac{\partial p}{\partial x} + \frac{1}{Re}\nabla^2 u - \frac{\partial \overline{u'^2}}{\partial x} - \frac{\partial \overline{u'v'}}{\partial y}, \tag{1b}$$

$$u\frac{\partial v}{\partial x} + v\frac{\partial v}{\partial y} = -\frac{\partial p}{\partial y} + \frac{1}{Re}\nabla^2 v - \frac{\partial \overline{u'v'}}{\partial x} - \frac{\partial \overline{v'^2}}{\partial y}. \tag{1c}$$

Herein $\nabla^2 = \partial^2/\partial x^2 + \partial^2/\partial y^2$, and (x, y), (u, v), $(u'v')$, $-\overline{u'^2}$, $-\overline{u'v'}$, $-\overline{v'^2}$, and p are Cartesian coordinates measuring the distance along and perpendicular to the wall, the corresponding time mean velocity components, the corresponding velocity fluctuations, the components of the Reynolds stress tensor, and the pressure, respectively. The Reynolds number is defined by $Re := \tilde{U}\tilde{L}/\tilde{\nu}$, where $\tilde{\nu}$ is the (constant) kinematic viscosity. Equation (1) describe flows past flat walls. Effects of wall curvature can be incorporated without difficulty but are beyond the scope of the present analysis.

When it comes down to the solution of the simplified version of these equations provided by asymptotic theory in the limit $Re \to \infty$, one is, of course, again faced with the closure problem. The point, however, is that these equations and the underlying structure represent closure independent basic physical mechanisms characterising various flow regions identified by asymptotic reasoning. This has been shown first in the outstanding papers [5, 8, 10, 31], and more recently and in considerable more depth and breath, in [24, 30] for the case of small-defect boundary layers, which will be considered in Sect. 2. Boundary layers exhibiting a slightly larger, i.e. a moderately large, velocity defect are treated in Sect. 3. Finally, Sect. 4 deals with situations where the velocity defect is of $O(1)$ rather than small.

2 Classical Theory of Turbulent Small-Defect Turbulent Boundary Layers

We first outline the basic ideas underlying an asymptotic description of turbulent boundary layers.

2.1 Preliminaries

Based on dimensional reasoning put forward by L. Prandtl and Th. von Kármán, a self-consistent time-mean description of firmly attached fully developed turbulent boundary layers holding in the limit of large Reynolds numbers Re, i.e. for $Re \to \infty$, has been proposed first in the aforementioned studies [5, 8, 10, 31]. One of the main goals of the present investigation is to show that this rational formulation can be derived from a minimum of assumptions:

(a) All the velocity fluctuations are of the same order of magnitude in the limit $Re \to \infty$, so that all Reynolds stress components are equally scaled by a single velocity scale u_{ref}, non-dimensional with a global reference velocity (hypothesis of locally isotropic turbulence);
(b) The pressure gradient does not enter the flow description of the viscous wall layer to leading order (assumption of firmly attached flow);
(c) The results for the outer predominantly inviscid flow region can be matched directly with those obtained for the viscous wall layer (assumption of "simplest possible" flow structure).

In the seminal studies [5,8,10,31], u_{ref} is taken to be of the same order of magnitude in the fully turbulent main portion of the boundary layer and in the viscous wall layer and, hence, equal to the skin-friction velocity

$$u_\tau := \left[Re^{-1}(\partial u/\partial y)_{y=0}\right]^{1/2}. \tag{2}$$

This in turn leads to the classical picture, according to which (i) the streamwise velocity defect with respect to the external impressed flow is small and of $O(u_\tau)$ across most of the boundary layer, while (ii) the streamwise velocity is itself small and of $O(u_\tau)$ inside the (exponentially thin) wall layer, and (iii) $u_\tau/U_e = O(1/\ln Re)$. Furthermore, then (iv) the celebrated universal logarithmic velocity distribution

$$u/u_\tau \sim \kappa^{-1} \ln y^+ + C^+, \quad y^+ := y\, u_\tau Re \to \infty. \tag{3}$$

is found to hold in the overlap of the outer (small-defect) and inner (viscous wall) layer. Here κ denotes the von Kármán constant; in this connection we note the currently accepted empirical values $\kappa \approx 0.384$, $C^+ \approx 4.1$, which refer to the case of a perfectly smooth surface, see [16] and have been obtained for a zero pressure gradient.

This might be considered to yield a stringent derivation of the logarithmic law of the wall (3), anticipating the existence of an asymptotic state and universality of the wall layer flow as $Re \to \infty$; a view which, however, has been increasingly challenged in more recent publications (e.g. [2–4]). It thus appears that – as expressed by Walker, see [30] – "...although many arguments have been put forward over the years to justify the logarithmic behaviour, non are entirely satisfactory as a proof, ...". As a result, one has to accept that matching (of inner and outer expansions), while ensuring self-consistency, is not sufficient to uniquely determine (3). In the following, from the viewpoint of the time-averaged flow description the logarithmic behaviour (3), therefore, will be taken to represent an experimentally rather than strictly theoretically based result holding in situations where the assumption (b) applies. The description of the boundary layer in the limit $Re \to \infty$ can then readily be formalised. In passing, we mention that in the classical derivations, see [5, 8, 10, 31], the assumption (b) is not adopted and (3) results from matching, rather than in the present study where it is imposed.

2.2 Leading-Order Approximation

Inside the wall layer where $y^+ = y\,u_\tau Re = O(1)$ the streamwise velocity component u, the Reynolds shear stress $\tau := -\overline{u'v'}$ and the pressure p are expanded in the form

$$u \sim u_\tau(x; Re)\, u^+(y^+) + \cdots, \tag{4a}$$

$$\tau \sim u_\tau^2(x; Re)\, t^+(y^+) + \cdots, \tag{4b}$$

$$p \sim p_0(x) + \cdots, \tag{4c}$$

where u^+ exhibits the limiting behaviour implied by (3):

$$u^+(y^+) \sim \kappa^{-1} \ln y^+ + C^+, \quad y^+ \to \infty. \tag{5}$$

Assumption (c), quoted in Sect. 2.1, then uniquely determines the asymptotic expansions of, respectively, u, τ, and p further away from the wall where the Reynolds stress τ is predominant. Let $\delta_0(x; Re)$ characterise the thickness of this outer main layer, i.e. of the overall boundary layer. In turn, one obtains

$$u \sim U_e(x) - u_\tau(x; Re)\, F_1'(x, \eta) + \cdots, \tag{6a}$$

$$\tau \sim u_\tau^2(x; Re)\, T_1(x, \eta) + \cdots, \tag{6b}$$

$$p \sim p_e(x) + \cdots, \tag{6c}$$

where $\eta := y/\delta_0$. Here and in the following primes denote differentiation with respect to η. Furthermore, U_e and p_e stand for the velocity and the pressure, respectively, at the outer edge $\eta = 1$ of the boundary layer (here taken as a sharp line with sufficient asymptotic accuracy) imposed by the external irrotational flow.

Matching of the expansions (4) and (6) by taking into account (5) forces a logarithmic behaviour of F_1',

$$F_1' \sim -\kappa^{-1} \ln \eta + C_0(x), \quad \eta \to 0, \tag{7}$$

yields $p_0(x) = p_e(x)$, and is achieved provided $\gamma := u_\tau / U_e$ satisfies the skin-friction relationship

$$\kappa/\gamma \sim \ln(Re\gamma\delta_0 U_e) + \kappa(C^+ + C_0) + O(\gamma). \tag{8}$$

From substituting (4) into the x-component (1b) of the Reynolds equations (1) one obtains the well-known result that the total stress inside the wall layer is constant to leading order,

$$du^+/dy^+ + t^+ = 1. \tag{9}$$

Moreover, the expansions (6) lead to a linearisation of the convective terms in the outer layer, so that there Bernoulli's law holds to leading order,

$$dp_e/dx = -U_e\, dU_e/dx. \tag{10}$$

The necessary balance with the gradient of the Reynolds shear stress then determines the magnitude of the thickness of the outer layer, i.e. $\delta_0 = O(u_\tau)$. As a consequence, the expansions (6) are supplemented with

$$\delta_0 \sim \gamma\, \Delta_1(x) + \cdots, \tag{11}$$

which in turn gives rise to the leading-order outer-layer streamwise momentum equation. After integration with respect to η and and employing the matching condition $T_1(x, 0) = 1$, the latter is conveniently written as

$$(E + 2\beta_0)\eta F_1' - EF_1 - \Delta_1 F_{1,e}\, F_{1x} = F_{1,e}(T_1 - 1), \tag{12a}$$

$$F_{1,e} := F_1(x, 1), \quad E := 1 - \Delta_1 \frac{dF_{1,e}}{dx}, \quad \beta_0 := -\Delta_1 F_{1,e} \frac{U_{ex}}{U_e}. \tag{12b}$$

From here on, the subscript x means differentiation with respect to x. The boundary layer equation (12a) is unclosed, and in order to complete the flow description, turbulence models for t^+ and T_1 have to be adopted. Integration of (12) then provides the velocity distribution in the outer layer and determines the yet unknown function $C_0(x)$ entering (7) and the skin-friction relationship (8), which completes the leading-order analysis.

As a main result, inversion of (8) with the aid of (11) yields

$$\gamma \sim \kappa\sigma[1 - 2\sigma \ln \sigma + O(\sigma)], \quad \sigma := 1/\ln Re, \quad \partial\gamma/\partial x = O(\gamma^2). \tag{13}$$

The skin-friction law (13) implies the scaling law (iii), already mentioned in Sect. 2.1, which is characteristic of classical small-defect flows.

2.3 Second-Order Outer Problem

Similar to the description of the leading-order boundary layer behaviour, the investigation of higher-order effects is started by considering the wall layer first. Substitution of (4a), (4b), (8) into (1b) yields upon integration (cf. [30]),

$$\frac{1}{Re}\frac{\partial u}{\partial y} + \tau \sim \gamma^2 U_e^2 - \frac{U_e U_{ex}}{\gamma Re}\, y^+ + \frac{\gamma U_e U_{ex}}{Re} \int_0^{y^+} u^{+2}\, dy^+ + \cdots. \tag{14}$$

Here the second and third terms on the right-hand side account, respectively, for the effects of the (imposed) pressure gradient, c.f. (10), and convective terms, which have been neglected so far. By using (5) and (12), the asymptotic behaviour of τ as $y^+ \to \infty$ can easily be obtained (e.g. [30]). Rewritten in terms of the outer-layer variable η, it is found to be described by

$$\frac{\tau}{U_e^2} \sim \gamma^2\left[1 + 2\frac{\Delta_0 U_{ex}}{\kappa U_e}\eta \ln \eta + \cdots\right] + \gamma^3\left[\frac{\Delta_0 U_{ex}}{\kappa^2 U_e}\eta(\ln \eta)^2 + \cdots\right] + \cdots, \quad (15)$$

as $\eta \to 0$, which immediately suggests the appropriate generalisation of the small-defect expansions (6a), (6b), (11):

$$u/U_e \sim 1 - \gamma F_1' - \gamma^2 F_2' + \cdots, \quad (16a)$$

$$\tau/U_e^2 \sim \gamma^2 T_1 + \gamma^3 T_2 + \cdots, \quad (16b)$$

$$\delta_0 \sim \gamma\, \Delta_1(x) + \gamma^2 \Delta_2(x) + \cdots. \quad (16c)$$

Here matching with the wall layer is achieved if

$$F_1' \sim -\kappa^{-1} \ln \eta + C_0(x)\,, \quad F_2' \sim C_1(x), \quad (17a)$$

$$T_1 \sim 1 + 2\frac{\Delta_0 U_{ex}}{\kappa U_e}\eta \ln \eta\,, \quad T_2 \sim \frac{\Delta_0 U_{ex}}{\kappa^2 U_e}\eta(\ln \eta)^2, \quad (17b)$$

as $\eta \to 0$, provided that the skin-friction relationship (8) is modified to explicitly include an additional term of $O(\gamma)$,

$$\kappa/\gamma \sim \ln(Re\gamma\delta_0 U_e) + \kappa(C^+ + C_0 + \gamma C_1) + \cdots. \quad (18)$$

Similar to $C_0(x)$, the function $C_1(x)$ depends on the specific turbulence model adopted, as well as the upstream history of the boundary layer.

2.4 Can Classical Small-Defect Theory Describe Boundary Layer Separation?

An estimate of the thickness δ_w of the viscous wall layer is readily obtained from the definition of y^+, see (3), and the (inverted) skin-friction relationship (13): $\delta_w = O[\gamma^{-1}\exp(-\kappa/\gamma)]$. In the limit $Re \to \infty$, therefore, the low-momentum region close to the wall is exponentially thin as compared to the outer layer, where Reynolds stresses cause a small $O(\gamma)$-reduction of the fluid velocity with respect to the mainstream velocity $U_e(x)$. This theoretical picture of a fully attached turbulent small-defect boundary layer has been confirmed by numerous comparisons with experimental data for flows of this type (e.g. [1, 14, 30]). However, it also indicates that attempts based on this picture to describe the phenomenon of boundary layer separation, frequently encountered in engineering applications, will face serious difficulties. Since the momentum flux in the outer layer, which comprises most of the boundary layer, differs only slightly from that in the external flow region, an $O(1)$-pressure rise almost large enough to cause flow reversal even there appears to be required to generate negative wall shear, which hardly can be considered as flow separation. This crude estimate is confirmed by a more detailed analysis dealing with the response of a turbulent small-defect boundary layer to a surface-mounted obstacle, carried out, among others, in [28]. Moreover, to date no self-consistent theory of flow separation compatible with the classical concept of a turbulent small-defect boundary layer has been formulated.

The above considerations strikingly contrast the case of laminar boundary layer separation, where the velocity defect is of $O(1)$ across the whole boundary layer and the associated pressure increase tends to zero as $Re \to \infty$. It, however, also indicates that a turbulent boundary layer may become more prone to separation by increasing the velocity defect. That this is indeed a realistic scenario can be inferred by seeking self-preserving solutions of (12), i.e. by investigating equilibrium boundary layers. Such solutions, where the functions F_1, T_1, characterising the velocity deficit and the Reynolds shear stress in the outer layer, respectively, solely depend on η, exist if the parameter β_0 in the outer-layer momentum equation (12a) is constant, i.e. independent of x. Equation (12a) then assumes the form

$$(1 + 2\beta_0)\eta F_1' - F_1 = F_{1,e}(T_1 - 1), \tag{19}$$

where

$$U_e \propto (x - x_{\mathrm{v}})^m \,, \quad m = -\beta_0/(1 + 3\beta_0) \,, \quad \Delta_1 F_{1,e} = (1 + 3\beta_0)(x - x_{\mathrm{v}}). \tag{20}$$

Herein $x = x_{\mathrm{v}}$ denotes the virtual origin of the boundary layer flow. In the present context flows associated with large values of β_0 are of most interest. By introducing suitably (re)scaled quantities in the form $F_1 = \beta_0^{1/2} \hat{F}(\hat{\eta})$, $T_1 = \beta_0 \hat{T}(\hat{\eta})$, $\eta = \beta_0^{1/2} \hat{\eta}$, the momentum equation (19) reduces to

$$2\hat{\eta}\hat{F}' = \hat{F}_e \hat{T}, \quad \hat{F}_e := \hat{F}(1) \tag{21}$$

in the limit $\beta_0 \to \infty$. Solutions of (21) describing turbulent boundary layers having a velocity deficit measured by $u_{\mathrm{ref}} := \beta_0^{1/2} u_\tau \gg u_\tau$ have been obtained first in [11]. Unfortunately, however, it was not realised that this increase of the velocity defect no longer allows for a direct match of the flow quantities in the outer and inner layer, which has significant consequences, to be elucidated below.

We note that in general $\beta_0(x)$ can be regarded as the leading-order contribution to the so-called Rotta–Clauser pressure-gradient parameter (e.g. [24]),

$$\beta := -U_e U_{ex} \delta^* / u_\tau^2 \,, \quad \delta^* := \delta_0 \int_0^\infty (1 - u/U_e)\, \mathrm{d}\eta. \tag{22}$$

As already mentioned in [11], this quantity allows for the appealing physical interpretation that u_{ref} is independent of the wall shear stress u_τ^2 for $\beta_0 \gg 1$.

3 Moderately Large Velocity Defect

Following the considerations summarised in the preceding section, we now seek solutions of (1) describing a relative velocity defect of $O(\varepsilon)$, where the newly introduced perturbation parameter ε is large compared to γ but still small compared to one: $\gamma \ll \varepsilon \ll 1$. From assumption (a), see Sect. 2.1, we then have $-\overline{u'v'} \sim \varepsilon^2$, and the linearised x-momentum equation immediately yields the estimate $\delta_0 = \varepsilon\Delta$, where $\Delta = O(1)$, for the boundary layer thickness. However, since $-\overline{u'v'} \sim \varepsilon^2$ with $\varepsilon^2 \gg u_\tau^2$, the solution describing the flow behaviour in the outer velocity defect region no longer matches with the solution for the universal wall layer as in the classical case. As a consequence, the leading-order approximation to the Reynolds shear stress must vanish in the limit $\eta = y/\delta_0 \to 0$. This indicates that the flow having a velocity defect of $O(\varepsilon)$ in the outer main part of the boundary layer exhibits a wake-type behaviour, leading to a finite wall slip velocity at its base and, therefore, forces the emergence of a sublayer, termed intermediate layer, where the magnitude of $-\overline{u'v'}$ reduces to $O(u_\tau^2)$, being compatible with the wall layer scaling.

3.1 Intermediate Layer

Here the streamwise velocity component u is expanded about its value at the base $\eta = 0$ of the outer defect region: $u/U_e \sim 1 - \varepsilon W - \gamma U_\mathrm{i} + \cdots$, so that the quantities W, U_i, assumed to be of $O(1)$, account, respectively, for the wall slip velocity, given by $u = U_e(1 - \varepsilon W)$ with $W > 0$, and the dominant contribution to u that varies with distance y from the wall. Integration of the x-momentum balance then shows that the Reynolds shear stress increases linearly with distance y for $y/\delta_0 \ll 1$:

$$\tau \sim \tau_w - \varepsilon(U_e^2 W)_x y\,, \quad y/\delta_\mathrm{i} = O(1). \tag{23}$$

Herein δ_i denotes the thickness of the intermediate layer and τ assumes its near-wall value τ_w as $y/\delta_\mathrm{i} \to 0$. Matching with the wall layer then requires that $\tau_w \sim u_\tau^2$, which, by taking into account (22), yields the estimate $\delta_\mathrm{i}/\delta_0 = O(\beta^{-1})$. Also, since $\tau_w \sim u_\tau^2$, we infer that $\delta_\mathrm{i} = O(u_\tau^2/\varepsilon)$ and, in turn, recover the relationship $\varepsilon \sim u_\tau \beta^{1/2}$, already suggested by the final considerations of Sect. 2.4. Formal expansions of u and $-\overline{u'v'}$ in the intermediate layer, therefore, are written as

$$u/U_e \sim 1 - \varepsilon W(x;\varepsilon,\gamma) - \gamma U_\mathrm{i}(x,\zeta), \tag{24a}$$

$$-\overline{u'v'}/(\gamma U_e)^2 \sim T_\mathrm{i}(x,\zeta;\varepsilon,\gamma) \sim 1 + \lambda\zeta, \tag{24b}$$

where $\zeta := y/\delta_\mathrm{i} = y\varepsilon/(\Delta\gamma^2)$ and $\lambda := -\Delta(U_e^2 W)_x/U_e^2$.

To close the problem for U_i, we adopt the common mixing length concept,

$$-\overline{u'v'} := \ell^2 \frac{\partial u}{\partial y}\left|\frac{\partial u}{\partial y}\right|, \tag{25}$$

by assuming that the mixing length ℓ behaves as $\ell \sim \kappa y$ for $y = O(\delta_{\mathrm{i}})$, which is the simplest form allowing for a match with the adjacent layers. Integration of (24b), supplemented with (25), then yields

$$\kappa U_{\mathrm{i}} = -\ln\zeta + 2\ln\left[(1+\lambda\zeta)^{1/2}+1\right] - 2(1+\lambda\zeta)^{1/2}, \tag{26}$$

from which the limiting forms

$$\kappa U_{\mathrm{i}} \sim -2(\lambda\zeta)^{1/2} + (\lambda\zeta)^{-1/2} + O(\zeta^{-3/2}), \quad \zeta \to \infty, \tag{27a}$$

$$\kappa U_{\mathrm{i}} \sim -\ln(\lambda\zeta/4) - 2 - \lambda\zeta/2 + O(\zeta^2), \quad \zeta \to 0, \tag{27b}$$

can readily be inferred. The behaviour (27a) holding at the base of the outer defect layer is recognised as the square-root law deduced first by Townsend in his study [29] of turbulent boundary layers exhibiting vanishingly small wall shear stress; the outermost layer so to speak "anticipates" the approach to separation as the velocity defect increases to a level larger than u_τ. We remark that Townsend in [29] identified the intermediate region as the so-called "equilibrium layer", where convective terms in (1b) are (erroneously within the framework of asymptotic high-Reynolds-number theory) considered to be negligibly small. Equation (27b) provides the logarithmic variation of U_{i} as $\zeta \to 0$, required by the match with the wall layer, which gives rise to the generalised skin-friction relationship

$$\frac{\kappa}{\gamma} \sim \ln\left(\frac{Re\gamma^2 U_e^3}{\beta_0^{1/2}}\right) + \beta_0\kappa W + O(\gamma\beta_0) \sim (1+\varepsilon W)\ln Re. \tag{28}$$

Note that (28) reduces to (8) when $\beta_0 = O(1)$.

Having demonstrated that classical theory of turbulent boundary layers in the limit of large Reynolds number can – in a self-consistent manner – be extended to situations where the velocity defect is asymptotically large as compared to u_τ but still $o(1)$, we now consider the flow behaviour in the outer wake-type region in more detail.

3.2 Outer Defect Region

Following the arguments put forward at the beginning of Sect. 3, we write, by making use of the stream function ψ, the flow quantities in the outer layer in the form

$$p \sim p_e(x) + \varepsilon^2 P(x,\eta;\varepsilon,\gamma), \tag{29a}$$

$$\psi/U_e \sim y - \varepsilon\delta_o F(x,\eta;\varepsilon,\gamma), \tag{29b}$$

$$-\left[\overline{u'^2}, \overline{v'^2}, \overline{u'v'}\right] \sim U_e^2\varepsilon^2[R,S,T](x,\eta;\varepsilon,\gamma). \tag{29c}$$

As before, here $\eta = y/\delta_0$ and we accordingly expand

$$Q \sim Q_1 + \varepsilon Q_2 + \cdots, \quad Q := F, P, R, S, T, W, \tag{30a}$$

$$\delta \sim \varepsilon\Delta_1 + \varepsilon^2\Delta_2 + \cdots, \tag{30b}$$

$$\beta/\beta_v \sim B_0(x) + \varepsilon B_1(x) + \cdots, \quad \beta_v \to \infty, \tag{30c}$$

where we require (without any loss of generality) that β_v equals β_0 at $x = x_v$, so that $\beta_0 = \beta_v B_0$ and $B_0(x_v) = 1$, $B_i(x_v) = 0$, $i = 1, 2, \ldots$. In analogy to (12), the first-order problem then reads

$$\frac{1}{U_e}\frac{\mathrm{d}(U_e\Delta_1)}{\mathrm{d}x}\eta F_1' - \frac{1}{U_e^3}\frac{\partial(U_e^3\Delta_1 F_1)}{\partial x} = T_1, \tag{31a}$$

$$F_1(x,0) = F_1'(x,1) = F_1''(x,1) = T_1(x,1) = 0, \tag{31b}$$

$$\eta \to 0: \quad T \sim (\kappa\eta F_1'')^2, \quad F_1' \sim W_1(x) - (2/\kappa)(\lambda\eta)^{1/2}. \tag{31c}$$

In the following we concentrate on solutions which are self-similar up to second order, i.e. $\partial F_1/\partial x \equiv \partial T_1/\partial x \equiv 0$ and $\partial F_2/\partial x \equiv \partial T_2/\partial x \equiv 0$. By again adopting the notations $F_1 = \hat{F}(\eta)$, $T_1 = \hat{T}(\eta)$, and setting $\Delta_1 = \hat{\Delta}(x)$, $U_e = \hat{U}(x)$, we recover the requirements (20), (21) for self-similarity at first order resulting from classical small-defect theory in the limit of large values of β_v, in agreement with (30b) and the definition of β provided by (22):

$$B_0 \equiv 1, \quad \hat{\Delta}\hat{F}_e = 3(x - x_v), \quad \hat{U} = (C/3)^{1/3}(x - x_v)^{-1/3}, \tag{32}$$

with a constant C, and

$$2\eta\hat{F}' = \hat{F}_e\hat{T}, \quad \hat{F}(0) = \hat{T}(0) = \hat{F}'(1) = \hat{F}''(1) = \hat{T}(1) = 0. \tag{33}$$

If, as in the discussion of the flow behaviour in the intermediate layer, a mixing length model $\hat{T} = [\ell(\eta)\hat{F}''(\eta)]^2$ in accordance with (25) is chosen to close the problem, integration of (33) yields the analytical expressions

$$\hat{F}'(\eta) = \frac{1}{2\hat{F}_e}\left[\int_\eta^1 \frac{z^{1/2}}{\ell(z)}\,\mathrm{d}z\right]^2, \quad \hat{F}_e = \left\{\frac{1}{2}\int_0^1\left[\int_\eta^1 \frac{z^{1/2}}{\ell(z)}\,\mathrm{d}z\right]^2\mathrm{d}\eta\right\}^{1/2}. \tag{34}$$

Equations (34) have been evaluated numerically by using a slightly generalised version of the mixing length closure originally suggested in [13],

$$\ell = c_\ell I(\eta)^{1/2}\tanh(\kappa\eta/c_\ell), \quad I = 1/(1 + 5.5\eta^6), \quad c_\ell = 0.085. \tag{35}$$

Herein $I(\eta)$ represents the well-known Klebanoff's intermittency factor proposed in [9]. One then obtains $W_1 = \hat{F}'(0) \doteq 13.868$, $\hat{F}_e \doteq 5.682$, and $\mathrm{d}\hat{\Delta}/\mathrm{d}x \doteq 0.528$, cf. (32). As seen in Fig. 1a, both $\hat{F}'$ and $\hat{T}$ vanish quadratically as $\eta \to 1$ as a result of the boundary conditions $\hat{T}(1) = \hat{T}'(1) = 0$, cf. (33). Also, note that $\hat{F}'$ exhibits the square-root behaviour required from the match with the intermediate layer as $\eta \to 0$.

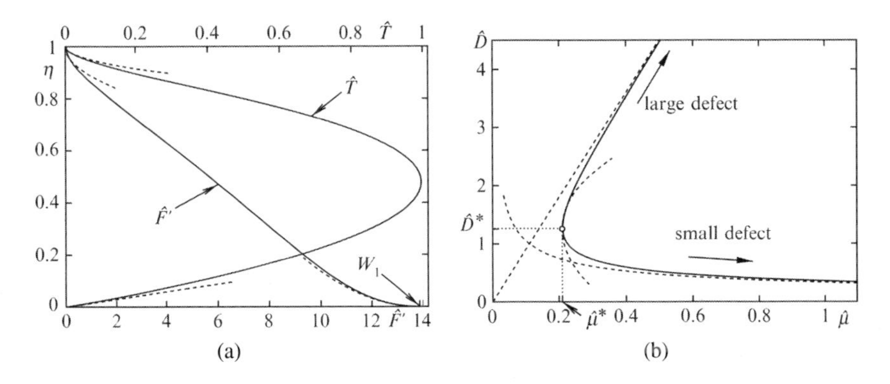

Fig. 1 Quasi-equilibrium flows: (**a**) $\hat{F}'(\eta)$, $\hat{T}(\eta)$, *dashed*: asymptotes found from (31b), (31c); (**b**) canonical representation (37), *dashed*: asymptotes (see last paragraph of Sect. 3) and parabola approximating the curve in the apex to leading order

Turning now to the second-order problem, we consider the most general case that the wall shear enters the description of the flow in the outer layer at this level of approximation (principle of least degeneracy). Therefore, we require $\varepsilon^3 T_2(0) \sim \gamma^2$, which finally determines the yet unknown magnitude of ε relative to γ, namely that $\varepsilon \sim \gamma^{2/3}$. Since, as pointed out before, $\varepsilon \sim \gamma \beta_0^{1/2}$, this implies that $\varepsilon\beta_0 = \Gamma = O(1)$. Inspection of the resulting second-order problem indicates that self similar solutions exist only if the external velocity distribution (32) predicted by classical theory is slightly modified in the form

$$\hat{U}(x) = (C/3)^{1/3}(x - x_v)^{-1/3+\mu}, \quad \mu \sim \gamma^{2/3}\mu_1 + \cdots, \tag{36}$$

where the $O(1)$-parameter μ_1 satisfies a solvability condition that represents the integral momentum balance obtained from integrating the second-order x-momentum equation from $\eta = 0$ to $\eta = 1$. It can be cast into the canonical form

$$9\hat{D}^2\hat{\mu} = 1 + \hat{D}^3. \tag{37}$$

Herein $\hat{D} = r^{1/3}\Gamma^{1/3}$, $\hat{\mu} = r^{-2/3}\mu_1$, and

$$r = \hat{F}_e^{-1}\int_0^1 (\hat{F}'^2 - \hat{R} + \hat{S})\,\mathrm{d}\eta\,. \tag{38}$$

A graph of the relationship (37) which represents the key result of the analysis dealing with quasi-equilibrium boundary layers having a moderately large velocity defect is shown in Fig. 1b. Most interesting, it is found that solutions describing flows of this type exist for $\hat{\mu} \geq \hat{\mu}^* = 2^{1/3}/6$ only and form two branches, associated with non-uniqueness of the quantity $\hat{D}$, which serves as a measure of velocity defect, for a specific value of the pressure gradient. Along the lower branch, $\hat{D} \leq \hat{D}^* = 2^{1/3}$ and decreases with increasing values of $\hat{\mu}$, so that classical small-defect theory is recovered in the limit $\hat{\mu} \to \infty$, where $\hat{D} \sim (9\hat{\mu})^{-1/2}$. In contrast, this limit leads to an unbounded growth of values $\hat{D} \geq \hat{D}^*$ associated with the upper branch: $\hat{D} \sim 9\hat{\mu}$ as $\hat{\mu} \to \infty$. This immediately raises the question if it is possible to formulate a general necessarily nonlinear theory which describes turbulent boundary layers having a finite velocity defect in the limit of infinite Reynolds number. We also note that the early experimental observations made by Clauser, see [6], seem to strongly point to this type of non-uniqueness.

4 Large Velocity Deficit

As in the cases of small and moderately small velocity defect we require the boundary layer to be slender. However, in contrast to the considerations of Sects. 2 and 3, the validity of this requirement can no longer be inferred from assumption (a) and the balance between convective and Reynolds stress gradient terms in the outer predominately inviscid region of the boundary layer which now yields $\partial\tau/\partial y = O(1)$, rather than $\partial\tau/\partial y \ll 1$ as earlier. A hint how this difficulty can be overcome is provided by the observation that the transition from a small to a moderately large velocity defect is accompanied with the emergence of a wake-type flow in this outer layer. One expects that this effect will become more pronounced as the velocity defect increases further, suggesting in turn that the outer part of the boundary layer, having a velocity defect of $O(1)$, essentially behaves as a turbulent free shear layer. An attractive strategy then is to combine the asymptotic treatment of such flows (e.g. [25]) in which the experimentally observed slenderness is enforced through the introduction of a Reynolds-number-independent parameter $\alpha \ll 1$ with the asymptotic theory of turbulent wall bounded flows.

4.1 Outer Wake Region

Let the parameter $\alpha \ll 1$ measure the lateral extent of the outer wake region, so that $\bar{y} := y/\alpha = O(1)$. Appropriate expansions of the various field quantities then are

$$p \sim p_e(x) + O(\alpha), \tag{39a}$$

$$q \sim \alpha\, q_0(x, \bar{y}) + o(\alpha), \tag{39b}$$

where q stands for Δ, ψ, $\tau := -\overline{u'v'}$, $\sigma_{(x)} := -\overline{u'^2}$, $\sigma_{(y)} := -\overline{v'^2}$. From substitution into (1b–1c) the leading order outer wake problem is found to be

$$\frac{\partial\psi_0}{\partial\bar{y}}\frac{\partial^2\psi_0}{\partial\bar{y}\partial x} - \frac{\partial\psi_0}{\partial x}\frac{\partial^2\psi_0}{\partial\bar{y}^2} = -U_e U_{ex} + \frac{\partial\tau_0}{\partial\bar{y}}, \tag{40a}$$

$$\bar{y} = 0: \quad \psi_0 = \tau_0 = 0, \tag{40b}$$

$$\bar{y} = \Delta_0(x): \quad \partial\psi_0/\partial\bar{y} = U_e, \quad \tau_0 = 0. \tag{40c}$$

As in the case of a moderately large velocity defect, we expect a finite wall slip $U_s(x) := \partial\psi_0/\partial\bar{y}$ at the base $\bar{y} = 0$ of this outer layer, which yields the limiting behaviour

$$\partial\psi_0/\partial\bar{y} \sim U_s(x) + O(\bar{y}^{3/2}), \quad \tau_0 \sim \Lambda_0\bar{y} + O(\bar{y}^{3/2}), \tag{41}$$

with $\Lambda_0 := U_s U_{sx} - U_e U_{ex} > 0$.

It is easily verified that the various layers introduced so far in the description of turbulent boundary layers share the property that their lateral extent is of the order of the mixing length ℓ characteristic for the respective layer. In contrast, the scalings given by (39) imply that ℓ is much smaller than the thickness of the outer wake region: $\ell \sim \alpha^{3/2} \ll \alpha$. This is a characteristic feature of free shear layers, of course, but also indicates that the outer wake region "starts to feel" the presence of the confining wall at distances $y \sim \alpha^{3/2}$, which in turn causes the emergence of an inner wake region.

4.2 Inner Wake Region

By introducing the stretched wall distance $Y = y/\alpha^{3/2} = O(1)$, inspection of (41) suggests the expansions

$$\psi \sim \alpha^{3/2}U_s(x) + \alpha^{9/4}\bar{\psi}(x, Y) + \cdots, \tag{42a}$$

$$\tau \sim \alpha^{3/2}\bar{T}(x, Y) + \cdots, \quad \ell \sim \alpha^{3/2}\bar{L}(x, Y) + \cdots, \tag{42b}$$

which leads to

$$\bar{T} = \Lambda_0 Y. \tag{43}$$

Furthermore, $\bar{T}$ and $\bar{\psi}$ are subject to the boundary conditions

$$T(x, 0) = \bar{\psi}(x, 0) = 0, \tag{44a}$$

$$\bar{\psi}_Y \sim \frac{2}{3}\frac{\Lambda_0^{1/2}}{\bar{L}_0}Y^{3/2}, \quad Y \to \infty, \quad \bar{L}_0 = \lim_{Y\to\infty}\bar{L}. \tag{44b}$$

The solution of the inner wake problem posed by (43), (44) can be obtained in closed form. It exhibits the expected square-root behaviour of $\bar{\psi}_Y$,

$$\bar{\psi}_Y \sim \bar{U}_s(x) + 2\frac{(\Lambda_0 Y)^{1/2}}{\chi(x)}, \quad \bar{L} \sim \chi(x)\bar{Y}, \quad Y \to 0. \tag{45}$$

Here $\bar{U}_s(x)$ denotes the correction of the slip velocity $U_s(x)$ caused by the inner wake region,

$$u_s \sim U_s(x) + \alpha^{3/4}\bar{U}_s(x) + \cdots, \tag{46a}$$

$$\bar{U}_s(x) = -\int_0^\infty \left(\frac{1}{\bar{L}} - \frac{1}{\bar{L}_0}\right)(\Lambda_0 Y)^{1/2}\,\mathrm{d}Y. \tag{46b}$$

At this point it is important to recall the basic assumption made at the beginning of this section, namely, that the slenderness parameter α is independent of Re, or more generally, asymptotes to a small but finite value as $Re \to \infty$. As a consequence, the outer and inner wake regions provide a complete description of the boundary layer flow in the formal limit $Re^{-1} = 0$. If, however, $0 < 1/Re \ll 1$ an additional sublayer forms at the base of the inner wake region. This sublayer plays a similar role as the intermediate layer discussed in Sect. 3.1: there the magnitude of the Reynolds shear stress, still varying linearly with distance from the wall, is reduced to $O(u_\tau^2)$, which is necessary to provide the square-root behaviour expressed in (45) and, finally, to allow for the match with the universal wall layer, see [19].

4.3 Numerical Solution of the Leading-Order Outer-Wake Problem

As earlier, a slightly modified version of the mixing length model proposed in [13] will be adopted to close the outer wake problem posed by (40). Numerical calculations were carried out for a family of retarded external flows controlled by two parameters m_s, k, with $m_s < 0$, $0 \leq k < 1$:

$$U_e(x; m_s, k) = (1 + x)^{m(x;m_s,k)}, \tag{47a}$$

$$\frac{m}{m_s} = 1 + \frac{k}{1-k}\,\Theta(2 - x)\big[1 - (1 - x)^2\big]^3. \tag{47b}$$

Herein Θ denotes the Heaviside step function. Self-similar solutions of the form $\psi_0 = \Delta_0 F(\xi)$, $\xi := Y/\Delta_0$, $\Delta_0 = b(1 + x)$, where $b = \text{const}$ and the position $x = -1$ defines the virtual origin of the flow, exist for $k = 0$ if $m_s > -1/3$ and are used to provide initial conditions at $x = 0$ for the downstream integration of (40) with U_e given by (47). As a specific example, we consider the case $F'(0) = 0.95$ of a relatively small velocity defect, imposed at $x = 0$, for which the requirement of self-similarity for $-1 < x < 0$ yields $b \doteq 0.3656$ and $m_s \doteq -0.3292$. The key results which are representative for the responding boundary layer and, most important, indicate that the present theory is capable of describing the approach to separation are displayed in Fig. 2. If k is sufficiently small, the distribution of the wall slip velocity U_s is smooth and $U_s > 0$ throughout. However, when k reaches a critical value $k_M \doteq 0.84258$, the slip velocity U_s is found to vanish at a single location $x = x_M$, but is positive elsewhere. A further increase of k provokes a breakdown of the calculations, accompanied with the formation of a weak singularity slightly upstream of x_M at $x = x_G$. A similar behaviour is observed for the boundary layer thickness Δ_0, which is smooth in the subcritical case $k < k_M$,

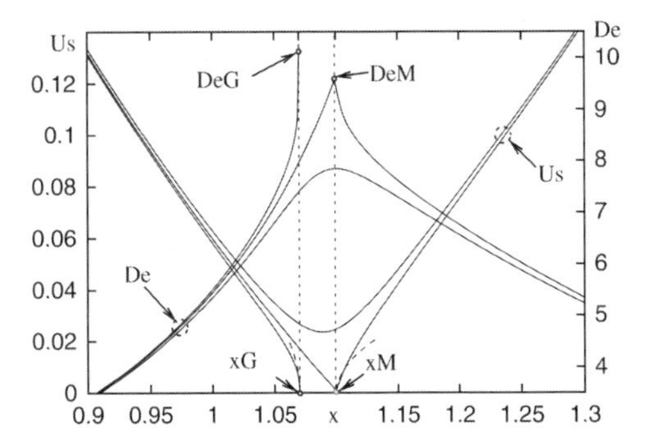

Fig. 2 Solutions of (40) for $|x - x_M| \ll 1$, $|k - k_M| \ll 1$, *dashed*: asymptotes expressed by (48b), (49)

exhibits a rather sharp peak $\Delta_{0,M}$ for $k = k_M$ at $x = x_M$, and approaches a finite limit $\Delta_{0,G}$ in an apparently singular manner in the supercritical case $k > k_M$.

Following the qualitatively similar behaviour of the wall shear stress that replaces U_s in the case of laminar boundary layers, see [17, 18, 27], the critical solution with $k = k_M$ is termed a marginally separating boundary layer solution. However, in vivid contrast to its laminar counterpart, is is clearly seen to be locally asymmetric with respect to $x = x_M$ where it is singular. This numerical finding is supported by a local analysis of the flow behaviour near $x = x_M$, carried out in [20]: it indicates that U_s decreases linearly with x upstream of $x = x_M$ but exhibits a square-root singularity as $x - x_M \to 0_+$,

$$U_s/P_{00}^{1/2} \sim -B(x - x_M), \qquad x - x_M \to 0_-, \tag{48a}$$

$$U_s/P_{00}^{1/2} \sim U_+(x - x_M)^{1/2}, \quad x - x_M \to 0_+, \tag{48b}$$

where $P_{00} = (\mathrm{d}p_e/\mathrm{d}x)(x_M)$. It is found that $U_+ \doteq 1.1835$, whereas the constant B remains arbitrary in the local investigation and has to be determined by comparison with the numerical results for $x \leq x_M$.

This local analysis also shows that a square-root singularity forms at a position $x = x_G < x_M$ for $k > k_M$,

$$U_s/P_{00}^{1/2} \sim U_-(x_G - x)^{1/2}, \quad x - x_G \to 0_-, \tag{49}$$

with some U_- to be determined numerically, and that the solution cannot be extended further downstream. This behaviour, which has been described first in [12], is reminiscent of the Goldstein singularity well-known from the theory of laminar boundary layers and, therefore, will be termed the turbulent Goldstein singularity. As shown in the next section, the bifurcating behaviour of the solutions for $k - k_M \to 0$ is associated with the occurrence of marginally separating flow.

4.4 Marginal Separation

According to the original boundary layer concept, pressure disturbances caused by the displacement of the external inviscid flow due to the momentum deficit, which is associated with the reduced velocities close to the wall, represent a higher order effect. Accordingly, higher-order corrections to the leading-oder approximation of the flow quantities inside and outside the boundary layer can be calculated in subsequent steps. However, as found first for laminar flows, this so-called hierarchical structure of the perturbation scheme breaks down in regions where the displacement thickness changes so rapidly that the resulting pressure response is large enough to affect the lowest-order boundary layer approximation (e.g. [26]). A similar situation is encountered for the type of turbulent flows discussed in the preceding section. Indeed, the slope discontinuity of Δ_0 and, in turn, of the displacement thickness forces a singularity in the response pressure, indicating a breakdown of the hierarchical approach to boundary layer theory. As for laminar flows, see [17, 18, 27], this deficiency can be overcome by adopting a local interaction strategy, so that the induced pressure disturbances enter the description of the flow in leading rather than higher order.

Again, similar to laminar flows, three layers (decks) characterising regions of different flow behaviour have to be distinguished inside the local interaction region, see Fig. 3. Effects of Reynolds stresses are found to be confined to the lower deck region (LD), having a streamwise and lateral extent of $O(\alpha^{3/5})$ and $O(\alpha^{6/5})$, respectively. Here the flow is governed by equations of the form (40). The majority of the boundary layer, i.e. the main deck (MD), behaves passively in the sense that it transfers displacement effects caused by the lower deck region unchanged to the external flow region taking part in the interaction process, the so-called upper deck (UD), and transfers the resulting pressure response unchanged to the lower deck. Solutions to the leading-order main and upper deck problems can be obtained in closed form which finally leads to the fundamental lower deck problem. By using suitably stretched variables, it can be written in terms of a stream function $\hat{\psi}(\hat{X}, \hat{Y})$ as (see [20])

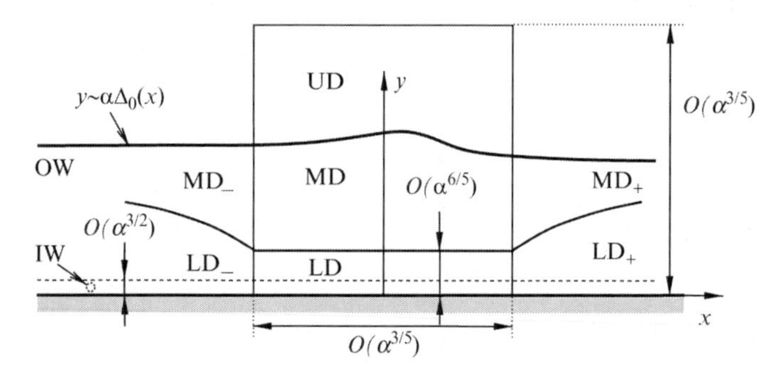

Fig. 3 Triple-deck structure, for captions see text, subscripts "−" and "+" refer to the continuation of flow regions up- and downstream of the local interaction zone, *dashed* line indicates inner wake

$$\frac{\partial \hat{\psi}}{\partial \hat{Y}} \frac{\partial^2 \hat{\psi}}{\partial \hat{Y} \partial \hat{X}} - \frac{\partial \hat{\psi}}{\partial \hat{X}} \frac{\partial^2 \hat{\psi}}{\partial \hat{Y}^2} = -1 - \Lambda(\Gamma)\, \hat{P}'(\hat{X}) + \frac{\partial \hat{T}}{\partial \hat{Y}}, \tag{50a}$$

$$\hat{T} = \frac{\partial^2 \hat{\psi}}{\partial \hat{Y}^2} \left| \frac{\partial^2 \hat{\psi}}{\partial \hat{Y}^2} \right|, \tag{50b}$$

$$\hat{P}(\hat{X}) = \frac{1}{\pi} \oint_{-\infty}^{\infty} \frac{\hat{A}'(\hat{S})}{\hat{X} - \hat{S}}\, \mathrm{d}\hat{S} \tag{50c}$$

$$\hat{Y} = 0: \quad \hat{\psi} = \hat{T} = 0, \tag{50d}$$

$$\hat{Y} \to \infty: \quad \hat{T} - \hat{Y} \to \hat{A}(\hat{X}), \tag{50e}$$

$$\hat{X} \to -\infty: \quad \hat{\psi} \to (4/15)\hat{Y}^{5/2} + \Gamma \hat{Y}\,, \quad 0 \le \Gamma \le 1, \tag{50f}$$

$$\hat{X} \to \infty: \quad \hat{\psi} \to \hat{X}^{5/6} F_+(\hat{\eta}), \quad \hat{\eta} := \hat{Y}/\hat{X}^{1/3}. \tag{50g}$$

The first and second term on the right-hand side of (50a) account for the imposed and induced pressure, respectively. The latter is given by the Hilbert integral (50c), where $\hat{A}$ characterises the displacement effect exerted by the lower deck region. The far-field condition (50e) expresses the passive character of the main deck mentioned before, whereas the conditions (50f), (50g) follow from the match with regions LD_-, LD_+ immediately upstream and downstream of the local interaction zone. The analysis of region LD_+ determines the function $F_+(\hat{\eta})$. Finally, the parameter Γ measures the intensity of the interaction process as the monotonically increasing but otherwise arbitrary function $\Lambda(\Gamma)$ expresses the magnitude of the induced pressure gradient.

As a representative example of flows encountering separation, the distributions of $\hat{A}$, $\hat{P}$, and the wall slip $\hat{U}_s := (\partial\hat{\psi}/\partial\hat{Y})(\hat{X}, \hat{Y} = 0)$, obtained by numerical solution of the triple-deck problem (50) for $\Gamma = 0.019$, $\Lambda = 3$, are depicted in Fig. 4a. Here the dot-and-dash lines indicate the upstream and downstream asymptotes, obtained from the analysis of the flow behaviour in the pre- and post-interaction regions (subscripts "–" and "+" in Fig. 3), while $\hat{X}_D$ and $\hat{X}_R$ denote the positions of, respectively, detachment and reattachment. It is interesting to note that the pas-

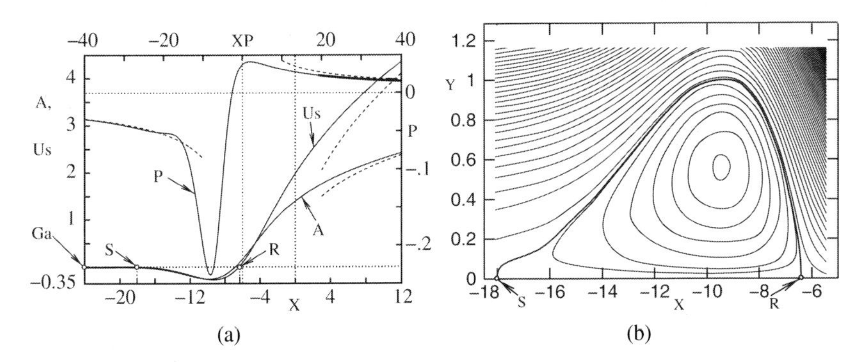

Fig. 4 Specific solution of (50), separation in $\mathcal{S}$, reattachment in $\mathcal{R}$: (**a**) key quantities, *dashed*: asymptotes found analytically; (**b**) streamlines

sage of $\hat{U}_s$ into the reverse-flow region where $\hat{U}_s < 0$ causes the interaction pressure $\hat{P}$ to drop initially before it rises sharply, overshoots and finally tends to zero in the limit $\hat{X} \to \infty$. This is in striking contrast to laminar flows, where flow separation always is triggered by an initial pressure rise, and reflects the fact that – in the case of turbulent flows considered here – the streamwise velocity component at the base $\hat{Y} = 0$ of the lower deck region is allowed to take on finite values, whereas the no-slip condition is enforced in its laminar counterpart.

Streamlines inside the lower-deck region are displayed in Fig. 4b which clearly shows the formation of a recirculating eddy. Also, we draw attention to the increasing density of streamlines further away from the wall and downstream of reattachment, associated with the strong acceleration of the fluid there as evident from the rapid increase of $\hat{U}_s$.

The interaction process outlined so far describes the behaviour or marginally separated turbulent flows in the limit $1/Re = 0$. As in the case of conventional, i.e. hierarchical, boundary layers having a velocity of defect of $O(1)$, additional sublayers form closer to the wall if $1/Re \ll 1$ but finite. Their analysis, outlined in [19], provides the skin-friction relationship in generalised form to include the effects of vanishing and negative wall shear – treated first in a systematic way in [24] – but also shows that these layers behave passively insofar as the lower deck problem (50) remains intact.

5 Conclusions and Outlook

In this study an attempt has been made to derive the classical two-layer structure of a turbulent small-defect boundary layer from a minimum of assumptions. As in [30], but in contrast to earlier investigations (e.g. [10]), the (logarithmic) law of the wall is taken basically as an empirical observation rather than a consequence of matching inner and outer layers, as the latter is not felt rich enough to provide a stringent foundation of this important relationship reflecting the dynamics of the flow close to the wall, which is not understood in full at present. Probably the first successful model that describes essential aspects of this dynamics is provided by Prandtl's mixing length concept, proposed more than 50 years before the advent of asymptotic theories in fluid mechanics. Significant progress has been achieved in more recent years and, in particular, by the pioneering work of Walker (e.g. [30]), whose untimely death ended a line of thought which certainly ought to be taken up again.

Following the brief outline of the classical small-defect theory, it is shown how a description of turbulent boundary layers having a slightly larger (i.e. moderately large) velocity defect, where the outer predominately inviscid layer starts to develop a wake-type behaviour, can be formulated. Further increase of the velocity defect to values of $O(1)$ causes the wake region to become even more pronounced and is seen to allow for the occurrence of reverse-flow regions close to the wall, resulting in what we believe to be the first fully self-consistent theory of marginally separated turbulent flows.

Unfortunately, however, this success seemingly does not shed light on the phenomenon of global or gross separation associated with flows past (more-or-less) blunt bodies or, to put it more precisely, flows which start at a stagnation point rather than a sharp leading edge. Indeed, a recent careful numerical investigation for the canonical case of a circular cylinder, presented, among others, in [22, 23], undoubtedly indicates that the boundary layer approaching separation exhibits a small rather than a large velocity defect, leading in turn to the dilemma addressed in Sect. 2.4. The accompanying asymptotic analysis based on the turbulence intensity gauge model introduced in [15], however, strongly suggests that a boundary layer forming on a body of finite extent and originating in a front stagnation point does not reach a fully developed turbulent state, even in the limit $Re \to \infty$. Specifically, it is found that the boundary layer thickness and the Reynolds shear stress are slightly smaller than predicted by classical small-defect theory, while the velocity defect in the outer region, and, most important, the thickness of the wall layer are slightly larger. As a consequence, the outer large-momentum region does not penetrate to distances from the wall which are transcendentally small. In turn, this situation opens the possibility to formulate a local interaction mechanism that describes the detachment of the boundary layer from the solid wall within the framework of free-streamline theory at pressure levels which are compatible with experimental observation. This is a topic of intense current investigations.

References

1. N. Afzal. Scaling of Power Law Velocity Profile in Wall-Bounded Turbulent Shear Flows. Meeting paper 2005-109, AIAA, 2005.
2. G. I. Barenblatt and A. J. Chorin. A note on the intermediate region in turbulent boundary layers. *Phys. Fluids*, 12(9):2159–2161, 2000. Letter.
3. G. I. Barenblatt, A. J. Chorin, and V. M. Prototoskin. Self-similar intermediate structures in turbulent boundary layers at large Reynolds numbers. *J. Fluid Mech.*, 410:263–283, 2000.
4. G. I. Barenblatt and N. Goldenfeld. Does fully developed turbulence exist? Reynolds number independence versus asymptotic covariance. *Phys. Fluids*, 7(12):3078–3082, 1995.
5. W. B. Bush and F. E. Fendell. Asymptotic analysis of turbulent channel and boundary-layer flow. *J. Fluid Mech.*, 56(4):657–681, 1972.
6. F. H. Clauser. The Turbulent Boundary Layer. In H. L. Dryden and Th. von Kármán, editors, *Advances in Applied Mechanics*, volume 4, pages 1–51. Academic, New York, 1956.
7. E. Deriat and J.-P. Guiraud. On the asymptotic description of turbulent boundary layers. *J. Theor. Appl. Mech.*, 109–140, 1986. Special Issue on Asymptotic Modelling of Fluid Flows. Original French article in *J. Méc. Théor. Appl.*
8. F. E. Fendell. Singular Perturbation and Turbulent Shear Flow near Walls. *J. Astronaut. Sci.*, 20(11):129–165, 1972.
9. P. S. Klebanoff. Characteristics of turbulence in a boundary layer with zero pressure gradient. NACA Report 1247, NASA – Langley Research Center, Hampton, VA, 1955. See also NACA Technical Note 3178, 1954.
10. G. L. Mellor. The Large Reynolds Number, Asymptotic Theory of Turbulent Boundary Layers. *Int. J. Eng. Sci.*, 10(10):851–873, 1972.

11. G. L. Mellor and D. M. Gibson. Equilibrium turbulent boundary layers. *J. Fluid Mech.*, 24(2):225–253, 1966.
12. R. E. Melnik. An Asymptotic Theory of Turbulent Separation. *Comput. Fluids*, 17(1):165–184, 1989.
13. R. Michel, C. Quémard, and R. Durant. Hypothesis on the mixing length and application to the calculation of the turbulent boundary layers. In S. J. Kline, M. V. Morkovin, G. Sovran, and D. J. Cockrell, editors, *Proceedings of Computation of Turbulent Boundary Layers – 1968 AFOSR-IFP-Stanford Conference*, volume 1, pages 195–207. Stanford, CA, 1969. Stanford University.
14. P. A. Monkewitz, K. A. Chauhan, and H. M. Nagib. Self-consistent high-Reynolds-number asymptotics for zero-pressure-gradient turbulent boundary layers. *Phys. Fluids*, 19(11):115101-1–115101-12, 2007.
15. A. Neish and F. T. Smith. On turbulent separation in the flow past a bluff body. *J. Fluid Mech.*, 241:443–467, 1992.
16. J. M. Österlund, A. V. Johansson, H. M. Nagib, and M. Hites. A note on the overlap region in turbulent boundary layers. *Phys. Fluids*, 12(1):1–4, 2000. Letter.
17. A. I. Ruban. Singular solutions of the boundary layer equations which can be extended continuously through the point of zero surface friction. *Fluid Dyn.*, 16(6):835–843, 1981. Original Russian article in *Izvestiya Akademii Nauk SSSR, Mekhanika Zhidkosti i Gaza* (6), 42–52, 1981.
18. A. I. Ruban. Asymptotic theory of short separation regions on the leading edge of a slender airfoil. *Fluid Dyn.*, 17(1):33–41, 1982. Original Russian article in *Izvestiya Akademii Nauk SSSR, Mekhanika Zhidkosti i Gaza* (1), 42–51, 1982.
19. B. Scheichl and A. Kluwick. On turbulent marginal boundary layer separation: how the half-power law supersedes the logarithmic law of the wall. *Int. J. Comput. Sci. Math. (IJCSM)*, 1(2/3/4):343–359, 2007. Special Issue on Problems Exhibiting Boundary and Interior Layers.
20. B. Scheichl and A. Kluwick. Turbulent Marginal Separation and the Turbulent Goldstein Problem. *AIAA J.*, 45(1):20–36, 2007. See also *AIAA Meeting paper 2005–4936*.
21. B. Scheichl and A. Kluwick. Asymptotic theory of bluff-body separation: a novel shear-layer scaling deduced from an investigation of the unsteady motion. *J. Fluids Struct.*, 24(8):1326–1338, 2008. Special Issue on the IUTAM Symposium on Separated Flows and their Control.
22. B. Scheichl, A. Kluwick, and M. Alletto. "How turbulent" is the boundary layer separating from a bluff body for arbitrarily large Reynolds numbers? *Acta Mech.*, 201(1–4):131–151, 2008. Special Issue dedicated to Professor Wilhelm Schneider on the occasion of his 70th birthday.
23. B. Scheichl and A. Kluwick. Level of Turbulence Intensities Associated with Bluff-Body Separation for Large Values of the Reynolds Number. Meeting paper 2008–4348, AIAA, 2008.
24. H. Schlichting and K. Gersten. *Boundary-Layer Theory*. Springer, Berlin, 8th edition, 2000.
25. W. Schneider. Boundary-Layer Theory of free turbulent shear flows. *Z. Flugwiss. Weltraumforsch. (J. Flight Sci. Space Res.)*, 15(3):143–158, 1991.
26. K. Stewartson. Multistructured Boundary Layers on Flat Plates and Related Bodies. In Y. Chia-Sun, editor, *Advances in Applied Mechanics*, volume 14, pages 145–239. Academic, New York, 1974.
27. K. Stewartson, F. T. Smith, and K. Kaups. Marginal Separation. *Studies in Applied Mathematics*, 67(1):45–61, 1982.
28. R. I. Sykes. An asymptotic theory of incompressible turbulent boundary-layer flow over a small hump. *J. Fluid Mech.*, 101(3):631–646, 1980.
29. A. A. Townsend. Equilibrium layers and wall turbulence. *J. Fluid Mech.*, 11(1):97–120, 1961.
30. J. D. A. Walker. Turbulent Boundary Layers II: Further Developments. In A. Kluwick, editor, *Recent Advances in Boundary Layer Theory*, volume 390 of CISM Courses and Lectures, pages 145–230. Springer, Vienna, 1998.
31. K. S. Yajnik. Asymptotic theory of turbulent shear flows. *J. Fluid Mech.*, 42(2):411–427, 1970.

A Deterministic Multiscale Approach for Simulating Dilute Polymeric Fluids

David J. Knezevic and Endre Süli

Abstract We introduce a numerical method for solving the coupled Navier–Stokes–Fokker–Planck model (i.e. a micro–macro model) for dilute polymeric fluids where polymer molecules are modelled as FENE dumbbells. The Fokker–Planck equation is posed on a high-dimensional domain and is therefore challenging from a computational point of view. We summarise analytical results for a Galerkin spectral method for the Fokker–Planck equation in *configuration space*, before combining this method with a finite element scheme in *physical space* to obtain an alternating-direction method for the high-dimensional Fokker–Planck equation. Alternating-direction methods have been considered previously in the literature for this problem (e.g. by Chauvière & Lozinski); we present an alternative framework here that is underpinned by rigorous numerical analysis, and numerical results demonstrating the effectiveness of our approach. The algorithm is well suited to implementation on a parallel computer, and we exploit this fact to make large-scale computations feasible.

1 Introduction

In this paper we introduce a computational framework for solving the Navier–Stokes–Fokker–Planck system of partial differential equations (also known as the *micro–macro model*) that governs the evolution of a dilute suspension of dumbbells in a Newtonian solvent, which is a well-studied model of dilute polymeric fluids [3, 23]. We refer to the approach of directly solving the coupled Navier–Stokes–Fokker–Planck system as the *deterministic multiscale* method; this approach has recently been used successfully in a number of papers by Lozinski, Chauvière

E. Süli (✉)
OUCL, University of Oxford, Parks Road, Oxford, OX1 3QD, UK,
E-mail: endre.suli@comlab.ox.ac.uk

A.F. Hegarty et al. (eds.), *BAIL 2008 - Boundary and Interior Layers*. Lecture Notes in Computational Science and Engineering,
DOI: 10.1007/978-3-642-00605-0,

and collaborators (see [4, 5, 19]), although those authors did not consider rigorous numerical analysis of their algorithm – such analysis is a major emphasis in the present paper as well as in [14, 15]. It is worth highlighting at the outset that there is an extensive literature on numerical methods for this problem, but most of the previous work on the subject addresses either fully macroscopic models (such as the Olroyd-B model) in order to circumvent the multiscale nature of the Navier–Stokes–Fokker–Planck system (see the text [23] for an overview of this field) or uses a stochastic approach in which the micro–macro system is treated using Monte–Carlo-type methods (cf. [22]). Compared to a fully macroscopic approach, the primary advantage of the deterministic multiscale method is that it does not involve "closure approximations"; the shortcomings of such approximations are well documented [11, 17, 26]. Also, a possible drawback of the stochastic approach is the presence of slowly decaying stochastic error terms. Variance reduction techniques have been developed to minimise the impact of this stochastic error in Monte–Carlo-type methods; nevertheless, circumventing this error completely is an important motivation for moving to fully deterministic micro–macro methods. The drawback of the deterministic multiscale approach, however, is that (as we shall see below) the Fokker–Planck equation is posed on a high-dimensional domain, and therefore solving it using deterministic methods is an imposing challenge from the computational point of view. Following Chauvière & Lozinski, our approach is to use an alternating-direction scheme to ameliorate the "curse of dimensionality", and we also use parallel computation to make large-scale simulations feasible in practice.

As indicated above, we are considering a dilute solution of microscopic dumbbells, i.e. two beads of small mass connected by a spring. The spring force law $\underset{\sim}{F}$ has a corresponding potential, $U : \mathbb{R}_{\geq 0} \to \mathbb{R}$, such that $\underset{\sim}{F}(\underset{\sim}{q}) = U'(\frac{1}{2}|\underset{\sim}{q}|^2)\underset{\sim}{q}$, where $\underset{\sim}{q} \in D$ is the *configuration* vector (or end-to-end vector) of a dumbbell. Here we consider the FENE force law [25], which, in non-dimensional form is:

$$U(\tfrac{1}{2}|\underset{\sim}{q}|^2) := -\frac{b}{2}\ln\left(1 - \frac{|\underset{\sim}{q}|^2}{b}\right), \qquad \underset{\sim}{F}(\underset{\sim}{q}) = \frac{\underset{\sim}{q}}{1 - |\underset{\sim}{q}|^2/b}, \tag{1}$$

where $D = B(0, \sqrt{b}) \subset \mathbb{R}^d$, $d = 2$ or 3. We assume that $b \in (2, \infty)$ (cf. [10] or Example 1.2 in [2]), and in practice b is typically chosen in the range [10, 100]. The theoretical results presented in this paper can be generalised to a broader class of FENE-like potentials that satisfy Hypotheses A and B from [15]. For simplicity of exposition, we restrict our attention to the FENE potential here.

Suppose the fluid is confined to a macroscopic physical domain Ω, assumed to be a bounded open set in $\mathbb{R}^d$. Let $\underset{\sim}{u} : (\underset{\sim}{x}, t) \in \Omega \times [0, T] \mapsto \underset{\sim}{u}(\underset{\sim}{x}, t) \in \mathbb{R}^d$ denote the macroscopic velocity field, and let $p : (\underset{\sim}{x}, t) \in \Omega \times [0, T] \mapsto p(\underset{\sim}{x}, t) \in \mathbb{R}$ denote the pressure. It is typical in this problem to let $\underset{\approx}{\kappa}$ denote the macroscopic velocity gradient, i.e. $\underset{\approx}{\kappa} := \nabla_{\underset{\sim}{x}} \underset{\sim}{u}$. Also, suppose the function $(\underset{\sim}{x}, \underset{\sim}{q}, t) \mapsto \psi(\underset{\sim}{x}, \underset{\sim}{q}, t)$ represents the probability, at time t, of finding a dumbbell with center of mass in the volume element $\underset{\sim}{x} + \mathrm{d}\underset{\sim}{x}$ and orientation vector in the element $\underset{\sim}{q} + \mathrm{d}\underset{\sim}{q}$. Then, for a suspension of FENE dumbbells, we have the following system (in non-dimensional form):

$$\frac{\partial \underset{\sim}{u}}{\partial t} + \underset{\sim}{u} \cdot \underset{\sim}{\nabla}_x \underset{\sim}{u} + \underset{\sim}{\nabla}_x p = \frac{\gamma}{\mathrm{Re}} \Delta_x \underset{\sim}{u} + \frac{b+d+2}{b} \frac{1-\gamma}{\mathrm{Re\,Wi}} \underset{\sim}{\nabla}_x \cdot \underset{\approx}{\tau}, \tag{2}$$

$$\underset{\sim}{\nabla}_x \cdot \underset{\sim}{u} = 0, \tag{3}$$

$$\underset{\approx}{\tau}(\underset{\sim}{x}, t) = \int_D \underset{\sim}{F} \otimes \underset{\sim}{q}\, \psi(\underset{\sim}{x}, \underset{\sim}{q}, t)\, \mathrm{d}\underset{\sim}{q}, \tag{4}$$

for $(\underset{\sim}{x}, t) \in \Omega \times (0, T]$, where ψ satisfies the Fokker–Planck equation:

$$\frac{\partial \psi}{\partial t} + \underset{\sim}{\nabla}_x \cdot (\underset{\sim}{u}\psi) + \underset{\sim}{\nabla}_q \cdot (\underset{\approx}{\kappa}\, \underset{\sim}{q}\, \psi) = \frac{1}{2\mathrm{Wi}} \underset{\sim}{\nabla}_q \cdot \left(M \underset{\sim}{\nabla}_q \left(\frac{\psi}{M} \right) \right), \tag{5}$$

or $(\underset{\sim}{x}, \underset{\sim}{q}, t) \in \Omega \times D \times (0, T]$. The system (2)–(5) is subject to the initial conditions:

$$\underset{\sim}{u}(\underset{\sim}{x}, 0) = \underset{\sim}{u}_0(\underset{\sim}{x}), \ \ \underset{\sim}{x} \in \Omega, \qquad \psi(\underset{\sim}{x}, \underset{\sim}{q}, 0) = \psi_0(\underset{\sim}{x}, \underset{\sim}{q}), \ \ (\underset{\sim}{x}, \underset{\sim}{q}) \in \Omega \times D. \tag{6}$$

In (2), Re is the Reynolds number, Wi is the Weissenberg number, which is the ratio of microscopic to macroscopic time-scales, and $\gamma \in (0, 1)$ is the ratio of solvent viscosity to total viscosity. In (5), M is the (normalised) FENE *Maxwellian* defined by

$$\underset{\sim}{q} \mapsto M(\underset{\sim}{q}) := \frac{1}{Z} \exp\left(-U(\tfrac{1}{2}|\underset{\sim}{q}|^2)\right) \in \mathrm{L}^1(D), \ Z := \int_D \exp\left(-U(\tfrac{1}{2}|\underset{\sim}{q}|^2)\right) \mathrm{d}\underset{\sim}{q},$$

which, in the case of the FENE model, is $M(\underset{\sim}{q}) := \frac{1}{Z}(1 - |\underset{\sim}{q}|^2/b)^{b/2}$. In fact, the form of the Fokker–Planck equation given in (5) uses a Kolmogorov symmetrisation [16]; it is equivalent to the 'standard' form of the equation:

$$\frac{\partial \psi}{\partial t} + \underset{\sim}{\nabla}_x \cdot (\underset{\sim}{u}\psi) + \underset{\sim}{\nabla}_q \cdot \left(\underset{\approx}{\kappa}\, \underset{\sim}{q}\, \psi - \frac{1}{2\mathrm{Wi}} \underset{\sim}{F}(\underset{\sim}{q})\psi \right) = \frac{1}{2\mathrm{Wi}} \Delta_q \psi, \tag{7}$$

but from our point of view the advantage of (5) is that the unbounded convection coefficient ($\underset{\sim}{F}$ in (7)) is absorbed into a weighted diffusion term, which is convenient from the point of view of analysis. It should be noted, however, that in [5] Lozinski & Chauvière proposed a numerical method based on (7) in which the substitution $\hat{\psi} := \psi / M^{2s/b}$ was used[1]; it was shown in Sect. 3.2 of [15] that with $b \geq 4s^2/(2s-1)$ and $s > 1/2$, this also leads to a well-posed problem and a stable semidiscretisation in any number of space dimensions, and hence all of the analytical results developed in this paper could also be developed based on the Lozinski–Chauvière substitution. Nevertheless, the symmetry of (5) simplifies analysis of the numerical methods we consider, and therefore we focus on

[1] Based on computational experience, Lozinski & Chauvière recommended $s = 2$ and $s = 2.5$ for $d = 2$ and $d = 3$, respectively.

the Maxwellian-transformed form of the Fokker–Planck equation in (5) for the remainder of this paper.

Since ψ is a probability density function (pdf) for each $\underset{\sim}{x} \in \Omega$, the initial datum should be non-negative:

$$\psi(\underset{\sim}{x}, \underset{\sim}{q}, 0) = \psi_0(\underset{\sim}{x}, \underset{\sim}{q}) \geq 0, \qquad \text{for a.e. } (\underset{\sim}{x}, \underset{\sim}{q}) \in \Omega \times D, \tag{8}$$

and should also satisfy the following normalisation property:

$$\int_D \psi_0(\underset{\sim}{x}, \underset{\sim}{q}) \, \mathrm{d}\underset{\sim}{q} = 1, \qquad \text{for a.e. } \underset{\sim}{x} \in \Omega. \tag{9}$$

It is crucial to note that (5) is posed in $2d$ spatial dimensions, plus time. Since the computational complexity of standard numerical methods for PDEs grows exponentially with the dimension of the spatial domain, the high-dimensionality of (5) represents a significant computational challenge. Therefore, in a coupled algorithm for (2)–(6), solving the Fokker–Planck equation is generally the bottleneck step and as a result the focus of this paper is on the analysis and implementation of efficient numerical methods for (5).

In the papers of Lozinski, Chauvière et al. [4, 5, 18–20] and Helzel & Otto [9], the authors decomposed the differential operator L from (5) by defining L_x and L_q acting in the $\underset{\sim}{x}$- and $\underset{\sim}{q}$-direction, respectively. They then used an alternating-direction numerical method (also referred to as an operator-splitting or dimension-splitting approach) based on these operators.[2] We pursue the same approach in this paper and we shall survey a number of stability and convergence results that we proved for our computational framework in the papers [14, 15].

Note that the splitting introduced above leads to a sequence of d-dimensional solves at each time step rather than a single $2d$-dimensional solve. Also, this splitting of L allows different numerical methods to be used in Ω and D (resulting in, what we call, a *heterogeneous* alternating-direction scheme). In Sect. 3 we consider heterogeneous alternating-direction numerical methods for the FENE Fokker–Planck equation on $\Omega \times D$ based on a finite element method in Ω and a single-domain Galerkin spectral method in D. These are appropriate choices because a finite element method is flexible enough to deal with the general domain Ω, whereas D is always a ball in $\mathbb{R}^d$, and therefore the L_q operator is well suited to a spectral discretisation via a polar or spherical coordinate transformation to a cartesian product domain.

The structure of this paper is as follows. We begin in Sect. 2 with an overview of the analysis and implementation of a Galerkin spectral method for the Maxwellian-transformed Fokker–Planck equation in configuration space. This spectral method is then integrated into an alternating-direction scheme for the full Fokker–Planck equation on $\Omega \times D$ in Sect. 3. Finally, we demonstrate the use of this alternating-direction scheme in an algorithm for the coupled Navier–Stokes–Fokker–Planck system for a channel flow problem of physical interest. We make concluding remarks in Sect. 5.

[2] These authors used (7), but the idea applies to (5) in the same way.

2 The Fokker–Planck Equation in Configuration Space

This section is concerned with the numerical approximation of the d-dimensional Fokker–Planck equation posed in configuration space:

$$\frac{\partial \psi}{\partial t} + \underset{\sim}{\nabla}_q \cdot (\underset{\approx}{\kappa}\, \underset{\sim}{q}\, \psi) = \frac{1}{2\mathrm{Wi}} \underset{\sim}{\nabla}_q \cdot \left(M \underset{\sim}{\nabla}_q \frac{\psi}{M} \right), \qquad (\underset{\sim}{q}, t) \in D \times (0, T], \tag{10}$$

where the $d \times d$ tensor $\underset{\approx}{\kappa}$ is assumed to belong to $(\mathrm{C}[0,T])^{d\times d}$ (i.e. it is independent of $\underset{\sim}{x}$) and is such that $\mathrm{tr}(\underset{\approx}{\kappa})(t) = 0$ for all $t \in [0,T]$. It will be assumed throughout that (10) is supplemented with the following initial and boundary conditions:

$$\psi(\underset{\sim}{q}, 0) = \psi_0(\underset{\sim}{q}), \qquad \text{for all } \underset{\sim}{q} \in D, \tag{11}$$

$$\psi(\underset{\sim}{q}, t) = o\left(\sqrt{M(\underset{\sim}{q})}\right), \qquad \text{as } \mathrm{dist}(\underset{\sim}{q}, \partial D) \to 0_+, \ \text{ for all } t \in (0,T]. \tag{12}$$

As in (8) and (9), the initial datum ψ_0 is such that $\psi_0 \geq 0$ and $\int_D \psi_0(\underset{\sim}{q})\, \mathrm{d}\underset{\sim}{q} = 1$.

The motivation for studying this subproblem is that, as indicated in Sect. 1, an efficient approach to the numerical solution of (5) in $2d + 1$ variables is based on operator-splitting with respect to $(\underset{\sim}{q}, t)$ and $(\underset{\sim}{x}, t)$. Thereby, the resulting time-dependent transport equation with respect to $(\underset{\sim}{x}, t)$ is completely standard, $\psi_t + \underset{\sim}{\nabla}_x \cdot (\underset{\sim}{u}(\underset{\sim}{x}, t)\psi) = 0$, while the transport-diffusion equation with respect to $(\underset{\sim}{q}, t)$ is (10).

2.1 Weak Formulation and Backward Euler Semidiscretisation

Following [15], let $\hat{\varphi} := \frac{\varphi}{\sqrt{M}}$ and $\underset{\sim}{\nabla}_M \hat{\varphi} := \sqrt{M}\, \underset{\sim}{\nabla}_q \left(\frac{\hat{\varphi}}{\sqrt{M}} \right)$, and define the function space $\mathrm{H}^1_0(D; M)$ to be the closure of $\mathrm{C}^\infty_0(D)$ in the norm of $\mathrm{H}^1(D; M)$, and

$$\mathrm{H}^1(D; M) := \left\{ \zeta \in \mathrm{L}^2(D) : \|\zeta\|^2_{\mathrm{H}^1(D;M)} := \int_D \left(|\zeta|^2 + |\underset{\sim}{\nabla}_M \zeta|^2 \right) \mathrm{d}\underset{\sim}{q} < \infty \right\}.$$

Then, (10) has the following weak formulation. Given $\hat{\psi}_0 := \psi_0 / \sqrt{M} \in \mathrm{L}^2(D)$, find $\hat{\psi} \in \mathrm{L}^\infty(0,T; \mathrm{L}^2(D)) \cap \mathrm{L}^2(0,T; \mathrm{H}^1_0(D; M))$ such that

$$\frac{\mathrm{d}}{\mathrm{d}t} \int_D \hat{\psi}\, \hat{\varphi}\, \mathrm{d}\underset{\sim}{q} - \int_D \underset{\approx}{\kappa}\, \underset{\sim}{q} \hat{\psi} \cdot \underset{\sim}{\nabla}_M \hat{\varphi}\, \mathrm{d}\underset{\sim}{q} + \frac{1}{2\mathrm{Wi}} \int_D \underset{\sim}{\nabla}_M \hat{\psi} \cdot \underset{\sim}{\nabla}_M \hat{\varphi}\, \mathrm{d}\underset{\sim}{q} = 0, \tag{13}$$

for all $\hat{\varphi} \in \mathrm{H}^1_0(D; M)$ in the sense of distributions on $(0,T)$, and $\hat{\psi}(\cdot, 0) = \hat{\psi}_0(\cdot)$. Notice that we solve for $\hat{\psi}$; ψ is recovered by setting $\psi := \sqrt{M}\, \hat{\psi}$. The Lozinski–Chauvière substitution introduced in Sect. 1 is identical to the substitution $\psi := \sqrt{M}\, \hat{\psi}$ in the case that $s = b/4$.

It is shown in Sect. 2 of [15] that $\mathrm{H}^1(D;M) = \mathrm{H}^1_0(D;M)$ and $\mathrm{H}^1_0(D) \subset \mathrm{H}^1_0(D;M)$.[3] The connection between $\mathrm{H}^1_0(D;M)$ and $\mathrm{H}^1_0(D)$ will prove helpful in the development of Galerkin methods for (13), since the construction of finite-dimensional subspaces of $\mathrm{H}^1_0(D)$ and the analysis of their approximation properties are well understood.

In [15], the following backward-Euler semidiscretisation of (13) was studied in detail: Let $N_T \geq 1$ be an integer, $\Delta t = T/N_T$, and $t^n = n\Delta t$, for $n = 0, 1, \ldots, N_T$. Discretising (13) in time using the backward Euler method yields the following semi-discrete numerical scheme.

Given $\hat{\psi}^0 := \hat{\psi}_0 = \psi_0/\sqrt{M} \in \mathrm{L}^2(D)$, find $\hat{\psi}^{n+1} \in \mathrm{H}^1_0(D;M)$, $n = 0, \ldots, N_T - 1$, such that

$$\int_D \frac{\hat{\psi}^{n+1} - \hat{\psi}^n}{\Delta t} \hat{\varphi} \, \mathrm{d}\underset{\sim}{q} - \int_D (\underset{\approx}{\kappa}^{n+1} \underset{\sim}{q} \, \hat{\psi}^{n+1}) \cdot \underset{\sim}{\nabla}_M \hat{\varphi} \, \mathrm{d}\underset{\sim}{q} + \frac{1}{2\mathrm{Wi}} \int_D \underset{\sim}{\nabla}_M \hat{\psi}^{n+1} \cdot \underset{\sim}{\nabla}_M \hat{\varphi} \, \mathrm{d}\underset{\sim}{q} = 0,$$

for all $\hat{\varphi} \in \mathrm{H}^1_0(D;M)$.

The following stability lemma for (14) was proved in Sect. 3 of [15].

Lemma 1. *Let* $\Delta t = T/N_T$, $N_T \geq 1$, $\underset{\approx}{\kappa} \in (\mathrm{C}[0,T])^{d\times d}$, $\hat{\psi}^0 \in \mathrm{L}^2(D)$, *and define* $c_0 := 1 + 4\mathrm{Wi}\, b \|\underset{\approx}{\kappa}\|^2_{\mathrm{L}^\infty(0,T)}$. *If* Δt *is such that* $0 < c_0 \Delta t \leq 1/2$, *then we have, for all* m *such that* $1 \leq m \leq N_T$,

$$\|\hat{\psi}^m\|^2 + \sum_{n=0}^{m-1} \Delta t \left\| \frac{\hat{\psi}^{n+1} - \hat{\psi}^n}{\sqrt{\Delta t}} \right\|^2 + \sum_{n=0}^{m-1} \frac{\Delta t}{2\mathrm{Wi}} \|\underset{\sim}{\nabla}_M \hat{\psi}^{n+1}\|^2 \leq \mathrm{e}^{2c_0 m \Delta t} \|\hat{\psi}^0\|^2.$$

Also, the existence and uniqueness of a weak solution of (13) was established in Theorem 3.2 of [15]. The proof makes use of the stability result in Lemma 1 in order to use compactness results for the bounded sequence of solutions to (14) as $\Delta t \to 0_+$.

2.2 Fully-Discrete Spectral Method

Let $\mathcal{P}_N(D)$ be a finite-dimensional subspace of $\mathrm{H}^1_0(D;M)$, to be chosen below, and let $\hat{\psi}^n_N \in \mathcal{P}_N(D)$ be the solution at time level n of our fully-discrete Galerkin method:

$$\int_D \frac{\hat{\psi}^{n+1}_N - \hat{\psi}^n_N}{\Delta t} \hat{\varphi} \, \mathrm{d}\underset{\sim}{q} - \int_D (\underset{\approx}{\kappa}^{n+1} \underset{\sim}{q} \, \hat{\psi}^{n+1}_N) \cdot \underset{\sim}{\nabla}_M \hat{\varphi} \, \mathrm{d}\underset{\sim}{q} + \frac{1}{2\mathrm{Wi}} \int_D \underset{\sim}{\nabla}_M \hat{\psi}^{n+1}_N \cdot \underset{\sim}{\nabla}_M \hat{\varphi} \, \mathrm{d}\underset{\sim}{q} = 0 \quad \forall \hat{\varphi} \in \mathcal{P}_N(D),\ n = 0, \ldots, N_T - 1, \tag{14}$$

$$\hat{\psi}^0_N(\cdot) := \text{the } \mathrm{L}^2(D) \text{ orthogonal projection of } \hat{\psi}_0(\cdot) = \hat{\psi}(\cdot, 0) \text{ onto } \mathcal{P}_N(D). \tag{15}$$

[3] In fact, these results hold for all FENE-like potentials, cf. Sect. 1.

The case $D \subset \mathbb{R}^2$ was considered in detail in [15]. Suppose we transform D into the rectangle $(r, \theta) \in R := (0,1) \times (0, 2\pi)$ using the polar coordinate transformation $\underset{\sim}{q} = (q_1, q_2) = (\sqrt{b}r\cos\theta, \sqrt{b}r\sin\theta)$. Also, suppose that $\hat{\psi} \in \mathrm{H}_0^1(D)$ and let $\tilde{\psi}(r,\theta) := \hat{\psi}(q_1, q_2)$. It was proved in Lemma 5.2 of [15] that $\tilde{\psi}$ can be written in polar coordinates as follows:

$$\tilde{\psi}(r,\theta) = \tilde{\psi}_1(r) + r\tilde{\psi}_2(r,\theta), \qquad (r,\theta) \in R = (0,1) \times (0, 2\pi). \tag{16}$$

Using the structure in (16), we defined in [15] the spectral basis $\mathcal{A}$ as $\mathcal{A} := \mathcal{A}_1 \cup \mathcal{A}_2$ where:

$$\begin{aligned}
\mathcal{A}_1 &:= \{(1-r)P_k(r) : k = 0, \ldots, N_r - 1\}, \\
\mathcal{A}_2 &:= \{r(1-r)P_k(r)\Phi_{il}(\theta) : k = 0, \ldots, N_r - 1;\ \ i = 0, 1;\ \ l = 1, \ldots, N_\theta\}.
\end{aligned}$$

P_k is a polynomial of degree k in $r \in [0,1]$ and $\Phi_{il}(\theta) = (1-i)\cos(2l\theta) + i\sin(2l\theta)$, $\theta \in [0, \pi]$. Notice that the polynomials in both $\mathcal{A}_1$ and $\mathcal{A}_2$ contain the factor $(1-r)$ in order to impose the homogeneous Dirichlet boundary condition on ∂D. Basis $\mathcal{A}$ is defined in order to mimic the decomposition (16) of the weak solution $\tilde{\psi}$ in polar coordinates: the role of span$(\mathcal{A}_1)$ is to approximate $\tilde{\psi}_1$ while span$(\mathcal{A}_2)$ is meant to approximate $r\tilde{\psi}_2$.

Now, let $\mathcal{P}_N(D)$ be span$(\mathcal{A})$ mapped from R to D. Approximation results were derived for this discrete space in Sect. 5 of [15], which enabled the derivation of the following optimal order spectral convergence estimate for the fully-discrete spectral method (14)–(15): for $\hat{\psi} \in \mathcal{H}^{k+1,l+1}(D)$ with $k, l \geq 1$ we have,

$$\begin{aligned}
&\|\hat{\psi} - \hat{\psi}_N\|_{\ell^\infty(0,T;\mathrm{L}^2(D))} + \|\underset{\sim}{\nabla}_M(\hat{\psi} - \hat{\psi}_N)\|_{\ell^2(0,T;\mathrm{L}^2(D))} \\
&\quad \leq C_1 N_r^{-k}\left(\|\hat{\psi}\|_{\ell^\infty(0,T;\mathcal{H}_r^k(D))} + \|\hat{\psi}\|_{\ell^2(0,T;\mathcal{H}_r^{k+1}(D))} + \left\|\frac{\partial\hat{\psi}}{\partial t}\right\|_{\mathrm{L}^2(0,T;\mathcal{H}_r^k(D))}\right) \\
&\quad + C_2 N_\theta^{-l}\left(\|\hat{\psi}\|_{\ell^\infty(0,T;\mathcal{H}_\theta^l(D))} + \|\hat{\psi}\|_{\ell^2(0,T;\mathcal{H}_\theta^{l+1}(D))} + \left\|\frac{\partial\hat{\psi}}{\partial t}\right\|_{\mathrm{L}^2(0,T;\mathcal{H}_\theta^l(D))}\right) \\
&\quad + C_3\Delta t\left\|\frac{\partial^2\hat{\psi}}{\partial t^2}\right\|_{\mathrm{L}^2(0,T;\mathrm{L}^2(D))},
\end{aligned} \tag{17}$$

(see Sect. 5 of [15] for definitions of the non-standard Sobolev spaces $\mathcal{H}^{k+1,l+1}(D)$, $\mathcal{H}_r^k(D)$ and $\mathcal{H}_\theta^l(D)$).

Note that we also considered a second basis, $\mathcal{B}$, in [15], proposed by Matsushima & Marcus [21] and Verkley [24], which satisfies the full pole condition on D (cf. [7]), and therefore the space defined by $\mathcal{B}$ is contained in $\mathrm{C}^\infty(\overline{D}) \cap \mathrm{C}_0(D)$. The numerical method based on $\mathcal{B}$ was found to be more efficient in practice than the one based on $\mathcal{A}$ for the FENE Fokker–Planck equation on D since $\hat{\psi}$ is typically very smooth. Finally, we considered a basis in [14] in the case of $d = 3$, referred to as basis $\mathcal{C}$, which, following [4], was defined as follows:

$$\mathcal{C} := \{Y_{lm}^{ik} : 0 \le k \le N_r - 1,\ i \in \{0,1\},\ l \in \{0,2,4,\ldots,N_{\mathrm{sph}}\} \text{ and } i \le m \le l\},$$

where $Y_{lm}^{ik}(r,\theta,\phi) := (1-r)Q_k(r)S_{l,m}^i(\theta,\phi)$, and the $S_{l,m}^i$ are spherical harmonics: $S_{l,m}^i(\theta,\phi) := C(l,m)\,P_l^m(\cos\phi)((1-i)\cos(m\theta) + i\,\sin(m\theta))$. Note that we showed in [14] that a splitting of the form (16) is not required in the case of $d = 3$.

A range of numerical results for spectral methods based on $\mathcal{A}$ and $\mathcal{B}$ in the case of $d = 2$ were presented in Sect. 7 of [15], and the convergence behaviour we obtained in practice was consistent with (17). The numerical method based on $\mathcal{C}$ is completely analogous, and it was shown in Sect. 2.6.3 of [13] that the convergence behaviour of this method in three dimensions is essentially the same as for methods $\mathcal{A}$ and $\mathcal{B}$ in two dimensions.

3 An Alternating-Direction Scheme for the Full Fokker–Planck Equation

In this section, we describe numerical methods for the Maxwellian-transformed Fokker–Planck equation posed on $\Omega \times D \times (0,T]$. Here we assume that $\underset{\sim}{u}$ is an *a priori* defined velocity field. Once the numerical scheme for the Fokker–Planck equation with a given $\underset{\sim}{u}$ is understood, it is straightforward to couple to the Navier–Stokes equations. These methods build upon the $\underset{\sim}{q}$-direction spectral method introduced in Sect. 2. In this case, the weak formulation is as follows: Given $\hat{\psi}_0 \in \mathrm{L}^2(\Omega \times D)$, find $\hat{\psi} \in \mathrm{L}^\infty(0,T;\mathrm{L}^2(\Omega \times D)) \cap \mathrm{L}^2(0,T;\mathcal{X})$ such that

$$\hat{\psi}(\underset{\sim}{x},\underset{\sim}{q},0) = \hat{\psi}_0(\underset{\sim}{x},\underset{\sim}{q}),\ (\underset{\sim}{x},\underset{\sim}{q}) \in \Omega \times D,$$

$$\frac{\mathrm{d}}{\mathrm{d}t}(\hat{\psi},\zeta) + \left(\underset{\sim}{u}\cdot\underset{\sim}{\nabla}_x\hat{\psi}\,,\,\zeta\right) - \left(\underset{\approx}{\kappa}\,\underset{\sim}{q}\hat{\psi}\,,\,\underset{\sim}{\nabla}_M\zeta\right) + \frac{1}{2\mathrm{Wi}}\left(\underset{\sim}{\nabla}_M\hat{\psi}\,,\,\underset{\sim}{\nabla}_M\zeta\right) = 0 \qquad \forall\,\zeta \in \mathcal{X},$$

in the sense of distributions on $(0,T)$, and again ψ is recovered by multiplying $\hat{\psi}$ by $\sqrt{M}$. Following Sect. 2, we impose a zero Dirichlet boundary condition on $\Omega \times \partial D$ for $t \in (0,T]$. See [14] for the hypotheses on $\underset{\sim}{u}$ and for the definition of the space $\mathcal{X}$.

The alternating-direction method under consideration here is nonstandard in the sense that we consider d-dimensional cross-sections (rather than one-dimensional cross-sections) of $\Omega \times D$. This leads to a formidable computational challenge because we typically need to solve a large number of problems posed in d spatial dimensions in each time-step. However, the method is extremely well suited to implementation on a parallel architecture since the $\underset{\sim}{q}$-direction solves are completely independent from one another, and similarly the $\underset{\sim}{x}$-direction solves are decoupled also. Our computational results in Sect. 4 were obtained using a parallel implementation of the alternating-direction methods described here.

3.1 The Alternating-Direction Methods

We now introduce the alternating-direction Galerkin methods for the weak formulation given above. These algorithms combine a classical Douglas–Dupont-type alternating-direction scheme [6] in the $\underset{\sim}{x}$-direction, with a new quadrature-based scheme in the $\underset{\sim}{q}$-direction.

First of all, define the bases

$$\{Y_k \in \mathcal{P}_N(D) : 1 \leq k \leq N_D\} \quad \text{and} \quad \{X_i \in V_h : 1 \leq i \leq N_\Omega\}, \tag{18}$$

such that $\text{span}(\{Y_k\}_{1 \leq k \leq N_D}) = \mathcal{P}_N(D)$ and $\text{span}(\{X_i\}_{1 \leq i \leq N_\Omega}) = V_h$, where V_h is an $\mathrm{H}^1(\Omega)$-conforming finite element space based on a mesh $\mathcal{T}_h$ of $\overline{\Omega}$. Let $\hat{\psi}_{h,N}$ denote our discrete solution, such that $\hat{\psi}_{h,N} \in V_h \otimes \mathcal{P}_N(D)$.

Also, we need to specify a quadrature rule on Ω. Let $\{(\underset{\sim}{x}_m, w_m), w_m > 0, \underset{\sim}{x}_m \in \overline{\Omega}, m = 1, \ldots, Q_\Omega\}$ define an element-based quadrature rule on $\mathcal{T}_h$, where the $\underset{\sim}{x}_m$ are the quadrature points and the w_m are the corresponding weights. Therefore, for functions $f, g \in \mathrm{C}^0(\overline{\Omega})$, the quadrature sum is evaluated element-wise as follows,

$$\sum_{m=1}^{Q_\Omega} w_m f(\underset{\sim}{x}_m) g(\underset{\sim}{x}_m) = \sum_{K \in \mathcal{T}_h} \sum_{l=1}^{Q_K} w_l^K f(\underset{\sim}{x}_l^K) g(\underset{\sim}{x}_l^K), \tag{19}$$

where Q_K is the number of quadrature points in element K. In [14], we introduced hypotheses on this quadrature rule that are necessary for our numerical analysis; we refer the reader to that paper for more details. The idea of using this quadrature rule in the context of the alternating-direction scheme is that by performing the $\underset{\sim}{q}$-direction solves at quadrature points $\underset{\sim}{x}_m$ we are able to recover a Galerkin formulation for the numerical method on $\Omega \times D$.

Noting that $\hat{\psi}_{h,N}$ can be written in terms of the coefficients $\{\hat{\psi}_{ik}\}$ as $\hat{\psi}_{h,N} := \sum_{i=1}^{N_\Omega} \sum_{k=1}^{N_D} \hat{\psi}_{ik} X_i Y_k \in V_h \otimes \mathcal{P}_N(D)$, we define the *line functions*, $\hat{\psi}_k$, for $k = 1, \ldots, N_D$, by $\hat{\psi}_k := \sum_{i=1}^{N_\Omega} \hat{\psi}_{ik} X_i \in V_h$. Then we have $\hat{\psi}_{h,N}(\underset{\sim}{x}, \underset{\sim}{q}) = \sum_{k=1}^{N_D} \hat{\psi}_k(\underset{\sim}{x}) Y_k(\underset{\sim}{q})$. These formulas shall be useful in the discussion of the alternating-direction methods below.

We now define two alternating-direction methods, referred to as method I and method II. The distinction between these schemes is that method I uses a semi-implicit spectral method in the $\underset{\sim}{q}$-direction (i.e. the term containing $\underset{\approx}{\kappa}$ is treated explicitly in time) whereas method II uses a fully-implicit temporal discretisation.

Method I: Semi-implicit scheme. Method I is initialised by computing the $\mathrm{L}^2(\Omega \times D)$ projection of the initial datum $\hat{\psi}_0 \in \mathrm{L}^2(\Omega \times D)$ onto $V_h \otimes \mathcal{P}_N(D)$, so that $\hat{\psi}_{h,N}^0 \in V_h \otimes \mathcal{P}_N(D)$ satisfies

$$\left(\hat{\psi}_0, \zeta\right) = \left(\hat{\psi}_{h,N}^0, \zeta\right) \qquad \text{for all } \zeta \in V_h \otimes \mathcal{P}_N(D). \tag{20}$$

Then, the alternating-direction method consists of two stages at each time-step: the $\underset{\sim}{q}$-direction stage and the $\underset{\sim}{x}$-direction stage. We begin with the $\underset{\sim}{q}$-direction stage, which essentially uses the Galerkin spectral method in D from Sect. 2.

Suppose $\hat{\psi}^n_{h,N} \in V_h \otimes \mathcal{P}_N(D)$. Then, in the $\underset{\sim}{q}$-direction stage we compute $\hat{\psi}^{n*}_{h,N}(\underset{\sim}{x}_m,\cdot) \in \mathcal{P}_N(D)$ for each $m = 1,\dots,Q_\Omega$ satisfying

$$\begin{aligned}
&\int_D \frac{\hat{\psi}^{n*}_{h,N}(\underset{\sim}{x}_m,\underset{\sim}{q}) - \hat{\psi}^{n}_{h,N}(\underset{\sim}{x}_m,\underset{\sim}{q})}{\Delta t} Y_l(\underset{\sim}{q})\, \mathrm{d}\underset{\sim}{q} \\
&\quad + \frac{1}{2\mathrm{Wi}} \int_D \underset{\sim}{\nabla}_M \hat{\psi}^{n*}_{h,N}(\underset{\sim}{x}_m,\underset{\sim}{q}) \cdot \underset{\sim}{\nabla}_M Y_l(\underset{\sim}{q})\, \mathrm{d}\underset{\sim}{q} \\
&\quad\quad = \int_D (\underset{\approx}{\kappa}^n(\underset{\sim}{x}_m)\, \underset{\sim}{q}\, \hat{\psi}^{n}_{h,N}(\underset{\sim}{x}_m,\underset{\sim}{q})) \cdot \underset{\sim}{\nabla}_M Y_l(\underset{\sim}{q})\, \mathrm{d}\underset{\sim}{q},
\end{aligned} \tag{21}$$

for $l = 1,\dots,N_D$. In order to separate out the $\underset{\sim}{x}$- and $\underset{\sim}{q}$-direction dependencies more clearly, we rewrite this equation in terms of line functions, i.e.:

$$\begin{aligned}
&\sum_{k=1}^{N_D} \hat{\psi}^{n*}_k(\underset{\sim}{x}_m) \left(\int_D Y_k(\underset{\sim}{q})\, Y_l(\underset{\sim}{q})\, \mathrm{d}\underset{\sim}{q} + \frac{\Delta t}{2\mathrm{Wi}} \int_D \underset{\sim}{\nabla}_M Y_k(\underset{\sim}{q}) \cdot \underset{\sim}{\nabla}_M Y_l(\underset{\sim}{q})\, \mathrm{d}\underset{\sim}{q} \right) \\
&= \sum_{k=1}^{N_D} \hat{\psi}^{n}_k(\underset{\sim}{x}_m) \left(\int_D Y_k(\underset{\sim}{q})\, Y_l(\underset{\sim}{q})\, \mathrm{d}\underset{\sim}{q} + \Delta t \int_D (\underset{\approx}{\kappa}^n(\underset{\sim}{x}_m)\, \underset{\sim}{q}\, Y_k(\underset{\sim}{q})) \cdot \underset{\sim}{\nabla}_M Y_l(\underset{\sim}{q})\, \mathrm{d}\underset{\sim}{q} \right),
\end{aligned} \tag{22}$$

for $l = 1,\dots,N_D$. This system is solved at each quadrature point $\underset{\sim}{x}_m$, $m = 1,\dots,Q_\Omega$, and the linear solves are completely independent from one another. This independence enables parallel computation to be used very effectively in this context.

The $\underset{\sim}{q}$-direction stage is complete once the values $\psi^{n*}_k(\underset{\sim}{x}_m)$, $k = 1,\dots,N_D$, $m = 1,\dots,Q_\Omega$, have been computed, and then we can begin solving in the $\underset{\sim}{x}$-direction. In the $\underset{\sim}{x}$-direction stage, we use a finite element discretisation of the transport equation, $\psi_t + \underset{\sim}{\nabla}_x \cdot (\underset{\sim}{u}(\underset{\sim}{x},t)\psi) = 0$, to update the output data from the $\underset{\sim}{q}$-direction stage. That is, for a given k, we find $\hat{\psi}^{n+1}_k \in V_h$, satisfying:

$$\int_\Omega \hat{\psi}^{n+1}_k X_i\, \mathrm{d}\underset{\sim}{x} + \Delta t \int_\Omega \left(\underset{\sim}{u}^{n+1} \cdot \underset{\sim}{\nabla}_x \hat{\psi}^{n+1}_k \right) X_i\, \mathrm{d}\underset{\sim}{x} = \sum_{m=1}^{Q_\Omega} w_m\, \hat{\psi}^{n*}_k(\underset{\sim}{x}_m)\, X_i(\underset{\sim}{x}_m), \tag{23}$$

for $i = 1,\dots,N_\Omega$, and, just as in the $\underset{\sim}{q}$-direction, these computations are decoupled from one another.

Once the $\underset{\sim}{x}$-direction computations are complete, we have the numerical solution at time level $n+1$: $\hat{\psi}^{n+1}_{h,N} = \sum_{k=1}^{N_D} \hat{\psi}^{n+1}_k Y_k \in V_h \otimes \mathcal{P}_N(D)$. Hence method I is defined by the initialisation (20), the $\underset{\sim}{q}$-direction spectral method (22) and the $\underset{\sim}{x}$-direction finite element method (23). In Lemma 3.2 of [14] we show that method I is equivalent to a one-step Galerkin formulation on $\Omega \times D$. This equivalent one-step formulation allows standard tools of numerical analysis to be applied to explore the stability and convergence properties of this method.

Method II: Fully-implicit scheme. Method II is very similar to method I, the sole difference being that the term containing $\underset{\approx}{\kappa}$ is now treated implicitly in time.

Using the line function notation of (22), the $\underset{\sim}{q}$-direction numerical method is defined as follows: Given the line functions $\hat{\psi}_k^n \in V_h$, $k = 1, \ldots, N_D$, determine the values $\hat{\psi}_k^{n*}(\underset{\sim}{x}_m)$ satisfying

$$\sum_{k=1}^{N_D} \hat{\psi}_k^{n*}(\underset{\sim}{x}_m) \Bigg(\int_D Y_k(\underset{\sim}{q})\, Y_l(\underset{\sim}{q})\, \mathrm{d}\underset{\sim}{q} + \frac{\Delta t}{2\mathrm{Wi}} \int_D \underset{\sim}{\nabla}_M Y_k(\underset{\sim}{q}) \cdot \underset{\sim}{\nabla}_M Y_l(\underset{\sim}{q})\, \mathrm{d}\underset{\sim}{q}$$
$$- \Delta t \int_D (\underset{\approx}{\kappa}^{n+1}(\underset{\sim}{x}_m)\, \underset{\sim}{q}\, Y_k(\underset{\sim}{q})) \cdot \underset{\sim}{\nabla}_M Y_l(\underset{\sim}{q})\, \mathrm{d}\underset{\sim}{q} \Bigg) = \sum_{k=1}^{N_D} \hat{\psi}_k^n(\underset{\sim}{x}_m) \int_D Y_k(\underset{\sim}{q})\, Y_l(\underset{\sim}{q})\, \mathrm{d}\underset{\sim}{q}, \tag{24}$$

for all $l = 1, \ldots, N_D$, and for each quadrature point $\underset{\sim}{x}_m$, $m = 1, \ldots, Q_\Omega$.

The initialisation and $\underset{\sim}{x}$-direction stages for method II are identical to those given for method I, hence we omit them here.

Clearly methods I and II are closely related to one another. Note, however, that from a practical point of view there is a trade-off in computational efficiency between the two methods because, on the one hand, method I requires less computation per time-step, since the matrix for the $\underset{\sim}{q}$-direction linear systems can be pre-assembled and LU-factorised only once since it is independent of $\underset{\approx}{\kappa}$, whereas the $\underset{\sim}{q}$-direction matrix for method II must be reassembled at each quadrature point. On the other hand, however, the fully implicit temporal discretisation used by method II tends to be more tolerant of large time-step sizes and coarse spatial discretisations than the semi-implicit scheme of method I, especially for larger flow rates and Weissenberg numbers (e.g. see Sect. 2.6.2 of [13]).

An important difference between methods I and II from the analytical point of view is that there is no equivalent one-step formulation available for method II. In [14], we proved stability and convergence results for method I based on its equivalent one-step formulation. That is, with some assumptions on the $\underset{\sim}{x}$-direction quadrature rule, we established stability results of the form of Lemma 1 for method I and, supposing that the set of shape functions for each element in $\mathcal{T}_h$ contains all polynomials of degree less than $s + 1$, we then proved the following error estimate for method I:

$$\|\hat{\psi} - \hat{\psi}_{h,N}\|_{\ell^\infty(0,T;\mathrm{L}^2(\Omega\times D))} + \|\underset{\sim}{\nabla}_M(\hat{\psi} - \hat{\psi}_{h,N})\|_{\ell^2(0,T;\mathrm{L}^2(\Omega\times D))}$$
$$\leq C_1 h^s \Bigg(\|\hat{\psi}\|_{\ell^\infty(0,T;\mathrm{H}^s(\Omega;\mathrm{L}^2(D)))} + \left\| \frac{\partial \hat{\psi}}{\partial t} \right\|_{\mathrm{L}^2(0,T;\mathrm{H}^s(\Omega;\mathrm{L}^2(D)))}$$
$$+ \left\| \hat{\psi} \right\|_{\ell^2(0,T;\mathrm{H}^s(\Omega;\mathrm{H}_0^1(D;M)))} + \left\| \hat{\psi} \right\|_{\ell^2(0,T;\mathrm{H}^{s+1}(\Omega;\mathrm{L}^2(D)))} \Bigg)$$

$$
\begin{aligned}
&+ C_2 N_r^{-k}\Big(\|\hat{\psi}\|_{\ell^\infty(0,T;\mathrm{L}^2(\Omega;\mathcal{H}_r^k(D)))} + \left\|\frac{\partial\hat{\psi}}{\partial t}\right\|_{\mathrm{L}^2(0,T;\mathrm{L}^2(\Omega;\mathcal{H}_r^k(D)))} \\
&\quad + \|\hat{\psi}\|_{\ell^2(0,T;\mathrm{H}^1(\Omega;\mathcal{H}_r^k(D)))} + \|\hat{\psi}\|_{\ell^2(0,T;\mathrm{L}^2(\Omega;\mathcal{H}_r^{k+1}(D)))}\Big) \\
&+ C_3 N_\theta^{-l}\Big(\|\hat{\psi}\|_{\ell^\infty(0,T;\mathrm{L}^2(\Omega;\mathcal{H}_\theta^l(D)))} + \left\|\frac{\partial\hat{\psi}}{\partial t}\right\|_{\mathrm{L}^2(0,T;\mathrm{L}^2(\Omega;\mathcal{H}_\theta^l(D)))} \\
&\quad + \|\hat{\psi}\|_{\ell^2(0,T;\mathrm{H}^1(\Omega;\mathcal{H}_\theta^l(D)))} + \left\|\hat{\psi}\right\|_{\ell^2(0,T;\mathrm{L}^2(\Omega;\mathcal{H}_\theta^{l+1}(D)))}\Big) \\
&+ C_4\Delta t\Big(\|\hat{\psi}\|_{\ell^2(0,T;\mathrm{L}^2(\Omega\times D))} + \|\hat{\psi}\|_{\mathrm{H}^2(0,T;\mathrm{L}^2(\Omega\times D))} + \|\nabla_{\underset{\sim}{x}}\nabla_M\hat{\psi}\|_{\ell^2(0,T;\mathrm{L}^2(\Omega\times D))} \\
&\quad + N_r^{-k}\|\hat{\psi}\|_{\ell^2(0,T;\mathrm{H}^1(\Omega;\mathcal{H}_r^{k+1}(D)))} + N_\theta^{-l}\|\hat{\psi}\|_{\ell^2(0,T;\mathrm{H}^1(\Omega;\mathcal{H}_\theta^{l+1}(D)))}\Big).
\end{aligned}
\tag{25}
$$

This error bound assumes that basis $\mathcal{A}$ is used for the $\underset{\sim}{q}$-direction spectral method; it would be straightforward (but laborious) to extend (25) to bases $\mathcal{B}$ or $\mathcal{C}$ introduced in Sect. 2.

We could not apply the same convergence argument to method II due to the absence of an equivalent one-step formulation; nevertheless, in Lemma 3.4 of [14], we proved the unconditional stability of method II.

4 The Micro–Macro Model

We now present some numerical results for a channel flow problem using a coupled algorithm for the Navier–Stokes–Fokker–Planck system (2)–(6) (see [14] for other computational results using the same approach, including a computation in the $d = 3$ case). We implemented the Navier–Stokes solver using a Taylor–Hood mixed finite element method [8] in the free `C++` finite element library `libMesh` [12]. We used a finite element space of continuous piecewise quadratic functions for V_h, and V_h was also used as the velocity space in the Taylor–Hood method, hence $\underset{\sim}{u}_h$, the finite element approximation to $\underset{\sim}{u}$, belongs to $(V_h)^d$. The alternating-direction method was implemented for parallel computation; the $\underset{\sim}{q}$-direction spectral method was implemented in `PETSc` [1] and `libMesh` was used for the $\underset{\sim}{x}$-direction finite element method (see [14] for more details of the implementation).

We considered a planar flow around a cylindrical obstacle in a channel. This is a standard benchmark problem in the polymer fluids literature (cf. Chap. 9 of [23]) and was also considered using deterministic multiscale methods by Chauvière & Lozinski in [4, 5, 19]. In the computation presented here, $\mathcal{T}_h$ contained 1505 triangular finite elements and $Q_\Omega = 9030$. For the $\underset{\sim}{q}$-direction spectral method we used basis $\mathcal{A}$. We imposed a parabolic inflow velocity profile for $\underset{\sim}{u}$ on the left boundary of Ω with $U_{\max} = 1$, a Neumann condition on the right boundary, a no-slip condition ($\underset{\sim}{u} = \underset{\sim}{0}$) for the obstacle and top boundary, and a symmetry condition on the bottom boundary. We used the parameters $b = 12$, $\gamma = 0.59$, $\mathrm{Re} = 1$ and we considered two choices of the Weissenberg number, (1) $\mathrm{Wi} = 1$ and (2) $\mathrm{Wi} = 3$.

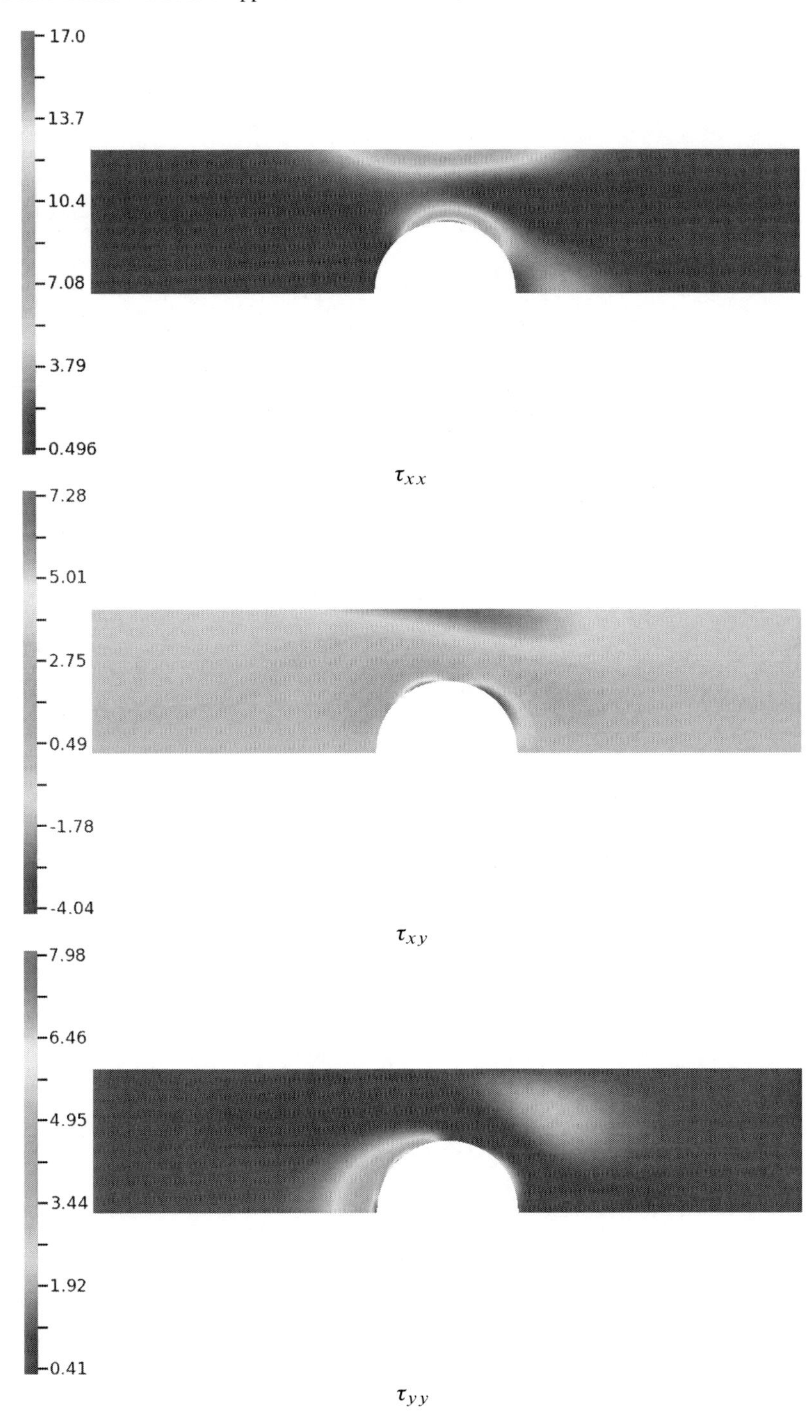

Fig. 1 The components of $\underset{\approx}{\tau}$ at $T = 5$ for the Wi $= 1$ case

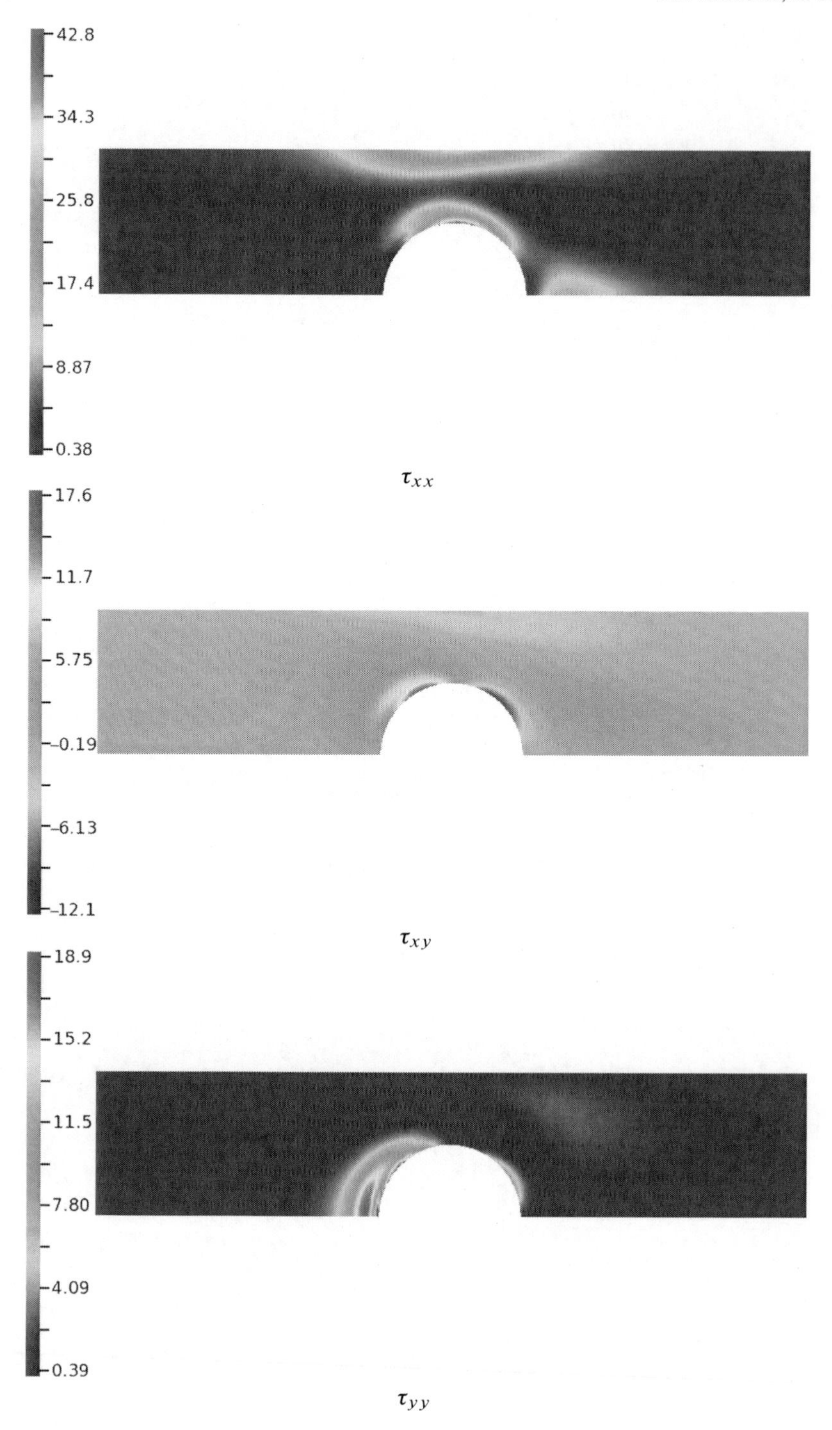

Fig. 2 The components of $\underset{\approx}{\tau}$ at $T = 5$ for the Wi $= 3$ case

Computational experimentation indicated that for both of these problems method II is significantly more efficient than method I (to the point where the semi-implicit method is computationally impractical), because, for the sake of stability, method I requires tighter restrictions on Δt and on the resolution of the discrete space $\mathcal{P}_N(D)$ (cf. Sect. 2.6.2 of [13]). Thus, we only present numerical results for the fully-implicit scheme here; for a detailed comparison of the two methods for a model problem with a milder velocity field, see Sect. 5.1 of [14].

We solved case (1) using method II with $(N_r, N_\theta) = (14, 14)$, so that $N_D = 406$ (recall that Q_Ω $\underset{\sim}{q}$-direction solves and N_D $\underset{\sim}{x}$-direction solves are performed in each time-step of the alternating-direction algorithm). More spectral modes were required to resolve the solution in case (2) due to the larger Weissenberg number and hence we used $(N_r, N_\theta) = (30, 30)$, i.e. $N_D = 1830$, in that case. We took 500 time-steps of size $\Delta t = 0.01$ and Figs. 1 and 2 show the components of $\underset{\approx}{\tau}$ at $T = 5$ in cases (1) and (2), respectively. These computations were performed on 80 processors of the Lonestar supercomputer at the Texas Advanced Computing Center (TACC), and took approximately 1.0 s per time-step in case (1) and 4.4 s per time-step in case (2) to perform.

5 Conclusions

We have summarised a range of results obtained in [14] and [15] for the analysis and implementation of numerical methods for solving the multiscale Navier–Stokes–Fokker–Planck system, which models the flow of dilute polymeric fluids. Most of our attention has been focused on the high-dimensional Fokker–Planck equation posed on the domain $\Omega \times D$ in $2d$ spatial dimensions. We developed an alternating-direction method for this equation that is efficient in practice and is also underpinned by rigorous numerical analysis.

We coupled this alternating-direction method to a mixed finite element method for the Navier–Stokes equations to obtain an algorithm for the coupled system (2)–(6). This algorithm was used to obtain computational results for a channel flow problem of physical interest. Parallel computation is particularly effective in the context of this problem because our alternating-direction solver for the high-dimensional Fokker–Planck equation is "embarrassingly parallel."

References

1. Balay, S., Buschelman, K., Eijkhout, V., Gropp, W. D., Kaushik, D., Knepley, M. G., McInnes, L. C., Smith, B. F., and Zhang, H. PETSc users manual. Tech. Rep. ANL-95/11 - Revision 2.1.5, Argonne National Laboratory, 2004.
2. Barrett, J. W., and Süli, E. Existence of global weak solutions to dumbbell models for dilute polymers with microscopic cut-off. *M3AS: Math. Model. Methods Appl. Sci. 18*, 6 (2008), 935–971.

3. Bird, R. B., Curtiss, C. F., Armstrong, R. C., and Hassager, O. *Dynamics of Polymeric Liquids, Volume 2, Kinetic Theory*, 2nd edn. Wiley, New York, 1987.
4. Chauvière, C., and Lozinski, A. Simulation of complex viscoelastic flows using Fokker–Planck equation: 3D FENE model. *J. Non-Newtonian Fluid Mech. 122* (2004), 201–214.
5. Chauvière, C., and Lozinski, A. Simulation of dilute polymer solutions using a Fokker–Planck equation. *Comput. Fluids 33* (2004), 687–696.
6. Douglas, J., and Dupont, T. Alternating-direction galerkin methods on rectangles. *Numerical Solution of Partial Differential Equations, II (SYNSPADE 1970)* (1971), 133–214.
7. Eisen, H., Heinrichs, W., and Witsch, K. Spectral collocation methods and polar coordinate singularities. *J. Comput. Phys. 96*, 2 (1991), 241–257.
8. Elman, H., Silvester, D., and Wathen, A. *Finite elements and fast iterative solvers*. Oxford Science Publications, Oxford, 2005.
9. Helzel, C., and Otto, F. Multiscale simulations of suspensions of rod-like molecules. *J. Comp. Phys. 216* (2006), 52–75.
10. Jourdain, B., Lelièvre, T., and Le Bris, C. Existence of solution for a micro-macro model of polymeric fluid: the FENE model. *J. Funct. Anal. 209*, 1 (2004), 162–193.
11. Keunings, R. On the Peterlin approximation for finitely extensible dumbbells. *J. Non-Newtonian Fluid Mech. 68* (1997), 85–100.
12. Kirk, B. S., Peterson, J. W., Stogner, R. M., and Carey, G. F. libmesh: A C++ library for parallel adaptive mesh refinement/coarsening simulations. *Eng. Comput. 23*, 3–4 (2006), 237–254.
13. Knezevic, D. J. Analysis and implementation of numerical methods for simulating dilute polymeric fluids. Ph.D. Thesis, University of Oxford, 2008. http://www.comlab.ox.ac.uk/people/David.Knezevic/.
14. Knezevic, D. J., and Süli, E. A heterogeneous alternating-direction method for a micro-macro model of dilute polymeric fluids. Submitted to M2AN, October 2008.
15. Knezevic, D. J., and Süli, E. Spectral galerkin approximation of Fokker–Planck equations with unbounded drift. *Accepted to M2AN, October 2008*. http://web.comlab.ox.ac.uk/people/Endre.Suli/biblio.html.
16. Kolmogorov, A. N. Über die analytischen Methoden in der Wahrscheinlichkeitsrechnung. *Math. Ann. 104* (1931).
17. Lielens, G., Halin, P., Jaumain, I., Keunings, R., and Legat, V. New closure approximations for the kinetic theory of finitely extensible dumbbells. *J. Non-Newtonian Fluid Mech. 76* (1998), 249–279.
18. Lozinski, A. Spectral methods for kinetic theory models of viscoelastic fluids. Ph.D. Thesis, École Polytechnique Fédérale de Lausanne, 2003.
19. Lozinski, A., and Chauvière, C. A fast solver for Fokker–Planck equation applied to viscoelastic flows calculation: 2D FENE model. *J. Comput. Phys. 189* (2003), 607–625.
20. Lozinski, A., Chauvière, C., Fang, J., and Owens, R. G. Fokker–Planck simulations of fast flows of melts and concentrated polymer solutions in complex geometries. *J. Rheology 47* (2003), 535–561.
21. Matsushima, T., and Marcus, P. S. A spectral method for polar coordinates. *J. Comput. Phys. 120* (1995), 365–374.
22. Öttinger, H. C. *Stochastic Processes in Polymeric Fluids*. Springer, Berlin, 1996.
23. Owens, R. G., and Phillips, T. N. *Computational Rheology*. Imperial College Press, London, 2002.
24. Verkley, W. T. M. A spectral model for two-dimensional incompressible fluid flow in a circular basin I. Mathematical formulation. *J. Comput. Phys. 136*, 1 (1997), 100–114.
25. Warner, H. R. Kinetic theory and rheology of dilute suspensions of finitely extendible dumbbells. *Ind. Eng. Chem. Fundamentals 11*, 3 (1972), 379–387.
26. Zhou, Q., and Akhavan, A. A comparison of FENE and FENE-P dumbbell and chain models in turbulent flow. *J. Non-Newtonian Fluid Mech. 109* (2003), 115–155.

Temperature Factor Effect on Separated Flow Features in Supersonic Gas Flow

V. Ya. Neyland, L.A. Sokolov, and V.V. Shvedchenko

Abstract The effect of the temperature factor (body temperature ratio to the stagnation temperature of external flow) on the separated flow features has been investigated in the supersonic gas flow near the concave angle. The strong effect of the temperature factor on the separated zone length and on the corresponding aerodynamic performances was revealed. It was shown that, if the angle is big enough, such flow cannot be described by free interaction theory, i.e. by triple deck theory.

Nomenclature

U	Velocity
p	Pressure
ρ	Density
T	Temperature
H	Total enthalpy
M	Mach number
Re	Reynolds number
Pr	Prandtl number
δ	Boundary layer thickness
ℓ	Boundary layer length
μ	Coefficient of viscosity
ω	Power in viscosity law
g_w	Temperature factor
γ	Specific heat ratio
x	Longitudinal coordinate
y	Normal coordinate
Δx	Separation zone length
θ	Flare angle

V.Y. Neyland (✉)
TsAGI, Zhukovsky, Moscow region, 140180, Russia, E-mail: Neyland@tsagi.ru

A.F. Hegarty et al. (eds.), *BAIL 2008 - Boundary and Interior Layers.* Lecture Notes in Computational Science and Engineering,
DOI: 10.1007/978-3-642-00605-0,

1 Introduction

The investigation of separated flow in viscous supersonic flow near a flat plate caused by the rear part of the flat plate deviating by an angle θ is an important task in the development of separated flow theory. It is also significant for the applications when temperature factor becomes small in the flight with high supersonic speed. It is important that, when testing models in the wind tunnel, the temperature factor may considerably differ from its flight value (Table 1). It can lead to the considerable deviations of the aerodynamic performances and heat fluxes in the wind tunnel from those in the actual flight. So far, the flow over the flat plate with deviated rear part was investigated in many theoretical and experimental works. A review of the results is given in [1–4].

During the first decades, theoretical investigations of these flows were divided into two directions. For the developed separated flows included the pressure "plateau" zone method with the criteria of Chapman–Korst [5, 6] was used. Later it was shown [7] that the criteria method of Chapman (for laminar flows) corresponds to the first approach of the strict asymptotic theory for Navier–Stokes equation. For small separated zones and zones of incipient separation without developed pressure "plateau" area, another approach based on the integral equations of the boundary layer was more appropriate.

After the development of free interaction theory [8–11] (usually called "triple deck" outside of Russia) the other multilayer solutions were obtained [7, 12].

Within asymptotic theory, the calculation of the flow near "compression corner" with an angle of $\theta \sim \mathrm{Re}^{-1/4}$ was performed by many authors [11–14]. Recently the author of [13] assumed that solution of this task within free interaction theory only exists up to some critical value of $\theta/\mathrm{Re}^{-1/4}$. Later similar calculations were performed in [14] more carefully and the authors showed that the conclusions of [13] were caused by a wrong calculation method. But they referred to the asymptotic reattachment theory developed in [7] which does not contain singularities. It should be noted that the applicability of the free interaction theory is a complicated matter, although the criticism of the numerical results of [13] by authors of [14] is, may be, correct.

Table 1 Temperature factor in flight ($T_w \sim 1{,}000$ K) as compared with that in wind tunnel

H, km	M = 10	15	20	25
40	0.193	0.0858	0.048	–
50	0.182	0.0807	0.046	–
60	0.196	0.0807	0.045	–
70	0.227	0.101	0.057	0.036
	M		T_0	g_w
	3–5		750	0.4
	6–10		1,075	0.279
	10, 12, 14, 18		2,600	0.115

The present article is based on two methods: qualitative analytical investigation of flow physical features and numerical investigation of the Navier–Stokes equations. The investigations assume that the flow is laminar everywhere.

2 Analytical Investigation

Let us consider supersonic flow over a flat plate at zero angle of attack. The rear part of the plate is deflected by an angle θ (Fig. 1). The angle θ, Mach number M and Reynolds number Re are so that the separated zone appears upstream of the angle.

Let us first consider the small separated zones and zones of arising separation. For this purposes it is convenient to use the method described in the monographs [15,16]. This approach was used in [8,9] for the free interaction theory development which later was proposed in [10] under the name of triple deck and using a slightly different way.

Thus, let us consider flow in a small vicinity of the separation point of the boundary layer (Fig. 1). Let a small pressure difference $\Delta p/p \ll 1$ be applied to the flow. In the major part of the boundary layer, where the longitudinal component of the velocity U is of the same order as the outer flow velocity Ue, we can use equation of the longitudinal momentum, state equation and relation $\rho_e u_e^2 \sim p$ to obtain

$$\rho U U_x \sim p_x, \quad \rho_e U_e \Delta U \sim \Delta p, \quad \frac{\Delta\rho}{\rho} \sim \frac{\Delta p}{p}.$$

Then, in this part of the boundary layer (area 2 in Fig. 1), because of the continuity equation, the disturbed streamline thickness assessment has the following form

$$\frac{\Delta\delta}{\delta_0} \sim \frac{\Delta p}{p},$$

where δ_0 is the typical value of boundary layer thickness upstream of the interaction area.

Near the wall, because of the boundary condition, in the undisturbed boundary layer there is always area 3 where the dynamic pressure will be of order Δp. It is

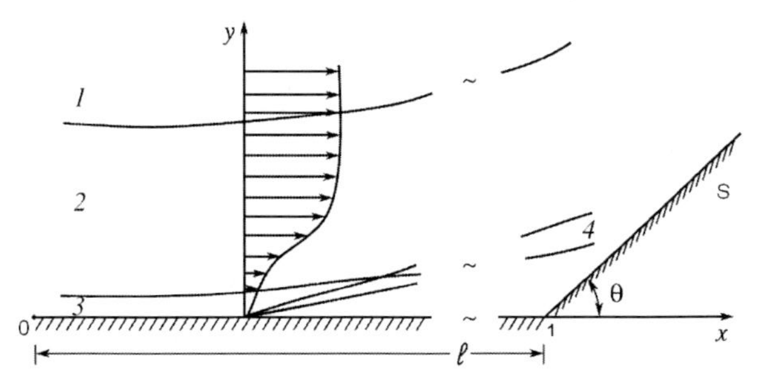

Fig. 1 Flow scheme

only true near the separation point, because far from it the Δp can be balanced by viscous forces and the task becomes linear.

Thus, for area 3 we obtain the assessment

$$\rho_3 U_3^2 \sim \Delta p. \tag{1}$$

In area 3, using the relationship for the velocity profile in the undisturbed boundary layer, we can obtain assessment

$$\frac{U_3}{U_e} \sim \frac{\delta_3}{\delta_0}. \tag{2}$$

In area 3 the flow upstream of the separation point is performed against inadvertent pressure difference $\Delta p > 0$ by viscous forces, i.e.

$$\frac{\rho U_3^2}{\Delta x} \sim \mu_3 \frac{U_3}{\delta_3^2}. \tag{3}$$

Due to (1) the thickness of area 3 varies as the undisturbed value, then (1) and (2) result in

$$\frac{\delta_3}{\delta_0} \sim \frac{U_3}{U_e} \sim \sqrt{\frac{\Delta p}{p}} \gg \frac{\Delta p}{p} \sim \frac{\Delta \delta_2}{\delta_0}.$$

Thus, in the first approach, the total variation of the boundary layer momentum thickness is produced by area 3.

This fact together with linear theory of supersonic flows (Ackeret formula) leads to the last estimate

$$(\mathrm{M}_e^2 - 1)^{1/2} \Delta p \sim \rho_e U_e^2 \frac{\delta_3}{\Delta x}. \tag{4}$$

To determine scales of the disturbed values Δx, Δp, U_3, δ_3 we get four equations (1)–(4). They give us estimates of all the required values

$$\Delta x \sim \ell \mathrm{Re}^{-3/8}, \quad \frac{\Delta p}{p} \sim \mathrm{Re}^{-1/4}, \quad \frac{U_3}{U_e} \sim \frac{\delta_3}{\delta_0} \sim \mathrm{Re}^{-1/8}. \tag{5}$$

The angle θ which produces the separation of the boundary layer has the order of the value $\mathrm{Re}^{-1/4}$ (ℓ is the boundary layer length upstream of the separation point).

Using the estimates (5) we can develop asymptotic theory of Navier–Stokes solution for the small separation zones at $\mathrm{Re} \to \infty$.

Let us consider this task for the flows with high supersonic speeds and small temperature factor using limit

$$\mathrm{Re} \to \infty, \ \mathrm{M} \to \infty, \ g_w = \frac{H_w}{H_e} \to 0,$$

where H is total enthalpy, indexes e and w correspond to the parameter values at the outer boundary of the boundary layer and the wall respectively.

Let us assume, that the interaction of the non-viscous flow with the boundary layer is small up to the separation point. Then

$$\delta_0/\ell \sim \left(\frac{\mu_0}{\rho_0 U_e \ell}\right)^{1/2}, \ \mathrm{M}_e \delta_0/\ell \ll 1.$$

In the major part of the boundary layer (area 2 in Fig. 1) the gas temperature will be of the order of the total temperature T_0 [17], $\rho_2 \sim \rho_0 \sim \rho_e/M_e^2$, $\mu_2 \sim \mu_0$, where ρ_0, μ_0 are the density and viscosity at $T = T_0$.

Friction and heat flux to the wall maintain their orders of the value in the whole boundary layer, i.e.

$$\left(\mu\frac{dU}{dY}\right)_{y\to 0} \sim \mu_0\frac{U_e}{\delta_0}; \quad \left(\mu\frac{dg}{dy}\right)_{y\to 0} \to \mu_0\frac{1}{\delta_0}.$$

Then the velocity and enthalpy profiles near the body surface will be (neglecting the inessential constants):

$$g \sim \left(g_w^{\omega+1} + \frac{y}{\delta_0}\right)^{1/(\omega+1)}, \ \frac{U}{U_e} \sim \left(g_w^{\omega+1} + \frac{y}{\delta_0}\right)^{1/(\omega+1)} - g_w. \tag{6}$$

Depending on the relationship of g_w and the disturbed pressure amplitude $\Delta p/p$, using (6) we can get profiles in the area 3

$$g_w^{\omega+1} \gg \frac{\delta_3}{\delta_0} \qquad g_3 \sim g_w, \ \frac{U_3}{U_e} \sim \frac{1}{g_w^{\omega}}\frac{\delta_3}{\delta_o} \tag{7}$$

$$g_w^{\omega+1} \ll \frac{\delta_3}{\delta_0} \qquad g_3 \sim \left(\frac{\delta_3}{\delta_0}\right)^{1/(\omega+1)}, \ \frac{U_3}{U_e} \sim \left(\frac{\delta_3}{\delta_0}\right)^{1/(\omega+1)}. \tag{8}$$

Let us consider regime (7). Near the separation point

$$\Delta p \sim \rho_3 U_3^2, \quad \frac{\delta_3}{\delta_0} \sim g_w^{(1+2\omega)/2} \cdot \left(\frac{\Delta p}{p}\right)^{1/2}, \ \frac{U_3}{U_e} \sim g_w^{1/2}\left(\frac{\Delta p}{p}\right)^{1/2}. \tag{9}$$

The thickness of the area with non-linear disturbances (area 3, $\Delta\delta_3 \sim \delta_3$) will be of a greater order than that of the area 2 $\Delta\delta_2$ if the following condition is valid

$$g_w^{(1+2\omega)/2}\sqrt{\frac{\Delta p}{p}} \gg \frac{\Delta p}{p} \to \frac{\Delta p}{p} \ll g_w^{1+2\omega}. \tag{10}$$

Then, using Ackeret formula (4) for the area 1 (the disturbed part of the external non-viscous flow) we get an estimate for the length of the disturbed flow Δx

$$\frac{\Delta x}{\ell} \sim \frac{\mathrm{M}_e \delta_0}{\ell} \cdot \sqrt{\frac{g_w^{1+2\omega}}{\Delta p/p}}. \tag{11}$$

It shows that as g_w decreases, $\Delta x/\ell$ decreases also.

Let us obtain an assessment for the critical pressure differential assuming, as usual, that the viscous and inertial members are of the same order in the Navier–Stokes equations.

$$\frac{\rho_3 U_3^2}{\Delta x} \sim \mu_3 \frac{U_3}{\delta_3^2}; \quad \frac{\Delta p}{p} \sim \left(\mathrm{M}_e \frac{\delta_0}{\ell}\right)^{1/2}. \tag{12}$$

Thus, when decreasing g_w at fixed deflection angle of the plate rear part, the separation zone length decreases (11).

Further decrease of g_w violates conditions (7) and (10). Now, let us assume that

$$\frac{\delta_3}{\delta_0} \sim \frac{\Delta\delta_2}{\delta_0} \sim \frac{\Delta p}{p} \sim g_w^{1+2\omega}. \tag{13}$$

Then, using Ackeret formula in the form of $\Delta p/p \sim \mathrm{M}_e\delta_2/\Delta x$ we get assessment for the action Δx

$$\Delta x \sim \mathrm{M}_e\delta_0. \tag{14}$$

The relations (9) are true as long as area 3 remains almost isothermal ($g_w^{\omega+1} \gg \delta_3/\delta_0$).

Having assessment (13), let us estimate the critical pressure differential at this regime, using the first estimate of (12)

$$\frac{\Delta p}{p} \sim \left(\mathrm{M}_e \frac{\delta_0}{\ell}\right)^{2/3} g_w^{-2(\omega+1)/3}. \tag{15}$$

The estimate shows that at fixed length of the disturbed area (14) decrease of g_w leads to the increase of the critical pressure difference (15). It means that at fixed angle of deflection of plate rear part the separated area length will also decrease.

And finally, if the isothermal condition of area 3 (7) is violated, estimate (14) for Δx will remain because $\Delta\delta_2 \gg \delta_3$. The estimate for g_3, U_3/U_e, and δ_3/δ_0 will have the form of (8). Then for the critical value of the pressure difference which produces separation initiation in the area 3 the following assessment is obtained, using equations (1) and (3)

$$\frac{\Delta p}{p} \sim (\mathrm{M}_e\delta_0/\ell)^{1/(2\omega+1)}.$$

This estimate is true for all small g_w when equation (8) is valid.

Now, let us investigate the effect of g_w on the separation zone length at a slightly higher value of θ when the zone appears with almost constant pressure but mixing layer at the outer boundary of the separated zone is still much thinner than the boundary layer separated from the body surface. To get the necessary assessment let us assume that the Korst–Chapman condition [5] or asymptotic attachment theory [7] is true.

In this case the following relationship must be valid $\theta_{inc} \ll \theta \ll 1$. The upper limitation provides fulfilment of the condition $\delta_4 \ll \delta_0$, where θ_{inc} is the angle at which mixing layer 4 leaves the body after separation.

At $\theta_{inc} \sim \theta$ we come back to the free interaction theory considered above (for example, at $\theta \sim \mathrm{Re}^{-1/4}$ and $g_w \sim 1$ the separation zone length Δx is determined by (5) according to free interaction theory (triple deck)).

At $\theta \gg \theta_{inc}$ friction forces acting on the gas along dividing streamline lead to the dynamic pressure increase thus providing the possibility to counteract pressure rise in the attachment area. The Korst–Chapman condition [5, 6] may be written as

$$\rho_4 U_4^2 \sim \Delta p, \quad \frac{\Delta p}{p} \sim \mathrm{M}_e \theta. \tag{16}$$

Here, index 4 designates parameters value at dividing streamline in the mixing layer. As $\delta_4 \ll \delta_0$ i.e. the separated zone is short, $\Delta x/L \ll 1$, so $\rho_4 \sim \rho_w$. Further, we must estimate the rate of $U_4(\Delta x)$ increase.

In the mixing layer 4 the acceleration occurs due to longitudinal momentum transfer when friction forces act on streamlines of the separated boundary layer. Thus, we can write down the following conditions

$$\frac{U_4}{\delta_4} \sim \frac{U_e}{g_w^{\omega}\delta_0}; \quad \frac{\rho_4 U_4^2}{\Delta x} \sim \mu_4 \frac{U_4}{\delta_4^2}. \tag{17}$$

In (17) the first condition corresponds to the conservation of friction stress value in area 4 to its value in the separated boundary layer, where $U \sim U_e$, thickness δ_0, and $\mu_4/\mu_0 \sim g_w^{\omega}$. The second condition in (17) is balance of orders of value of viscous and inertial members in the longitudinal momentum equation.

Resolving (17) we obtain the estimates

$$\frac{\delta_4}{\delta_0} \sim g_w^{(1+2\omega)/3} \left(\frac{\Delta x}{\ell}\right)^{1/3}, \quad \frac{U_4}{U_e} \sim g_w^{(1+2\omega)/3} \left(\frac{\Delta x}{\ell}\right)^{1/3}. \tag{18}$$

Here, ℓ is boundary layer length up to the separation point, while Δx is the mixing zone length from the separation point to the attachment point. At $g_w \sim 1$ (18) corresponds to the known selfsimilar solution of Prandtl equation for the mixing layer between external flow with shear profile and stagnation zone. Now, using condition in the attachment zone (16) we get dependence of the separation zone length on θ and g_w

$$\frac{\Delta x}{\ell} \sim (\mathrm{M}_e \theta)^{3/2} \cdot g_w^{(1+2\omega)/2}.$$

Thus, the separation zone length decreases with g_w decreasing in this regime also.

3 Numerical Investigation

Numerical investigation was performed with the use of computer codes packet of numerical integration of Navier–Stokes equations by time-dependent method developed in TSAGI [18–20].

The initial boundary-value problem was solved by the integro-interpolation method (finite volume method). Implicit monotonic scheme of the type of Godunov [21] scheme and the approximate method of Roe [22] of solving the Riemann problem on break-up of arbitrary discontinuity were used in the approximation of the convection component of flow vectors in half-integer nodes. The principle of minimal derivates [23] was used for raising the order of approximation to second one in the case of interpolation of dependent variables to the face of elementary cell. A difference scheme of the type of central differences of the second order of accuracy was used in the approximation of the diffusion component of flow vectors on the face of elementary cell. The modified Newton–Raphson method was used for solving the nonlinear finite-difference equations. The set of linear algebraic equations was solved using the GMRES(k) method of minimal residuals [24].

The flow field near the two-dimensional compression corner protruding into the supersonic flow (Fig. 1) has been calculated at following parameters: Reynolds number based on a plate length up to the corner point $\mathrm{Re} = 10^6$, Mach number $\mathrm{M} = 5$, Prandtl number $\mathrm{Pr} = 2/3$, temperature factor $g_w = 10^{-3}/1$, specific heat ratio $\gamma = 5/3$, viscosity law $\mu \sim T^{\omega}(\omega = 0.5)$.

In the investigated area the coordinate origin coincides with the beginning of the non-deflected part of the plate, deviation point is located at $x = 1$, the end of the investigated area is located at $x = 5$.

At the left boundary the undisturbed flow was chosen. Upper boundary of the computed area was chosen so that the boundary conditions were also undisturbed external flow. Right boundary of the computed area was chosen so that error in soft boundary conditions did not effect on the solution in the vicinity of the separation zone. The condition of no-slip were chosen at the body surface. The special grid thickening at the plate beginning was performed to correctly follow the abrupt pressure gradient at the leading edge. It should be noted that errors at the leading edge does not effect on the solution downstream and dissipate quickly with distance from leading edge if the separation zone is not located near the leading edge.

Following the method of analytical grid development [20] the grid thickening near the body surface was performed with line number about 20–40% of the total number in the direction of the boundary layer thickness. It allowed one ensure high resolution of boundary layer near the body surface. This method of analytical grid is appropriate for small $\theta < 10°$ when the separation zone dimensions are small. It allows one cover the separation zone with a grid of necessary density and to simulate actual flow pattern.

The grid resolution in the area of abrupt pressure gradient in the attachment zone also strongly effects the quantity of the obtained results. The additional grid thickening in this area both in longitudinal and transversal coordinates is required. In the rest of the computation area the grid is quasi-uniform. The grid resolution in the separation area does not strongly effect the flow pattern if separation does not start from the leading edge.

For angles $\theta = 10°, 20°$ the separation zone dimensions are much more than the boundary layer and the mixing layer thickness. In this case the method of analytical grid is not working. The issue may be in development of adaptive grids [25]. This method allows not only to get correct flow pattern but also to considerably decrease the required number of lines (for $\theta = 10°$). Sometimes ($\theta = 20°$) it is the only possible method to get solution.

With the task features in mind, the grids were used having one-dimension adaptation in the direction normal to the surface constructed by "equidistribution" method [26]. It allows one to fine solve the mixing layer and to check the solution on grids with different number of nodes. If the resolution and adaptation were correctly chosen the mixing layer position practically does not change when the number of grid nodes varies by 2–4 times. To get a final solution the adaptive iterations on the grids with small number of nodes were repeated many times until convergence was obtained. Then the solution was checked with number of grid nodes variation by 2–4 times. The solutions obtained with the use of analytical grids were unstable when the number of nodes increased. The maximum dimensions of the grids were $1{,}600 \times 200$ for $\theta = 10°, 20°$ and 800×200 for $\theta < 10°$. At this, the solutions were checked for convergence on different grids.

For big angles $\theta = 10°, 20°$ the grid resolution does not effect the mixing layer position but considerably changes flow pattern inside the separated zone.

4 Results

Figure 2 shows the pressure distribution along the x-axis on the corner surface. Pressure is normalized to $\rho_\infty U_\infty^2$ ($p_\infty/\rho_\infty U_\infty^2 = 1/\gamma \mathrm{M}^2 = 0.024$). On the plate surface ($\theta = 0$) at $x \sim 1$ the pressure becomes constant increasing as the temperature factor increases. For $\theta = 2.5°$ full attached flow occurs. For larger angles the separated flow occurs with the separated zone length increasing as both θ and temperature factor increase. When temperature factor increases the separation point moves to the left while the attachment point moves to the right. Temperature factor increase causes small pressure rise in the separated zone and increases pressure steps smoothness.

At $\theta = 2.5°$ separation starts near the leading edge, where the flow parameters vary considerably along the x-axis. The pressure steps smoothness decreases. Each value of the temperature factor corresponds to a certain value of pressure in the separation zone and to a certain x-coordinate of reaching maximum pressure. Inside the separation zone there were observed pressure oscillations caused by the development of vortices.

Figure 3 gives x-coordinates of the separation and attachment points where friction becomes zero. For small angles variation of full separation length Lx is caused (in equal proportions) by variation of its components Lx_1 and Lx_2 (upstream and

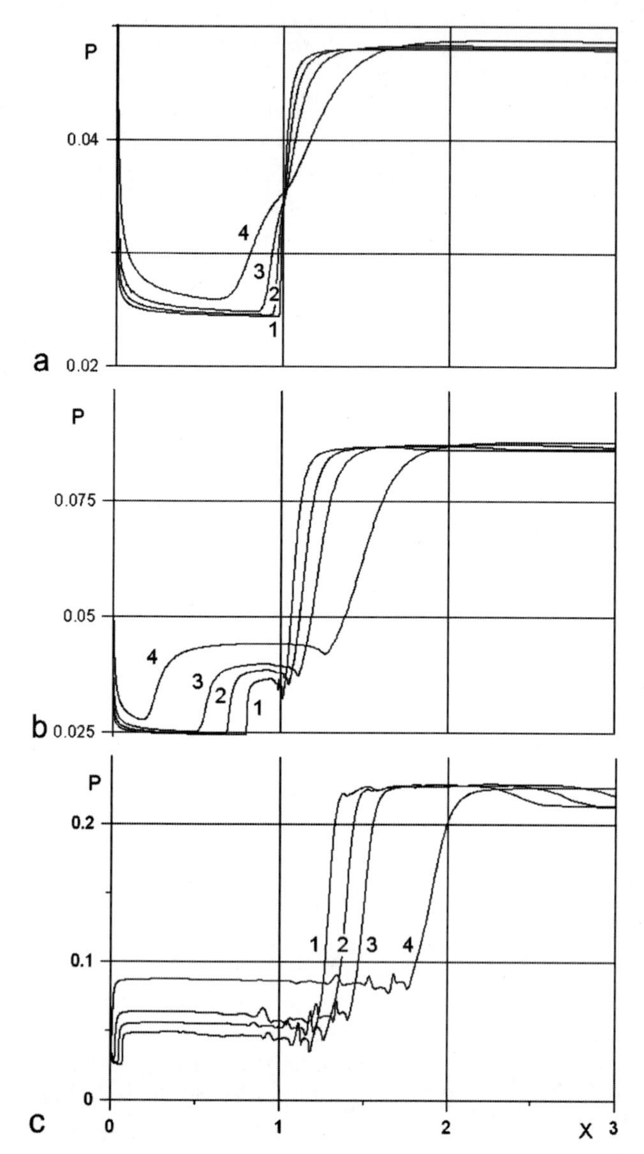

Fig. 2 Pressure distribution along x-coordinate of the corner $\theta^{\circ} = 5, 10, 20$ (**a, b, c**) at temperature factor $g_w = 0.001, 0.1, 0.3, 1$ (curves 1–4)

downstream of the corner correspondingly). For an angle $\theta = 20^{\circ}$ $Lx_1 \sim 1$ and major variation of Lx is caused by Lx_2 component, i.e. by considerable displacement of the attachment point. For θ values increase of temperature factor leads to the separation zone length increase.

For angles $\theta = 10, 15, 20^{\circ}$ the vortices were observed inside the separation zone (Fig. 4b–d) similar as to "bubble" in the work [27]. For angles $\theta = 5^{\circ}, 7.5^{\circ}$ (Fig. 4a) vortices were not discovered.

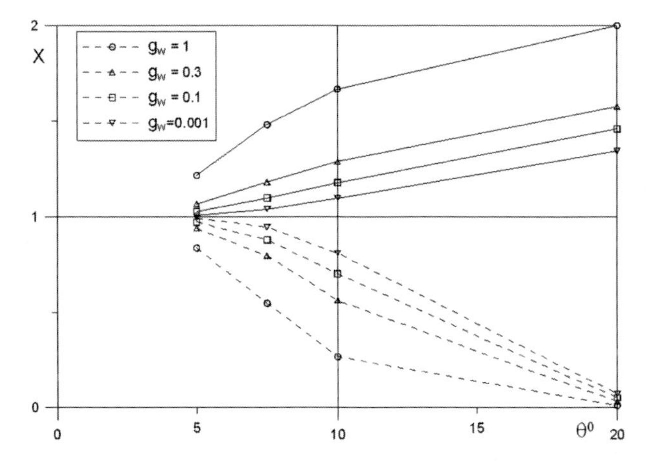

Fig. 3 Variation x-coordinates of separation (*dash line*) and attachment points (*solid line*) depending on angle θ and temperature factor $g_w = 10^{-3}, 0.1, 0.3, 1$

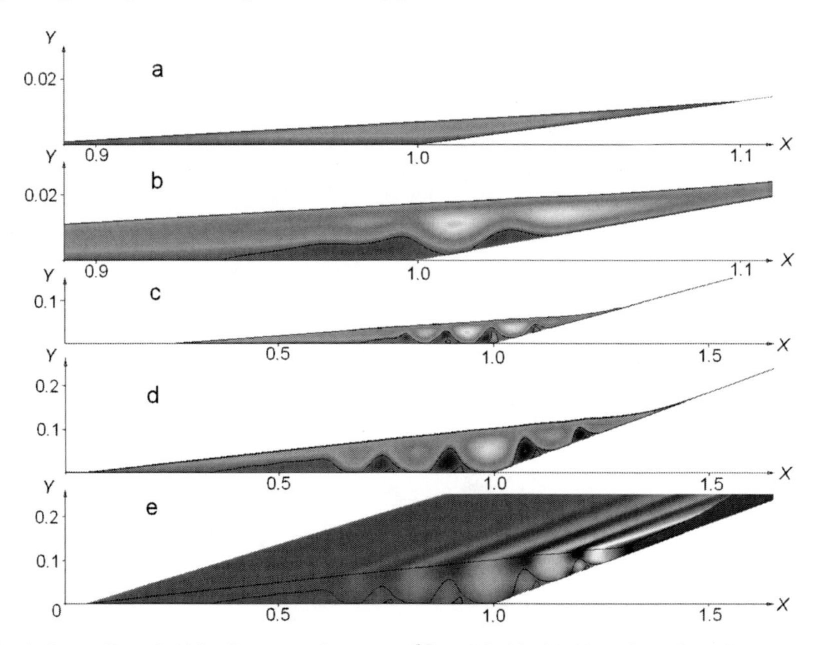

Fig. 4 Streamlines field for flow near the corner $\theta^\circ = 7.5, 10, 15, 20$ (**a**, **b**, **c**, **d**) and pressure (**e**) field $\theta^\circ = 20$ with temperature factor $g_w = 0.1$ (*solid line* corresponds to the zero streamline)

Investigation showed that, when the grid resolution was high enough, the large vortices sizes did not change. We should remark, that for $g_w = 1$ separation zone is large, thus, the grid step in the x-direction is 1.5–2 times greater than that for $g_w = 10^{-3}$–10^{-1}. That is why for $g_w = 1$ the vortices were only specified in details at a grid of 1,600 × 200 nodes. The grid of 800 × 200 did not give vortices details while at $g_w = 10^{-3} \cdots 10^{-1}$ vortices were seen quite clearly.

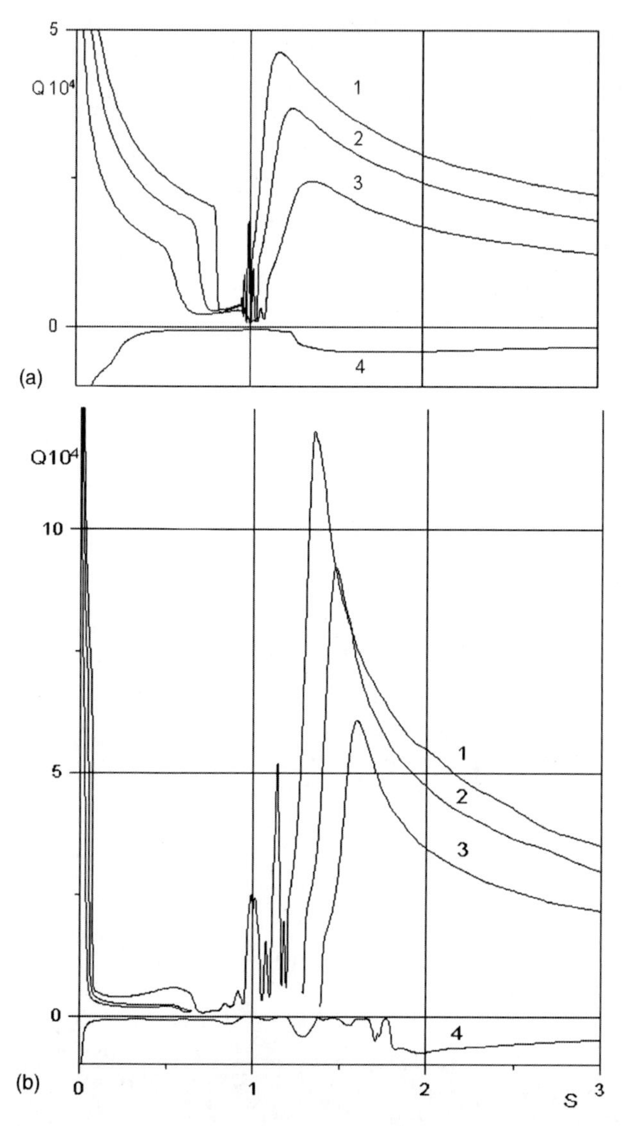

Fig. 5 Heat flux distribution along corner coordinate S for $\theta = \mathbf{10}$, 20 (**a**, **b**) at temperature factor $g_w = 0.001, 0.1, 0.3, 1$ (curves 1–4)

It is possible, that vortices development is caused by the separation of a viscous sublayer at the bottom of locally inviscid jet that flows out of the attachment zone of the main separation zone. In this case pressure across the separation zone becomes variable, i.e. $\partial p/\partial y \neq 0$ (Fig. 4b).

It is very interesting to investigate the effect of the temperature factor on the heat flux in the attachment zone (Fig. 5). The heat flux is normalized by $\rho_\infty U_\infty^3$. For all angles, when the temperature factor decreases maximum heat flux reaches its limit

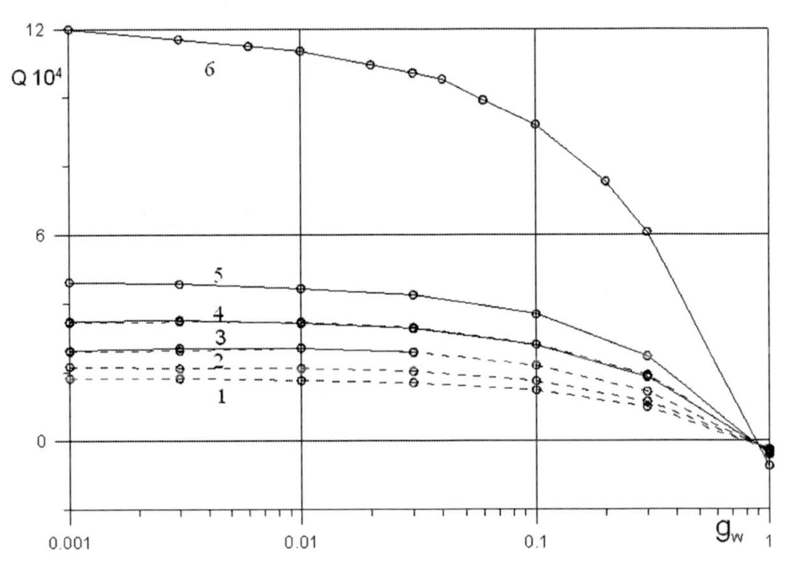

Fig. 6 Variation of maximum heat flux in the attachment zone depending on temperature factor g_w for angles $\theta° = \mathbf{0}$, 2.5, 5, 7.5, 10, 20 (curves 1–6). *Solid line* corresponds to the adaptive grid, *dash line* –to the analytical grid

value (Fig. 6). For angles $\theta = 5°$, $\theta = 7.5°$ results are shown obtained both on analytical and adaptive grids, while for angles $\theta = 10, 20$ - obtained on the adaptive grids only. For angles $\theta = 10,\ 20°$ (Fig. 5) there are heat flux splashes inside the separation zone (up to 40% of its maximum value at the attachment point) caused by vortices. For $\theta = 5°$ at $g_w \sim 10^{-2}$ and for $\theta = 7.5°$ at $g_w \sim 3 \times 10^{-3}$ there is weak maximum. It may be caused by calculation accuracy at small temperature factor when additional grid thickening is required because of the big density gradient near the surface. Such thickening, being small, does not effect on the global flow pattern in the separation zone, but is important for local heat flux modeling.

Big practical interest is the temperature factor effect on the effectiveness of flight controls of "ramp" type. Figure 7 gives the difference Δx between the pressure center locations in two cases: with g_w simulation and with pressure "step" obtained from inviscid corner flow. If there is a separation, we can specify two regions that effect the pressure center location: increased pressure zone inside the separation zone and increased pressure zone at the attachment point. For $\theta \leq 7.5°$ the input of the increased pressure zone inside the separation zone is practically balanced by the displacement to the right of the increased pressure zone at the attachment point.

So, there are two counteracting tendencies that determine the pressure center location depending on the temperature factor: (1) displacement to the left on the plate and (2) displacement of the attachment zone to the right. For small angles ($\theta < 7.5°$) displacement to the left on the plate overrides the displacement of the attachment zone to the right. For big angles ($\theta > 7.5°$) the input of high pressure

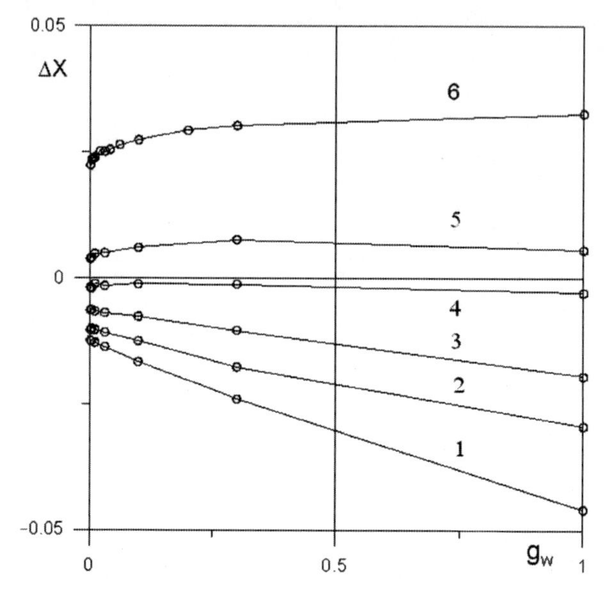

Fig. 7 Pressure center displacement ΔX as compared with position for inviscid flow near angle $\theta^\circ = 0, 2.5, 5, 7.5, 10, 20$ (curves 1–6) depending on temperature factor g_w

in the attachment zone on the pressure center position becomes significant and the pressure center moves to the right.

These results are given for the calculation area $x = [0,\ 5]$. For the ramp of actual geometry with small sizes similar effects may be observed, but it is a quite different task with different geometry.

5 Conclusion

The qualitative analytical investigation was carried out related to the temperature factor effect on separation flow physical features caused by a compression corner in the supersonic viscous flow. The numerical results of simulation the same flow based on the Navier–Stokes equations are also presented. It is shown that the separation zone length decreases as the temperature factor decreases. For high values of the compression corner in numerical investigations there were discovered vortices in the separation zone that were not observed before. These vortices effect considerably on heat exchange in the separation zones. The temperature factor effect on the pressure center position was investigated. It is shown that at small corner angles a temperature factor increase may deteriorate static stability of the vehicle, while at big angles it may improve static stability of the vehicle.

The work has been performed under the grant of RFBR (Russian Foundation for Basic Research) (projects No 07-08-000124 and 06-08-01558a)

References

1. Neiland, V. Ya., Kukanova, N. I., 1965, Investigation of flows with separation zones. Review of TsAGI, No. 129.
2. Lapin, Yu. V., Loytsianskii, L. G , Lun'kin, Yu. P., Neiland, V. Ya., Sychev, V. V., Tirskii, G. A., 1970, Mechanics of Viscous Liquids and Gases. Theory of Laminar and Turbulent Boundary Layers. Mechanics in USSR for Fifty Years. Vol. II, Nauka, Moscow.
3. Charwat, A. F., 1970, Supersonic Flows with Imbedded Separation Regions, in: Advances in Heat Transfer, Vol. 6. Academic, New York.
4. Chang, P. K., Separation of Flow. Pergamon, Oxford.
5. Chapman, D. R., 1951, An analysis of base pressure at supersonic velocities and comparison with experiment. NASA Rep., No 1051. pp. 23.
6. Korst, H. H., 1956, A theory for base pressure in transonic and supersonic flow. J. Appl. Mech., Vol. 23, No. 4, pp. 593–600.
7. Neiland, V. Ya., 1970, Asymptotic theory of plane steady supersonic flows with separation zones. Fluid Dynamics, No. 3, pp. 372.
8. Neiland, V. Ya., 1968, Supersonic viscous flow near a separation point. Abstracts of the 3-rd All-Union Congress on Theoretical and Applied Mathematics, Nauka, Moscow, pp. 224.
9. Neiland, V. Ya., 1969, Theory of laminar boundary layer separation in supersonic flow. Fluid Dynamics, No. 4, pp. 33.
10. Stewartson, K., Williams, P. G., 1969, Self-induced separation. Proc. Roy. Soc. London. Ser. A., Vol. 312, 1509, pp. 181–206.
11. Stewartson, K., 1970, On laminar boundary layers near corners//Quart J. Mech. Appl. Math., Vol. 23, No. 2. pp. 137–152.
12. Neiland, V. Ya., 1971, Flow behind the boundary layer separation point in a supersonic stream. Fluid Dynamics, No. 3, pp. 378.
13. Smith, F. T., Khorrami, A. F., 1991, The interactive breakdown in supersonic ramp flow. J. Fluid. Mech., Vol. 224, pp. 197–215.
14. Korolev, G. L., Gajjar, J. S. B., Ruban, A. I., 2002, Once again on the supersonic flow separation near a corner. J. Fluid Mech., Vol. 463, pp. 173–199.
15. Neiland, V. Ya., Bogolepov, V. V., Dudin, G. N., Lipatov, I. I., 2004, Asymptotic Theory of Supersonic Viscous Gas Flows. M.: Fizmatlit, pp. 456.
16. Neiland, V. Ya., Bogolepov, V. V., Dudin, G. N., Lipatov, I. I., 2007, Asymptotic Theory of Supersonic Viscous Gas Flows. Elsevier, Oxford, The Netherlands, pp. 536.
17. Hayes, W. D., and Probstein, R. F., 1959, Hypersonic Flow Theory. Academic, New York.
18. Egorov, I. V., Zaitsev, O. L., 1991, On an approach to the numerical solution of the two-dimensional Navier-Stokes equations by the shock capturing method. Journal of Computational Mathematics and Mathematical Physics. Vol. 31, No 2. pp. 286–299.
19. Babaev, I. Yu., Bashkin, V. A., Egorov, I. V., 1994, Numerical solution of the Navier-Stokes equations using variational iteration methods. Comp. Maths Math Phys., Vol. 34, No 11. pp. 1455–1462.
20. Bashkin, V. A., Egorov, I. V., Ivanov, D. V., 1997, Application of Newton's method to the calculation of internal supersonic separated flows. Zh. Prikl. Mekh. Tekhn. Fiz., Vol. 38, No 1, pp. 30–42.
21. Godunov, S. K., 1959, Mat. Sb., Vol. 47, pp. 271.
22. Roe, P. L., 1981, Aproximate Rieman Solvers, Parameter Vectors, and Difference Scheme. Journal Computation Physics, Vol. 43, pp. 357–372.
23. Kolgan, V. P. , 1972, Uch. Zap. Tsentr. Aerogidrodin. Inst., Vol. 3, No. 6, pp. 68.

24. Saad, Y., Shultz, M. H., 1986, GMRES: a generalized minimal residual algorithm for solving nonsymmetric linear systems. SIAM Journal of Scientific and Statistical Computing, No.7, pp. 856–869.
25. Gil'manov, A. N., 2000, Methods of Adaptive Meshes in Gas Dynamic Problems, I. Nauka, Fizmatlit, Moscow, pp. 247.
26. Anderson, D. A., Tannehill, J. C., Pletcher, R. H., 1984, Computational Fluid Mechanics And Heat Transfer. Hemisphere, New York.
27. Stemmer, C., Adams, N. A., 2004, Investigation of supersonic boundary layers by DNS, ECCOMAS 2004 Proceedings, Vol. II.

Recent Results on Local Projection Stabilization for Convection-Diffusion and Flow Problems

Lutz Tobiska

Abstract A survey of stabilization methods based on local projection is given. The class of steady problems considered covers scalar convection-diffusion equations, the Stokes problem and the linearized Navier–Stokes equations.

1 Introduction

It is well known that standard finite element discretizations applied to convection-diffusion or incompressible flow problems show spurious oscillations in the case of higher Reynolds numbers, owing to dominating convection. A first proposal to handle this instability for low-order finite element discretization has been the use of upwind finite elements [1]. Another idea, suitable also for higher-order finite elements, is the streamline upwind Petrov-Galerkin (SUPG) stabilization proposed in [2] and analyzed for a scalar convection-diffusion equation in [3]. The method is based on adding weighted residuals to the standard Galerkin method to enhance stability without losing consistency. The same idea is useful in circumventing the Babuška–Brezzi condition which restricts the set of possible finite element spaces that approximate velocity and pressure for incompressible flows. Such a pressure-stabilized Petrov–Galerkin (PSPG) method has been studied for low equal-order interpolations of the Stokes problem in [4]. A detailed error analysis of these SUPG/PSPG-type stabilizations applied to the incompressible Navier–Stokes equations, including both the case of inf–sup stable and equal-order interpolations, can be found in [5]. Recently, local projection stabilization (LPS) [6–8] methods have become quite popular, in particular because of their commutative properties in optimization problems [9] and stabilization properties similar to those of the SUPG

L. Tobiska
Institute for Analysis and Computational Mathematics, Otto von Guericke University Magdeburg, PF 4120, D-39016 Magdeburg, Germany, E-mail: tobiska@ovgu.de

A.F. Hegarty et al. (eds.), *BAIL 2008 - Boundary and Interior Layers.* Lecture Notes in Computational Science and Engineering,
DOI: 10.1007/978-3-642-00605-0,

method [10]. In the following we give an overview of recent developments for this class of stabilizations applied to various problems.

2 Convection-Diffusion Problem

2.1 Standard Galerkin and SUPG

We start with the convection-diffusion equation

$$-\varepsilon\Delta u + b\cdot\nabla u + \sigma u = f \quad \text{in } \Omega, \qquad u = 0 \quad \text{on } \Gamma \tag{1}$$

in a bounded domain $\Omega \subset \mathbb{R}^d$ with Lipschitz continuous boundary $\Gamma = \partial\Omega$. For simplicity we assume $\nabla\cdot b = 0$ and $\sigma > 0$ which guarantees a unique weak solution $u \in H_0^1(\Omega)$. Note that in the interesting case $0 < \varepsilon \ll 1$, the solution exhibits boundary and interior layers whose positions depend on the convection field b. Let $V_h \subset H_0^1(\Omega)$ be a finite element space with mesh size h. Then the discrete problem for the standard Galerkin approach is:

Find $u_h \in V_h$ such that for all $v_h \in V_h$

$$a(u_h, v_h) := \varepsilon(\nabla u_h, \nabla v_h) + (b\cdot\nabla u_h + \sigma u_h, v_h) = (f, v_h)$$

where $(\cdot,\cdot)$ denotes the inner product in L^2 and its vector-valued analogues. Stability and convergence for piecewise polynomials of degree $r \geq 1$ follow from the coercivity of the bilinear form $a(\cdot,\cdot)$ and the Lemma of Cea:

$$\begin{aligned} a(v,v) &\geq ||v||_{1,\varepsilon}^2 := \varepsilon|v|_1^2 + \sigma\|v\|_0^2 && \forall v \in V,\\ ||u-u_h||_{1,\varepsilon} &\leq C\, h^r |u|_{r+1}, && u \in H_0^1(\Omega)\cap H^{r+1}(\Omega). \end{aligned}$$

Nevertheless it is well-known that spurious oscillations appear if $\varepsilon \ll h$. This observation shows that the norm $\|\cdot\|_{1,\varepsilon}$ is too weak to suppress global oscillations.

The SUPG [2, 3] modifies the Galerkin method by adding weighted residuals of the strong form of the differential equations, resulting in:

Find $u_h \in V_h$ such that for all $v_h \in V_h$

$$\begin{aligned} &\varepsilon(\nabla u_h, \nabla v_h) + (b\cdot\nabla u_h + \sigma u_h, v_h)\\ &\quad + \sum_{K\in\mathcal{T}_h} \tau_K(-\varepsilon\Delta u_h + b\cdot\nabla u_h + \sigma u_h - f, b\cdot\nabla v_h)_K = (f, v_h) \end{aligned}$$

where $\mathcal{T}_h$ denotes a decomposition of Ω into cells $K \in \mathcal{T}_h$, $(\cdot,\cdot)_K$ is the inner product in $L^2(K)$, and τ_K is a user-chosen stabilization parameter. For $\tau_K \sim h_K$, stability follows again from coercivity of the associated bilinear form, but now with respect to the stronger norm

$$|||v|||_{SUPG} := \left(||v||_{1,\varepsilon}^2 + \sum_{K \in \mathcal{T}_h} \tau_K \|b \cdot \nabla v\|_{0,K}^2 \right)^{1/2},$$

which suppresses global oscillations. A clever estimation of the convection term uses integration by parts and the stability with respect to $||| \cdot |||_{SUPG}$:

$$\begin{aligned}
\left|(b \cdot \nabla (u - i_h u), v_h)\right| &\leq |(u - i_h u, b \cdot \nabla v_h)| + |(u - i_h u, v_h \nabla \cdot b)| \\
&\leq \sum_{K \in \mathcal{T}_h} \tau_K^{-1/2} \|u - i_h u\|_{0,K} \, \tau_K^{1/2} \|b \cdot \nabla v_h\|_{0,K} + C h^{r+1} |u|_{r+1} \, \|v_h\|_0 \\
&\leq C \left[\left(\sum_{K \in \mathcal{T}_h} \tau_K^{-1} h_K^{2r+2} |u|_{r+1,K}^2 \right)^{1/2} + h^{r+1} |u|_{r+1} \right] |||v_h|||_{SUPG}
\end{aligned}$$

resulting in the improved error estimate

$$|||u - u_h|||_{SUPG} \leq C \, (\varepsilon^{1/2} + h^{1/2}) \, h^r \, |u|_{r+1}$$

for P_r finite elements. Note that in boundary layers we usually have $|u|_{r+1} \sim \varepsilon^{-(r+1/2)}$, which means that the above error estimate becomes useless. Nevertheless local error estimates have been derived that support theoretically the good approximation properties away from layers observed in numerical computations; see, e.g., [11].

Thus the SUPG is a consistent method with improved stability and convergence properties compared to the standard Galerkin approach. However, consistency is obtained at the cost of computing several additional terms to assemble the coefficient matrix of the discrete system.

2.2 *Local Projection Stabilization (LPS)*

A detailed study of the stability and convergence analysis of the SUPG shows that in the discrete problem only the term

$$\sum_{K \in \mathcal{T}_h} \tau_K (b \cdot \nabla u_h, b \cdot \nabla v_h)_K$$

is responsible for improved stability properties. However, skipping all other terms in the SUPG leads to an inconsistent method for which the consistency error scales with τ_K. A remedy is to add a term that controls only the fluctuations of the derivatives in the streamline direction $b \cdot \nabla u_h$. Let $\mathcal{M}_h$ denote a decomposition of Ω into 'macro' cells $M \in \mathcal{M}_h$ of diameter h_M with $h_K \sim h_M$ for $\overline{K} \cap \overline{M} \neq \emptyset$, D_h a discontinuous projection space associated with the decomposition $\mathcal{M}_h$, $\pi_h : L^2(\Omega) \to D_h$ the L^2 projection, and $\kappa_h := id - \pi_h$ the fluctuation operator. Then our modified discrete problem is:

Find $u_h \in V_h$ such that for all $v_h \in V_h$

$$\varepsilon(\nabla u_h, \nabla v_h) + (b \cdot \nabla u_h + \sigma u_h, v_h) + \sum_{M \in \mathcal{M}_h} \tau_M (\kappa_h(b \cdot \nabla u_h), \kappa_h(b \cdot \nabla v_h))_M = (f, v_h).$$

The modified bilinear form associated with the left-hand side is coercive with respect to the mesh-dependent norm

$$|||v|||_{LPS} := \left(||v||_{1,\varepsilon}^2 + \sum_{M \in \mathcal{M}_h} \tau_M \|\kappa_h(b \cdot \nabla v)\|_{0,M}^2 \right)^{1/2}.$$

Now the consistency error depends on τ_M and the projection space D_h. If the discontinuous space of piecewise polynomials of degree at most $r-1$ is selected, which we write as $D_h = P_{r-1}^{\text{disc}}(\mathcal{M}_h)$, then for $\tau_M \sim h_M$ we get

$$\begin{aligned} &\left| \sum_{M \in \mathcal{M}_h} \tau_M \, (\kappa_h(b \cdot \nabla u), \kappa_h(b \cdot \nabla v_h))_M \right| \\ &\quad \leq \sum_{M \in \mathcal{M}_h} \tau_M^{1/2} h_M^r |b \cdot \nabla u|_{r,M} \, \tau_M^{1/2} \|\kappa_h(b \cdot \nabla v_h)\|_{0,M} \\ &\quad \leq C \left(\sum_{M \in \mathcal{M}_h} h_M^{2r+1} |b \cdot \nabla u|_{r,M}^2 \right)^{1/2} |||v_h|||_{LPS}. \end{aligned}$$

Using the L^2 stability of the fluctuation operator we see that

$$|||v_h|||_{LPS} \leq C \, |||v_h|||_{SUPG} \quad \forall v_h \in V_h$$

which means that the SUPG is at least as stable as the LPS. Having in mind only the coercivity of the bilinear forms with respect to $||| \cdot |||_{SUPG}$ and $||| \cdot |||_{LPS}$, respectively, one might think that the LPS is less stable compared to the SUPG. But in [10] an inf–sup condition for the LPS bilinear form in a stronger norm (which turns out to be equivalent to the SUPG norm) has been shown, i.e., the stability properties of LPS and SUPG are in fact comparable.

2.3 Basics in the Error Analysis of LPS

We assume that $Y_h \approx H^1(\Omega)$ is a finite element space associated with a decomposition of Ω into cells $K \in \mathcal{T}_h$ and $V_h = Y_h \cap H_0^1(\Omega)$ denotes the approximation space. Let the discontinuous projection space $D_h = \oplus_M D_h(M)$ live on a decomposition into macro cells $M \in \mathcal{M}_h$, where the case $\mathcal{T}_h = \mathcal{M}_h$ is assumed to be included. We will see that the key idea of the LPS lies in the existence of a special interpolant $j_h : H^2(\Omega) \to Y_h$ that displays the usual interpolation properties and satisfies in addition the orthogonality property

$$(w - j_h w, q_h) = 0 \qquad \forall w \in H^2(\Omega),\ \forall q_h \in D_h.$$

This orthogonality enables an estimation of the critical part of the convection term after integrating by parts for $\tau_M \sim h_M$ as follows:

$$\begin{aligned}
|(u - j_h u, b \cdot \nabla v_h)| &= |(u - j_h u), \kappa_h(b \cdot \nabla v_h))| \\
&\leq \sum_{M \in \mathcal{M}_h} \tau_M^{-1/2} \|u - j_h u\|_{0,M}\ \tau_M^{1/2} \|\kappa_h(b \cdot \nabla v_h)\|_{0,M} \\
&\leq C \left(\sum_{M \in \mathcal{M}_h} h_M^{2r+1} |u|^2_{r+1,M} \right)^{1/2} |||v_h|||_{LPS}.
\end{aligned}$$

Dealing with all other terms in the usual way, we end up with the error estimate

$$|||u - u_h|||_{LPS} \leq C\, (\varepsilon^{1/2} + h^{1/2})\, h^r\, |u|_{r+1} \tag{2}$$

for $\tau_M \sim h_M$ [6,7,12,13]. Now the question arises: under which conditions does an interpolation j_h with additional orthogonality properties exist? Examples have been given for the transport equation ($\varepsilon = 0$) in [12] and the Oseen equation in [6], where the two-level variant has been studied in which the decomposition into cells is generated from a macro mesh by certain refinement rules. We indicate this by writing $\mathcal{T}_h = \mathcal{M}_{h/2}$. In the general case we have

Theorem 1 ([7]). *Let the local inf–sup condition*

$$\inf_{q_h \in D_h(M)} \sup_{v_h \in Y_h(M)} \frac{(v_h, q_h)_M}{\|v_h\|_{0,M}\ \|q_h\|_{0,M}} \geq \beta_1 > 0, \qquad \forall M \in \mathcal{M}_h \tag{3}$$

with $Y_h(M) := \{w_h|_M : w_h \in Y_h,\ w_h = 0 \text{ on } \Omega \backslash M\}$ be satisfied. Then there is an interpolation $j_h : H^2(\Omega) \to Y_h$ with the usual interpolation error estimates and the additional orthogonality property

$$(w - j_h w, q_h) = 0, \quad \forall q_h \in D_h,\ \forall w \in H^2(\Omega).$$

In order to fulfil all assumptions of the convergence analysis, two different requirements for the pair (V_h, D_h) of approximation and projection space have to be reconciled:

- D_h has to be rich enough to guarantee a certain order of consistency
- D_h should be small enough w.r.t. V_h to guarantee $j_h u - u \perp D_h$

Two main approaches have been considered in the literature:

$$\text{one-level } (V_h^+, D_h) \quad \Leftrightarrow \quad \text{two-level } (V_h, D_{2h}).$$

In the one-level approach, a standard finite element space is chosen as the projection space D_h to guarantee the consistency order. Then, the approximation space $V_h = Y_h \cap H_0^1(\Omega)$ is (if necessary) enriched to V_h^+ such that the assumptions of Theorem 1 are fulfilled. In the two-level approach, a standard finite element space is chosen as the approximation space V_h and the projection space D_h is thinned out to a space D_{2h} on the next coarser mesh level.

In the following we give explicit examples satisfying all assumptions needed for the above error estimation, see [7] for details. Let b_K and $\tilde{b}_K$ denote the (mapped) bubble functions of lowest polynomial degree that vanish on the boundary ∂K of a simplex and hexahedron respectively. We introduce the enriched approximation spaces on triangles and quadrilaterals respectively:

$$P_r^+ := P_r + \bigoplus_{K \in \mathcal{T}_h} b_K \cdot P_{r-1}(K)$$

$$Q_r^+ := Q_r + \bigoplus_{K \in \mathcal{T}_h} \text{span}\,(\tilde{b}_K \cdot x_i^{r-1},\, i = 1, \ldots, d).$$

An overview of different variants is given in Table 1 and illustrated in the two-dimensional case $d = 2$ for $r = 1$ and $r = 2$ in Figs. 1–4.

One disadvantage of the one-level approach is the increasing number of degrees of freedom owing to the enrichments in particular in the case of simplices. However, this can be overcome by static condensation. In the two-level approach the stencil of the stabilizing term increases due to the larger support of $\kappa_h(b \cdot \nabla\varphi_i)$ compared with that of $b \cdot \nabla\varphi_i$ (for each basis function φ_i in V_h). This might not fit into the data structure of an available code.

So far we have only considered the case of boundary conditions of Dirichlet type. Mixed boundary conditions lead often to a limited regularity of the solution of a convection-diffusion problem. In [13], it is shown how the error analysis of the

Table 1 Possible space pairs in the LPS

one-level		two-level	
V_h^+	D_h	V_h	D_{2h}
P_r^+	P_{r-1}^{disc}	P_r	P_{r-1}^{disc}
Q_r^+	P_{r-1}^{disc}	Q_r	Q_{r-1}^{disc}

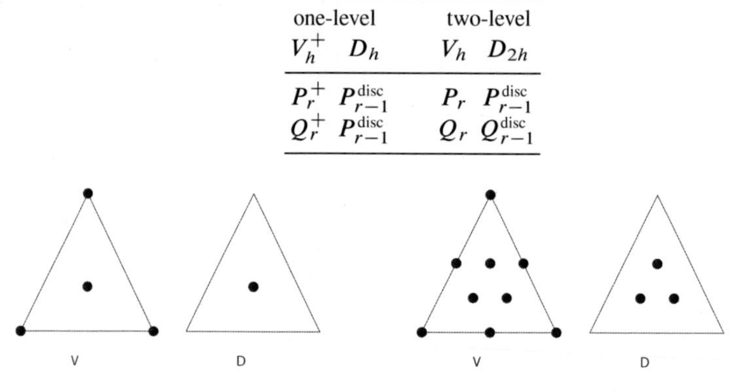

Fig. 1 Approximation and projection spaces on triangles (one-level approach)

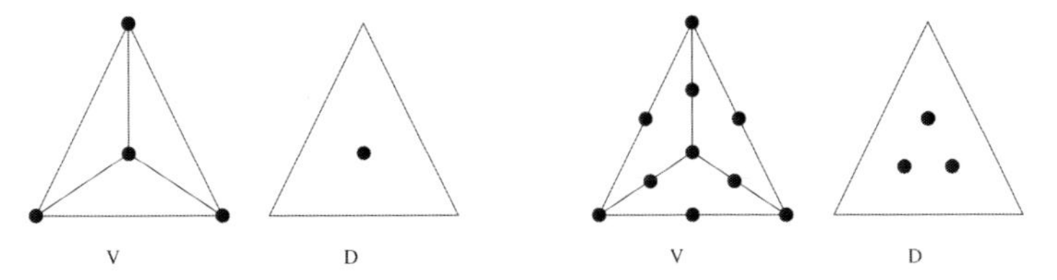

Fig. 2 Approximation and projection spaces on triangles (two-level approach)

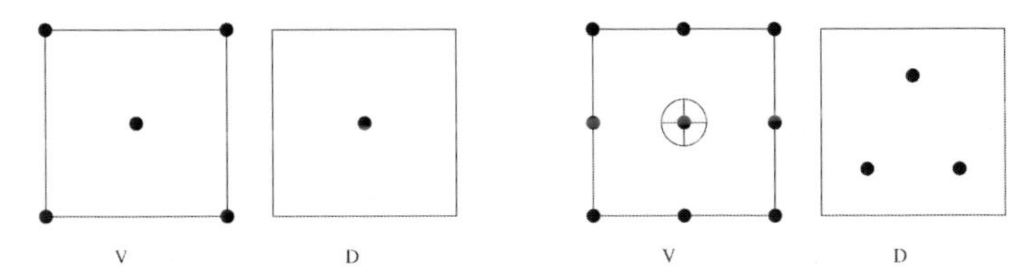

Fig. 3 Approximation and projection spaces on quadrilaterals (one-level approach)

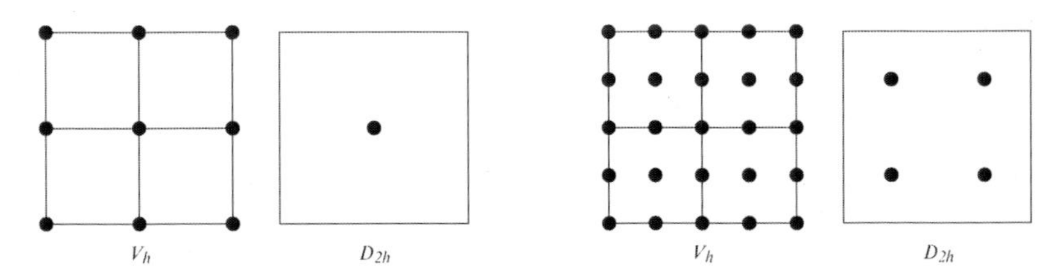

Fig. 4 Approximation and projection spaces on quadrilaterals (two-level approach)

one-level LPS can be extended to the case of boundary conditions of mixed Dirichlet and Neumann type.

2.4 Relationship to Other Stabilization Methods

The LPS is akin to but not exactly equal to the subgrid scale stabilization introduced by Guermond [14], who considered gradients of fluctuations instead of fluctuations of gradients. Thus the stabilizing term has the form

$$\sum_{K \in \mathcal{K}_h} \tau_K(\nabla(id - P_H)u_h, \nabla(id - P_H)v_h)_K$$

where $P_H : v_h \to V_H$ is a projection onto the (resolvable) coarse scales. This can be also interpreted as adding artificial viscosity only for the fine scales of the finite element space V_h. For certain scale separations of $V_h = V_H \oplus V_H^{\perp}$ both methods give spectrally equivalent stabilization terms (simplices) or even coincide (lowest order case on simplices). However, in the general case the stabilizing terms are not spectrally equivalent. For more details see [7].

We mention that a special variant of the LPS has been already introduced by Layton [15] as a mixed method combined with scale separation of the finite element space V_h. The projection space has been chosen as $D_H = \nabla V_H$, where V_H denotes the approximation space on a coarser mesh level. The analysis given in [15] does not use orthogonality of the interpolation and leads to the suboptimal convergence rate of $4/3$ instead of $3/2$ for the scaling $\tau \sim h^{4/3}$, $H \sim h^{2/3}$. In order to gain the full $3/2$-power of h, the orthogonality of the interpolation has been used in [12] for solving the transport equation ($\varepsilon = 0$) discretized by the two-level (Q_1, Q_0)-LPS.

There is also a close relation to the stabilization method using orthogonal subscales (OSS) proposed by Codina in [16, 17]. In the OSS the projection π_h is chosen as the L^2 projection into the finite element ansatz space without forcing boundary conditions, i.e., $D_h = Y_h$. Since this projection is no longer local, the stencil of the stabilizing term increases as in the two-level LPS approach or one has to solve a global system in $V_h \times Y_h$ to approximate u and $b \cdot \nabla u$. For details we refer to [16].

2.5 Choice of the Stabilization Parameter

A general strategy to select appropriate stabilization parameters τ_K is to equilibrate different terms in the *a priori* error estimates. In this way, the asymptotic behaviour of τ_K with respect to the meshsize and the polynomial degree of the finite element spaces can be fixed. For convection-diffusion equations in one space dimension, it is known that in the constant coefficient case with $c = 0$ and piecewise linear elements the stabilization parameter in the SUPG method can be chosen in such a way that the discrete solution becomes nodally exact.

It has been shown in [18] that in the one-dimensional, constant coefficient case (with $c = 0$), the one-level version of the $(P_r^+, P_{r-1}^{\text{disc}})$-LPS is equal to the P_{r+1}-differentiated residual method (DRM). Note that in 1D one has $P_r^+ = P_{r+1}$. Moreover, a successive elimination of the higher modes in the P_{r+1}-DRM by static condensation leads to the P_r-DRM, where the P_1-DRM coincides with the SUPG. These observations allow the derivation of explicit formulas for the stabilization parameter in the LPS and DRM such that the P_1 part of the corresponding discrete solutions is nodally exact. For more details, see [18]. The convergence properties of the DRM on arbitrary and on layer-adapted meshes are investigated in [19]. Finally, we mention that the DRM is also closely related to the variational multiscale method (VMS) studied in [20].

2.6 LPS on Layer Adapted Meshes

It has been mentioned already in Sect. 2.1 that $|u|_{r+1} \sim \varepsilon^{-(r+1/2)}$ in boundary layers and error estimates like (2) lose their value as $\varepsilon \to 0$. For the model problem

(1) in the unit square $\Omega = (0,1)^2$, two different types of layers can appear: for $b = (b_1, b_2)$ with $b_1, b_2 > 0$ we observe only exponential layers along the outflow part ($x = 1$ or $y = 1$) of the domain whereas for $b = (b_1, 0)$ with $b_1 > 0$ an exponential layer along $x = 1$ and two characteristic layers along $y = 0$ and $y = 1$ are present. The idea is to use special layer-adapted (so-called S-type) meshes and suitable enriched approximation spaces. Consider the following enrichment of the usual Q_r space of continuous, piecewise (mapped) polynomials of degree r in each variable

$$Q_r^{6+} := Q_r + \bigoplus_{K \in \mathcal{T}_h} Q_K^*(K),$$

$$Q_K^*(\widehat{K}) := \text{span}\left\{(1-\hat{x}_1^2)(1-\hat{x}_2^2)\hat{x}_i^{p-1},\ (1 \pm \hat{x}_{i+1})(1-\hat{x}_i^2)L_{r-1}(\hat{x}_i)\right\},$$

where $i \in \{0, 1\}$ modulo 2 and L_{r-1} denotes the Legendre polynomial of order $r-1$. The projection space is set to be $D_h = P_{r-1}^{\text{disc}}$. Note that $P_{r+1} \subset Q_r^{6+}$. The number of subintervals in each coordinate direction of the tensor product mesh on Ω will be denoted by N.

Theorem 2 ([21,22]). *Let $b_1, b_2 > 0$, $(Y_h, D_h) = (Q_r^{6+}, P_{r-1}^{disc})$, and the stabilization parameter be given by $\tau_K \sim N^{-2}$ on the coarse mesh and $\tau_K = 0$ on the fine mesh. Then, there is an interpolant u^I such that*

$$|||u^I - u^N|||_{LPS} \leq C\,(N^{-1}\log N)^{r+1},\ \|u - u^N\|_{1,\varepsilon} \leq C\,(N^{-1}\log N)^{r+1}$$

on a Shishkin mesh. For the characteristic layers case ($b_1 > 0$, $b_2 = 0$) an appropriate choice of the stabilization parameter in the characteristic layer region leads to the same estimate. Moreover, for $r = 1$ we have the supercloseness result for the spaces $(V_h, D_h) = (Q_1, P_0^{disc})$

$$|||u^I - u^N|||_{LPS} \leq C\,(N^{-1}\log N)^2,\ \|u - u^N\|_{1,\varepsilon} \leq C\,N^{-1}\log N.$$

Apart from the lowest-order case, we have to handle a considerable set of additional degrees of freedom because of the large enrichment of Q_r. Next, we consider a moderate enrichment of Q_r such that $P_{r+1} \not\subset Q_r^+$ and give a supercloseness result.

Theorem 3 ([23]). *Let the approximation space Y_h be enriched only on the coarse mesh part so that $Y_h = Q_r^+$ on the coarse and $Y_h = Q_r$ on the fine mesh part. Then on Shishkin and Bakhvalov–Shishkin type meshes the interpolant u^I is superclose, i.e.,*

$$|||u^I - u^N|||_{LPS} \leq C\,N^{-(r+1/2)}$$

whereas for the solution u one has only

$$\|u - u^N\|_{1,\varepsilon} \leq \begin{cases} C\,(N^{-1}\log N)^{-r} & \text{for a Shishkin mesh,} \\ C\,N^{-r} & \text{for a Bakhvalov-Shishkin mesh.} \end{cases}$$

Note that the enrichments in Theorem 3 consists of only two additional degrees of freedom per coarse mesh cell. Thus, compared to Theorem 2, a considerable reduction in the number of degrees of freedom has been achieved.

3 Stokes Problem

3.1 Standard Galerkin and PSPG

Now we consider the Stokes Problem

$$-\Delta u + \nabla p = f \quad \text{in}\ \Omega,\ \text{div}\, u = 0 \quad \text{in}\ \Omega,\ u = 0 \quad \text{on}\ \Gamma \tag{4}$$

in a bounded domain $\Omega \subset \mathbb{R}^d$ with Lipschitz continuous boundary $\Gamma = \partial\Omega$. There is a unique weak solution $(u, p) \in H_0^1(\Omega)^d \times L_0(\Omega)$. Let $V_h \subset H_0^1(\Omega)^d$ and $Q_h \subset L_0^2(\Omega)$ be finite element spaces with a mesh size h, approximating velocity and pressure respectively. Then the discrete problem of the standard Galerkin approach is:

Find $(u_h, p_h) \in V_h \times Q_h$ such that for all $(v_h, q_h) \in V_h \times Q_h$

$$(\nabla u_h, \nabla v_h) - (p_h, \text{div}\, v_h) + (q_h, \text{div}\, u_h) = (f, v_h).$$

It is well-known [24] that the Babuška–Brezzi condition

$$\exists \beta_0 > 0,\ \forall h: \qquad \inf_{q_h \in Q_h} \sup_{v_h \in V_h} \frac{(q_h, \text{div}\, v_h)}{\|q_h\|_0\, |v_h|_1} \geq \beta_0 \tag{5}$$

guarantees stability and convergence of a unique solution $(u_h, p_h) \in V_h \times Q_h$ of the Galerkin method. The condition (5) restricts the possible choices of approximation spaces V_h and Q_h; in particular, equal-order interpolations for velocity and pressure are excluded. One way to circumvent the inf–sup condition is to add weighted residuals of the strong form of the differential equation resulting in the stabilized formulation

Find $(u_h, p_h) \in V_h \times Q_h$ such that for all $(v_h, q_h) \in V_h \times Q_h$

$$A_{PSPG}((u_h, p_h); ((v_h, q_h)) = (f, v_h) + \sum_{K \in \mathcal{T}_h} \alpha_K (f, \nabla q_h)_K \tag{6}$$

with the discrete bilinear form A_h given by

$$\begin{aligned} A_{PSPG}((u, p); (v, q)) := {} & (\nabla u, \nabla v) - (p, \text{div}\, v) + (q, \text{div}\, u) \\ & + \sum_{K \in \mathcal{T}_h} \alpha_K (-\Delta u + \nabla p, \nabla q)_K. \end{aligned} \tag{7}$$

For continuous pressure approximations $Q_h \subset H^1(\Omega)$, the form A_{PSPG} is coercive on the product space $V_h \times Q_h$ with respect to the norm

$$|||(v,q)|||_{PSPG} := \left(|v|_1^2 + \sum_{K \in \mathcal{T}_h} \alpha_K |q|_{1,K}^2 \right)^{1/2}$$

provided that the stabilization parameter has been chosen as $\alpha_K = \alpha_0 h_K^2$ where the positive constant α_0 satisfies an certain upper bound. This residual-based stabilization technique proposed and analyzed in [4] is also known as the pressure stabilized Petrov–Galerkin (PSPG) approach [25]. Over the years it has been extended and combined with the SUPG for solving the (linearized) Navier–Stokes equations.

3.2 Local Projection Stabilization

Beside the residual-based approach, projection-based stabilization techniques have also been developed for the Stokes problem. A method based on the projection of the pressure gradient onto a continuous finite element space has been proposed in [26]. Although the method is consistent (in a certain sense) it is expensive due to the nonlocality of the projection. Becker and Braack proposed in [27] to project the pressure gradient onto a discontinuous finite element space living on a coarser mesh. This method is not consistent, but it is cheaper owing to the locality of the projection. Nevertheless, as a two-level approach the stabilizing term leads to an larger stencil which might not fit into the data structure of an available code.

A revision of the residual-based PSPG approach shows that the improved stability properties rely on adding the term

$$\sum_{K \in \mathcal{T}_h} \alpha_K (\nabla p, \nabla q)_K \quad \text{instead of} \quad \sum_{K \in \mathcal{T}_h} \alpha_K (-\Delta u + \nabla p - f, \nabla q)_K$$

to the Galerkin method. The other terms are only needed to preserve consistency. Now, replacing in the first term the pressure gradients by the fluctuations, we obtain the LPS for equal-order interpolations [28].

Let the approximation spaces for velocity and pressure be generated by a scalar finite element space $Y_h \approx H^1(\Omega)$, such that $V_h = (Y_h \cap H_0^1(\Omega))^d$ and $Q_h = Y_h \cap L_0^2(\Omega)$. We will consider for simplicity only the one-level approach, thus the discontinuous projection space D_h lives on the same decomposition $\mathcal{M}_h = \mathcal{T}_h$ as the approximation space Y_h. As above we introduce the fluctuation operator $\kappa_h := \text{id} - \pi_h$ with the L^2 projection $\pi_h : L^2(\Omega) \to D_h$. Now the stabilized discrete problem reads:

Find $(u_h, p_h) \in V_h \times Q_h$ such that for all $(v_h, q_h) \in V_h \times Q_h$

$$A_h\big((u_h,p_h);(v_h,q_h)\big) := (\nabla u_h, \nabla v_h) - (p_h, \operatorname{div} v_h) + (q_h, \operatorname{div} u_h) \\ + \sum_{K\in\mathcal{T}_h} \alpha_K \big(\kappa_h \nabla p_h, \kappa_h \nabla q_h\big)_K = (f, v_h). \tag{8}$$

As in the Galerkin method the bilinear form is not coercive on the product space $V_h \times Q_h$; indeed we have only

$$A_h((v_h,q_h);(v_h,q_h)) = |v_h|_1^2 + \sum_{K\in T_h} \alpha_K \|\kappa_h(\nabla q_h)\|_{0,K}^2$$

and the right-hand side vanishes for all $(v_h,q_h) = (0,q_h)$ with $\nabla q_h \in D_h$. Therefore, it is essential that an inf–sup condition can be proven in the mesh-dependent norm

$$|||(v,q)||| := \left(|v|_1^2 + \|q\|_0^2 + \sum_{K\in\mathcal{T}_h} \alpha_K \|\kappa_h(\nabla q)\|_{0,K}^2 \right)^{1/2}.$$

Lemma 1 ([28]). *Let (Y_h, D_h) satisfy the local inf–sup condition in Theorem 1 and let $h_K^2/\alpha_K \le C$. Then there is a positive constant $\beta > 0$ independent of h such that*

$$\inf_{(v_h,q_h)\in V_h\times Q_h} \sup_{(w_h,r_h)\in V_h\times Q_h} \frac{A_h\big((v_h,q_h);(w_h,r_h)\big)}{|||(v_h,q_h)|||\;|||(w_h,r_h)|||} \ge \beta.$$

Let us briefly discuss the different properties of PSPG and LPS. In the PSPG method A_{PSPG} is coercive on the product space $V_h \times Q_h$ for a restricted range of the stabilization parameter, more precisely $\alpha_K = \alpha_0 h_K^2$ with an upper bound for α_0 depending on the polynomial degree used in the definition of Y_h. In contrast to that the bilinear form A_h of the LPS satisfies an inf–sup condition on the product space $V_h \times Q_h$ for $\alpha_K = \alpha_0 h_K^2$ and any $\alpha_0 \in \mathbb{R}^+$. The theoretically larger range for α_0 can be also seen in computations [28].

3.3 Error Estimates

As in the case of a scalar convection-diffusion equation one has to balance two requirements: D_h has to be rich enough to guarantee a certain order of consistency and D_h has to be sparse enough to allow the existence of an interpolant $j_h : H^1(\Omega)^d \to Y_h^d$ (needed to prove Lemma 1) such that the interpolation error is perpendicular to D_h^d. The larger domain of definition ($H^1(\Omega)$ instead of $H^2(\Omega)$) is not a problem since interpolants of Scott–Zhang type can be used [29].

We briefly discuss the essential points in the error analysis. For the consistency error we get from the L^2 stability of κ_h the estimate ($\alpha_K \sim h_K^2$)

$$|A_h((u-u_h),(p-p_h);(w_h,r_h))| = \left| \sum_{K\in\mathcal{T}_h} \alpha_K(\kappa_h(\nabla p),\kappa_h(\nabla q_h)) \right|$$
$$\leq C \left(\sum_{K\in\mathcal{T}_h} h_K^{2r} |\nabla p|_{r-1}^2 \right)^{1/2} |||(w_h,r_h)|||$$

provided D_h comprises piecewise polynomials of degree $r-2$. Using Lemma 1 and the estimation of the consistency error it remains to estimate the approximation error. The most difficult part of it is the estimate

$$|(r_h, \operatorname{div}(u-j_h u))| = |(\nabla r_h, u-j_h u)| = |(\kappa_h(\nabla r_h), u-j_h u)|$$
$$\leq C\, h^r |u|_{r+1} \left(\sum_{K\in\mathcal{T}_h} \alpha_K \|\kappa_h \nabla r_h\|_{0,K}^2 \right)^{1/2}$$

in which we used the orthogonality property of the interpolant. Putting all pieces together we get the main theorem for the Stokes problem.

Theorem 4 ([28]). *Let the solution of* (4) *be smooth enough such that* $(u,p) \in \left(V \cap H^{r+1}(\Omega)^d\right) \times \left(Q \cap H^r(\Omega)\right)$ *and* $P_{r-2}^{disc} \subset D_h$. *Then, under the assumptions of Lemma 1 and* $\alpha_K \sim \alpha_0 h_K^2$, *there exists a positive constant C independent of h such that*

$$|||(u-u_h, p-p_h)||| \leq C\, h^r (\|u\|_{r+1} + \|p\|_r).$$

Moreover, if the Stokes problem is $H^2(\Omega)^d \times H^1(\Omega)$ *regular, there exists a positive constant C independent of h such that*

$$\|u-u_h\|_0 \leq C\, h^{r+1}(\|u\|_{r+1} + \|p\|_r).$$

Note that in contrast to the PSPG approach for the LPS scheme considered we did not require higher regularity of the pressure when using equal-order interpolations.

3.4 Examples

In the following we list approximation and projection spaces from [28] satisfying all assumptions needed for the error estimate in Theorem 4. It turns out that some known stabilization methods in the literature can be recovered as special cases of the one-level LPS.

3.4.1 Simplicial Elements, First-Order Methods

Let the solution and projection spaces be given by $(V_h, Q_h) = (P_1^d, P_1)$ and $D_h = \{0\}$, respectively. Then the fluctuation operator becomes the identity and we get the method proposed by Brezzi and Pitkäranta in [30]. Now, let as above

b_K denote the (mapped) bubble function that belongs to P_{d+1} and vanishes at the boundary ∂K. We enrich the space of continuous, piecewise linear function by adding the bubble functions on each cell, i.e.,

$$P_1^+ = P_1 + \bigoplus_{K \in \mathcal{T}_h} \text{span}\, b_K.$$

If we enrich only the velocity space so that $(V_h, Q_h, D_h) = ((P_1^+)^d, P_1, \{0\})$, no stabilization is needed since the pair $(V_h, Q_h) = ((P_1^+)^d, P_1)$, the so called 'Mini'-element [31], satisfies the inf–sup condition (5). However, enriching the spaces for approximating velocity and pressure, we get an equal-order interpolation and the LPS becomes necessary. A possible choice with optimal first-order convergence is $(V_h, Q_h, D_h) = ((P_1^+)^d, P_1^+, P_0)$ [28].

3.4.2 Simplicial Elements, Higher-Order Methods

Unlike Subsect. 2.3 we consider the (less) enriched approximation space

$$\widetilde{P}_r^+ := P_r + \bigoplus_{K \in \mathcal{T}_h} b_K \cdot P_{r-2}(K)$$

which fits the projection space $D_h = P_{r-2}^{\text{disc}}$ and the choice $\alpha_K \sim \alpha_0 h_K^2$. Then an LPS method with optimal convergence order $r \geq 2$ is generated by $(V_h, Q_h, D_h) = ((\widetilde{P}_r^+)^d, \widetilde{P}_r^+, P_{r-2}^{\text{disc}})$ [28].

3.4.3 Hexahedral Elements, First-Order Methods

We consider first the case where the approximation and projection spaces are given by $(V_h, Q_h) = (Q_1^d, Q_1)$ and $D_h = \{0\}$, respectively. The fluctuation operator is the identity and we end up again with the stabilization proposed by Brezzi and Pitkäranta in [30]. Now, by enriching only the velocity space, we can derive pairs of finite elements (V_h, Q_h) satisfying the inf–sup condition (5) such that no stabilization is needed. Although similar to the case of triangular elements, where two additional degrees of freedom per cell have been added, in the quadrilateral case $(d = 2)$ we have to add at least three additional degrees of freedom in the conforming and non-conforming case [32, 33]. Further examples of enrichments of the velocity space leading to inf–sup stable element pair (V_h^+, Q_1) have been studied in [34, 35]. Enriching both the velocity and the pressure space, we get an equal-order interpolation and the LPS becomes needed. Let us enrich the space of continuous, piecewise multi-linear functions by adding the bubble functions on each cell, i.e.,

$$Q_1^+ = Q_1 + \bigoplus_{K \in \mathcal{T}_h} \text{span}\, \tilde{b}_K.$$

Now a possible choice of spaces with a first-order convergence property is $(V_h, Q_h, D_h) = ((Q_1^+)^d, Q_1^+, Q_0)$ [28].

3.4.4 Hexahedral Elements, Higher-Order Methods

It turns out that for hexahedral elements and $r \geq 2$ the standard spaces Q_r are already rich enough that the pair $(Y_h, D_h) = (Q_r, Q_{r-2}^{\mathrm{disc}})$ satisfies the local inf–sup condition of Theorem 1. Thus an LPS method with optimal convergence order $r \geq 2$ is generated by $(V_h, Q_h, D_h) = (Q_r^d, Q_r, Q_{r-2}^{\mathrm{disc}})$ [28]. Note that the 'smallest' projection space that guarantees the consistency order $r \geq 2$ is the mapped or unmapped space P_{r-2}^{disc}. Since in both cases the inclusion $P_{r-2}^{disc} \subset Q_{r-2}^{disc}$ holds true, the local inf–sup condition of Theorem 1 is still satisfied and we obtain the optimal convergence order also for the choice $(V_h, Q_h, D_h) = (Q_r^d, Q_r, P_{r-2}^{\mathrm{disc}})$. For details and other pairs of finite element spaces we refer to [28].

3.5 Elimination of Enrichments

It has been shown by Bank and Welfert in [36] that the bubble part of the velocity components for the Mini element discretization of the Stokes problem, i.e., $(V_h, Q_h) = ((P_1^+)^d, P_1)$, can be locally eliminated and lead to a formulation equivalent to the stabilized method proposed by Hughes, Franca, and Balestra in [4]. Furthermore, in [37] special enrichments of both the velocity space $V_h = P_1^d$ and the pressure space $Q_h = P_1$ have been introduced and shown to lead by static condensation to a Galerkin least squares stabilized formulation of the Stokes problem. For this, on each cell of the triangulation the velocity components are enriched by two bubble functions and the pressure by a function that does not vanish at the cell boundaries.

Of course, the additional degrees of freedom introduced by the enrichments of the $(V_h, Q_h, D_h) = ((P_1^+)^d, P_1^+, P_0)$-LPS can also be eliminated locally by static condensation. The resulting scheme corresponds to the stabilized method of Hughes, Franca, and Balestra [4] with an additional grad/div stabilization which can be written as

Find $(u_L, p_L) \in V_L \times Q_L = P_1^d \times P_1$ such that for all $(v_L, q_L) \in V_L \times Q_L$

$$(\nabla u_L, \nabla v_L) - (p_L, \operatorname{div} v_L) + \sum_{K \in \mathcal{T}_h} \gamma_K (\operatorname{div} u_L, \operatorname{div} v_L)_K = (f, v_L),$$

$$(q_L, \operatorname{div} u_L) + \sum_{K \in \mathcal{T}_h} (-\Delta u_L + \nabla p_L, \tau_K \nabla q_L)_K = \sum_{K \in \mathcal{T}_h} (f, \tau_K \nabla q_L)_K.$$

Here the parameters γ_K and τ_K behave like [28]

$$\gamma_K \sim \frac{h_K^2}{\alpha_K} \sim 1, \qquad \tau_K(x) \sim h_K^2 b_K(x)$$

which is in agreement with the suggested choice in the literature. The result in [28, 37] demonstrates that pressure bubbles play a role in explaining the addition of the least-squares form of the continuity equation in stabilized methods for the Stokes problem.

4 Oseen Problem

4.1 Standard Galerkin and LPS

We consider finally the Oseen problem

$$-\varepsilon\Delta u + (b\cdot\nabla)u + \sigma u + \nabla p = f \text{ in } \Omega,\ \nabla\cdot u = 0 \quad \text{in } \Omega,\ u = 0 \quad \text{on } \Gamma,$$

which can be understood as a testbed for developing stable and accurate approximations of the incompressible Navier–Stokes equations. The reason for that is that this simpler problem (a unique solution exists for all $\varepsilon > 0$) already includes the two sources of instabilities: the instability due to dominant convection ($\varepsilon \ll 1$) and the instability caused by pairs of finite elements that are not inf–sup stable. The weak formulation of the Oseen problem reads

Find $(u, p) \in V \times Q$ such that for all $(v, q) \in V \times Q$

$$\begin{aligned} A\big((u,p);(v,q)\big) :&= \varepsilon(\nabla u, \nabla v) + \big((b\cdot\nabla)u, v\big) + \sigma(u, v) \\ &\quad - (p, \operatorname{div} v) + (q, \operatorname{div} u) = (f, v) \end{aligned}$$

where $V := H_0^1(\Omega)^d$, $Q := L_0^2(\Omega)$, $\varepsilon > 0$, $\sigma \geq 0$, $b \in W^{1,\infty}(\Omega)$, $\operatorname{div} b = 0$ have been assumed. Now, let us consider the case of equal order interpolation in which the velocity and the pressure space are generated by the same scalar finite element space $Y_h \approx H^1(\Omega)$, namely $V_h := Y_h^d \cap V$ and $Q_h := Y_h \cap Q$ [6,7,38]. Then the stabilized discrete problem is:

Find $(u_h, p_h) \in V_h \times Q_h$ such that

$$(A+S)\big((u_h,p_h);(v_h,q_h)\big) = (f, v_h) \qquad \forall (v_h, q_h) \in V_h \times Q_h$$

where the stabilization term is given by

$$\begin{aligned} S\big((u_h,p_h);(v_h,q_h)\big) := \sum_{M\in\mathcal{M}_h} \Big[&\tau_M\big(\kappa_h((b\cdot\nabla)u_h), \kappa_h((b\cdot\nabla)v_h)\big)_M \\ &+ \mu_M\big(\kappa_h(\operatorname{div} u_h), \kappa_h(\operatorname{div} v_h)\big)_M + \alpha_M\big(\kappa_h(\nabla p_h), \kappa_h(\nabla q_h)\big)_M\Big] \end{aligned}$$

with user-chosen parameters τ_M, μ_M, and α_M. Here $\mathcal{M}_h$ denotes a decomposition of Ω into macro cells needed to define the projection spaces D_h while the approximation spaces live on a decomposition $\mathcal{T}_h$ not necessary equal to $\mathcal{M}_h$. Furthermore, $\kappa_h = \mathrm{id} - \pi_h$ is the fluctuation operator and π_h the (vector-valued) L^2 projection

Table 2 Possible (mapped) spaces in the LPS for the Oseen problem

one-level			two-level		
V_h^+	Q_h^+	D_h	V_h	Q_h	D_{2h}
$(P_r^+)^d$	P_r^+	P_{r-1}^{disc}	$(P_r)^d$	P_r	P_{r-1}^{disc}
$(Q_r^+)^d$	Q_r^+	P_{r-1}^{disc}	$(Q_r)^d$	Q_r	Q_{r-1}^{disc}

into the discontinuous projection space D_h. An interesting option is an additional projection space for controlling the fluctuations of the divergence since that term is fully consistent [38]. Under reasonable assumptions the bilinear form $A+S$ satisfies an inf–sup condition on the spaces Y_h and D_h with respect to the mesh-dependent norm

$$|||(v,q)|||_{OSE} := \Big(\varepsilon|v|_1^2 + \sigma\|v\|_0^2 + \rho\|q\|_0^2 + S\big((v,q);(v,q)\big)\Big)^{1/2}$$

with $\rho > 0$ [7]. Moreover, the following error estimate holds true.

Theorem 5 ([7]). *Let $\alpha_M, \mu_M, \tau_M \sim h_M$, b piecewise smooth, $P_{r-1}^{disc} \subset D_h$ and (Y_h, D_h) satisfies the local inf–sup condition* (3). *Then there is a positive constant C independent of h such that*

$$|||(u-u_h, p-p_h)|||_{OSE} \le C(\varepsilon^{1/2} + h^{1/2})\, h^r \left(\|u\|_{r+1} + \|p\|_{r+1}\right).$$

We show in Table 2 examples of spaces that satisfy all assumptions which guarantee the stated error estimate. Note that in the two-level approach we divide a macro simplex M into $d+1$ simplices K by connecting the barycenter with the vertices. A macro hexahedron is subdivided into 2^d hexahedrons in the usual way. For more details we refer to [7].

4.2 LPS for Inf–Sup Stable Elements

The local projection stabilization has been also applied to inf–sup stable discretizations of the Oseen equation in [8, 39]. An interesting point is that for inf–sup stable finite element pairs one does not need an $H^1(\Omega)$ stable interpolation operator with additional orthogonality properties to prove stability of the discrete problem, unlike the case of equal-order interpolation. Consequently, one has much more flexibility in choosing the approximation and projection spaces [39]. We replace the stabilizing term above by

$$
\begin{aligned}
S((u_h, p_h), (v_h, q_h)) := \sum_{K \in \mathcal{T}_h} \Big(& \tau_K (\kappa_h^1 (b \cdot \nabla u_h), \kappa_h^1 (b \cdot \nabla v_h))_K \\
& + \mu_K (\kappa_h^2 (\operatorname{div} u_h), \kappa_h^2 (\operatorname{div} v_h))_K \Big) \\
& + \sum_{E \in \mathcal{E}_h} \gamma_E \, \langle [p_h]_E, [q_h]_E \rangle_E
\end{aligned}
$$

in order to handle both continuous ($\gamma_E = 0$) and discontinuous ($\gamma_E > 0$) pressure spaces Q_h. Note that a pressure $p \in H^1(\Omega)$ does not cause any consistency error and that we have introduced two projection spaces resulting in two fluctuation operators. Most of the known inf–sup stable elements approximate the velocity components by elements of order r and the pressure by elements of order $r-1$, which yields error estimates of order r, cf. [8], which in the convection-dominated case ($\varepsilon < h$) is half an order less than the LPS with equal-order interpolation. However, the same convergence order can be achieved in the one-level case by standard finite element spaces without any enrichments [39]; a possible variant is

$$
(V_h, Q_h, D_h^1, D_h^2) = ((Q_r)^d, P_{r-1}^{\text{disc}}, (Q_{r-2}^{\text{disc}})^d, P_{t-1}^{\text{disc}})
$$

with the parameter choice $\tau_K \sim h_K$, $\mu_K \sim 1$, and $\gamma_E \sim h_E$. Furthermore, there are inf–sup stable elements approximating both the velocity components and the pressure by elements of order r, which yield error estimates of order $r + 1/2$ in the convection-dominated case. For details see [11, 39].

4.3 LPS as an hp-Method

The a priori error analysis and the parameter design of the LPS have been extended to study the dependence of the error not only on the mesh size but also on the polynomial degree [8, 38]. As an example we give a result for the two-level variant of equal order interpolation, i.e., we assume $(Y_h, D_{2h}) = (P_r, P_{r-1})$ with $r \geq 1$.

Theorem 6 ([38]). *Let $\nu \leq h_M / r^2$ and let the stabilization parameters be chosen as $\tau_M \sim h_M/r^2$, $\mu_M \sim h_M/r^2$, and $\alpha_M \sim h_M/r^2$. Then there is a constant $C = C(\beta_1) > 0$ independent of h such that for $l \leq r$*

$$
|||(u - u_h, p - p_h)|||_{OSE} \leq C(\beta_1) \frac{h^{l+1/2}}{r^l} (\|u\|_{l+1} + \|p\|_{l+1}).
$$

Compared with the interpolation error, this estimate is optimal with respect to h but in general not with respect to r.

4.4 LPS on Anisotropic Meshes

In Sect. 2.6 we discussed the convergence properties of the LPS on layer-adapted meshes. Unfortunately, no precise information is known regarding how the derivatives of the solution of the Oseen problem behave in different parts of the domain. Thus, an important ingredient for the construction of layer-adapted meshes is missing. Nevertheless, highly anisotropic meshes are often used to resolve layers. In [40] an extension of the LPS has been proposed which uses different scalings for the fluctuations of the derivatives in x and y direction. For the two-level approach with equal-order interpolation, i.e., $(V_h, Q_h, D_{2h}) = ((Q_1)^d, Q_1, (Q_0)^d)$, optimal anisotropic error estimates have been established.

References

1. M. Tabata. A finite element approximation corresponding to the upwind differencing. *Memoirs of Numerical Mathematics*, 1:47–63, 1977.
2. A.N. Brooks and T.J.R. Hughes. Streamline upwind/Petrov-Galerkin formulations for convection dominated flows with particular emphasis on the incompressible Navier-Stokes equations. *Comput. Methods Appl. Mech. Engrg.*, 32:199–259, 1982.
3. U. Nävert. *A finite element method for convection-diffusion problems*. PhD thesis, Chalmers University of Technology, Göteborg, 1982.
4. T.J.R. Hughes, L.P. Franca, and M. Balestra. Errata: "A new finite element formulation for computational fluid dynamics. V. Circumventing the Babuška-Brezzi condition: a stable Petrov-Galerkin formulation of the Stokes problem accommodating equal-order interpolations". *Comput. Methods Appl. Mech. Eng.*, 62(1):111, 1987.
5. L. Tobiska and R. Verfürth. Analysis of a streamline diffusion finite element method for the Stokes and Navier–Stokes equations. *SIAM J. Numer. Anal.*, 33:107–127, 1996.
6. M. Braack and E. Burman. Local projection stabilization for the Oseen problem and its interpretation as a variational multiscale method. *SIAM J. Numer. Anal.*, 43:2544–2566, 2006.
7. G. Matthies, P. Skrzypacz, and L. Tobiska. A unified convergence analysis for local projection stabilisations applied to the Oseen problem. *Math. Model. Numer. Anal. M2AN*, 41(4):713–742, 2007.
8. G. Rapin, G. Lube, and J. Löwe. Applying local projection stabilization to inf-sup stable elements. In Karl Kunisch, Günther Of, and Olaf Steinbach, editors, *Numerical mathematics and advanced applications. Proceedings of the 7th European Conference (ENUMATH 2007) held in Graz, September 10–14, 2007*, pages 521–528, Springer, Berlin, 2008.
9. M. Braack and G. Lube. Finite elements with local projection stabilization for incompressible flow problems. *J. Comput. Math. (to appear)*, 2008.
10. P. Knobloch and L. Tobiska. On the stability of finite element discretizations of convection-diffusion-reaction equations. *Preprint 08-11, Faculty of Mathematics, University Magdeburg*, 2008.
11. H.-G. Roos, M. Stynes, and L. Tobiska. *Robust numerical methods for singularly perturbed differential equations. Convection-diffusion-reaction and flow problems*. Number 24 in SCM. Springer, Berlin, 2008.
12. R. Becker and M. Braack. A two-level stabilization scheme for the Navier-Stokes equations. In *Numerical mathematics and advanced applications*, pages 123–130. Springer, Berlin, 2004.
13. G. Matthies, P. Skrzypacz, and L. Tobiska. Stabilization of local projection type applied to convection-diffusion problems with mixed boundary conditions. *ETNA*, 32:90–105, 2008.

14. J.-L. Guermond. Stabilization of Galerkin approximations of transport equations by subgrid modeling. *M2AN Math. Model. Numer. Anal.*, 33:1293–1316, 1999.
15. W.J. Layton. A connection between subgrid scale eddy viscosity and mixed methods. *Appl. Math. Comput.*, 133:147–157, 2002.
16. R. Codina. Stabilization of incompressibility and convection through orthogonal sub-scales in finite element methods. *Comput. Methods Appl. Mech. Eng.*, 190:1579–1599, 2000.
17. R. Codina. Stabilized finite element approximation of transient incompressible flows using orthogonal subscales. *Comput. Methods Appl. Mech. Eng.*, 191:4295–4321, 2002.
18. L. Tobiska. On the relationship of local projection stabilization to other stabilized methods for one-dimensional advection-diffusion equations. *Comput. Methods Appl. Mech. Eng.*, 2008. doi:10.1016/j:cma.2008.10.016.
19. L. Tobiska. Analysis of a new stabilized higher order finite element method for advection-diffusion equations. *Comput. Methods Appl. Mech. Engrg.*, 196:538–550, 2006.
20. T.J.R. Hughes and G. Sangalli. Variational multiscale analysis: the fine-scale Green's function, projection, optimization, localization, and stabilized methods. *SIAM J. Numer. Anal.*, 45(2):539–557, 2007.
21. S. Franz and G. Matthies. Local projection stabilization on S-type meshes for convection-diffusion problems with characteristic layers. *Preprint MATH-NM-07-2008, TU Dresden*, 2008.
22. G. Matthies. Local projection stabilization for higher order discretizations of convection-difusion problems on Shishkin meshes. *Adv. Comput. Math.*, 2008. doi: 10.1007/s10444-008-9070-y.
23. G. Matthies. Local projection methods on layer adapted meshes for higher order discretizations of convection-difusion problems. *Preprint July 1, 2008, Ruhr University Bochum*, 2008.
24. V. Girault and P.-A. Raviart. *Finite element methods for Navier–Stokes equations*. Springer, Berlin, 1986.
25. T.E. Tezduyar, S. Mittal, S.E. Ray, and R. Shih. Incompressible flow computations with stabilized bilinear and linear equal order interpolation velocity pressure elements. *Comput. Methods Appl. Mech. Eng.*, 95:221–242, 1992.
26. R. Codina and J. Blasco. A finite element formulation for the Stokes problem allowing equal velocity-pressure interpolation. *Comput. Methods Appl. Mech. Engrg.*, 143: 373–391, 1997.
27. R. Becker and M. Braack. A finite element pressure gradient stabilization for the Stokes equations based on local projections. *Calcolo*, 38(4):173–199, 2001.
28. S. Ganesan, G. Matthies, and L. Tobiska. Local projection stabilization of equal order interpolation applied to the Stokes problem. *Math. Comp.*, 77(264):2039–2060, 2008.
29. L.R. Scott and S. Zhang. Finite element interpolation of nonsmooth functions satisfying boundary conditions. *Math. Comp.*, 54:483–493, 1990.
30. F. Brezzi and J. Pitkäranta. On the stabilization of finite element approximations of the stokes problem. In W. Hackbusch, editor, *Efficient solution of elliptic systems*, Notes on Numerical Fluid Mechanics, pages 11–19. Vieweg, 1984.
31. D.N. Arnold, F. Brezzi, and M. Fortin. A stable finite element for the Stokes equation. *CALCOLO*, 21:337–344, 1984.
32. W. Bai. The quadrilateral 'Mini' finite element for the Stokes equation. *Comput. Methods Appl. Mech. Engrg.*:41–47, 1997.
33. L.P. Franca, S.P. Oliviera, and M. Sarkis. Continuous Q1/Q1 Stokes element stabilized with non-conforming null average velocity functions. *Math. Models Meth. Appl. Sci. (M3AS)*, 17:439–459, 2007.
34. P. Knobloch and L. Tobiska. Stabilization methods of bubble type for the Q1/Q1-element applied to the incompressible Navier-Stokes equations. *Math. Model. Numer. Anal. M2AN*, 34(1):85–107, 2000.
35. P. Mons and G. Roge. L' élément q_1-bulle/q_1. *Math. Model. Numer. Anal. M2AN*, 26:507–521, 1992.

36. R.E. Bank and B.D. Welfert. A comparision between the mini-element and the Petrov-Galerkin formulation for the generalized Stokes problem. *Comput. Methods Appl. Mech. Engrg.*, 83(1):61–68, 1990.
37. L.P. Franca and S.P. Oliviera. Pressure bubbles stabilization features in the Stokes problem. *Comput. Methods Appl. Mech. Engrg.*, 192:1929–1937, 2003.
38. G. Lube, G. Rapin, and J. Löwe. Local projection stabilization of finite element methods for incompressible flows. In Karl Kunisch, Günther Of, and Olaf Steinbach, editors, *Numerical mathematics and advanced applications. Proceedings of the 7th European Conference (ENUMATH 2007) held in Graz, September 10–14, 2007*, pages 481–488, Springer, Berlin, 2008.
39. G. Matthies and L. Tobiska. Local projection type stabilisation applied to inf-sup stable discretisations of the Oseen problem. Preprint 07–47, Fakultät für Mathematik, University Magdeburg, 2007.
40. M. Braack. A stabilized finite element scheme for the Navier-Stokes equations on quadrilateral anisotropic meshes. *Math. Model. Numer. Anal. M2AN*, 42:903–924, 2008.

Part II
Contributed Papers

Numerical Simulation of the Towing Tank Problem Using High Order Schemes

L. Beneš, J. Fürst, and Ph. Fraunié

Abstract The article deals with the numerical simulation of 2D and 3D unsteady incompressible flows with stratifications. The mathematical model is based on the Boussinesq approximation of the Navier–Stokes equations. The flow field in the towing tank with a moving sphere is modelled for a wide range of Richardson numbers. The obstacle is modeled via appropriate source terms. The resulting set of partial differential equations is then solved by the fifth-order finite difference WENO scheme, or by the second-order finite volume AUSM MUSCL scheme. For the time integration, the second-order BDF method was used. Both schemes are combined with the artificial compressibility method in dual time.

1 Introduction

Modelling of Atmospheric Boundary Layer (ABL) flows plays a significant role in many industrial applications. It is well known that the influence of the stratification is significant in many processes in ABL flows (e.g., it affects the transport of pollutants, plays a significant role in determining the environmental and human consequences of accidents). Stratified flows in environmental applications are characterized by a variation of fluid density in the vertical direction that can result in qualitative and quantitative changes of the flow by buoyancy. Stable stratification generally suppresses any vertical mixing of mass and momentum. The present work was motivated by a desire to obtain a better understanding of these effects.

L. Beneš (✉)
Department of Technical Mathematics, Faculty of Mechanical Engineering,
Czech Technical University in Prague, Karlovo náměstí 13, 12135 Praha 2, Czech Republic,
E-mail: benes@marian.fsik.cvut.cz

A.F. Hegarty et al. (eds.), *BAIL 2008 - Boundary and Interior Layers.*
Lecture Notes in Computational Science and Engineering,
DOI: 10.1007/978-3-642-00605-0,

2 Mathematical Model

The flow in ABL can be usually assumed to be incompressible. Nevertheless, the density is not constant owing to temperature changes, gravity, etc. Thus an equation for the density must be considered. This type of flow is described by the Navier–Stokes equations for viscous incompressible flow with variable density; these equations are simplified by the Boussinesq approximation. Density and pressure are divided into two parts: a background part (with subscript $_0$) plus a perturbation (with superscript $'$). The background component is chosen to fulfill the hydrostatic balance equation $\partial p_0(z)/\partial z = -\rho_0(z)g$. The system of equations obtained is partly linearized around the average state ρ_*. The resulting set of equations can be written in the form

$$
\begin{aligned}
\frac{D\rho'}{Dt} &= -w\frac{d\rho_0}{dz}, \\
\frac{D\mathbf{u}}{Dt} + \frac{1}{\rho_*}\nabla p' &= \nu\Delta\mathbf{u} + \frac{\rho'}{\rho_*}\mathbf{g} + \frac{1}{\rho_*}\mathbf{f}, \\
\nabla\mathbf{u} &= 0,
\end{aligned}
\tag{1}
$$

where ρ is the density, $\mathbf{u} = (u, v, w)$ is the velocity, p is the pressure, ν is the viscosity, $\mathbf{g} = (0, 0, -g)$ is the gravity and $\mathbf{f}$ represents other forces (e.g., Coriolis force, source terms). We assume that $\rho_* = 1$ and we shall omit the primes above the density and pressure disturbances.

Equations (1) are rewritten in the vector conservative form

$$
PW_t + F(W)_x + G(W)_y + H(W)_z = S(W).
$$

Here $W = [\rho, u, v, w, p]^T$, $F = F^{in} - \nu F^v$, $G = G^{in} - \nu G^v$ and $H = H^{in} - \nu H^v$ contain the inviscid fluxes F^{in}, G^{in}, H^{in} and viscous fluxes F^v, G^v, H^v, while S is the gravity and source term and $P = diag(1, 1, 1, 1, 0)$. These fluxes and source term are

$$
\begin{aligned}
F^{in}(W) &= [\rho u, u^2 + p, uv, uw, u]^T, & G^{in}(W) &= [\rho v, uv, v^2 + p, vw, v]^T, \\
H^{in}(W) &= [\rho w, uw, vw, w^2 + p, w]^T, & S(W) &= [-v d\rho_0/dz, 0, 0, -\rho g, 0]^T
\end{aligned}
\tag{2}
$$

$$
\begin{aligned}
F^v(W) &= [0, u_x, v_x, w_x, 0]^T, & G^v(W) &= [0, u_y, v_y, w_y, 0]^T, \\
H^v(W) &= [0, u_z, v_z, w_z, 0]^T.
\end{aligned}
$$

3 Numerical Schemes

3.1 Spatial Discretization

Two different numerical schemes were used for the spatial discretization. We discretize only terms containing spatial derivatives. The system of ordinary differential equations (with respect to the time derivative) that is generated is solved by an appropriate ODE method; see [Bla01].

The first scheme is based on a flux-splitting method for incompressible flow and WENO-interpolation. The second method is the finite volume AUSM MUSCL scheme with the Hemker–Koren limiter.

3.1.1 Flux Splitting for Incompressible Flows

The discretization in space is achieved by standard fourth-order differences for viscous terms and by the following high-order flux-splitting method [Issa85]. Divide the inviscid flux $F^{in}(W)$ into two parts, the convective flux $F^c(W) = [\rho u, u^2, uv, uw, 0]^T$ and the pressure flux $F^p(W) = [0, p, 0, 0, \beta^2 u]^T$, then approximate the flux derivative by

$$F^{in}(W)_x\big|_i \approx \frac{1}{\Delta x}\left[F^c_{i+1/2} - F^c_{i-1/2}\right] + \frac{1}{\Delta x}\left[F^p_{i+1/2} - F^p_{i-1/2}\right]. \tag{3}$$

Here each subscript denotes the value at the corresponding point on the Cartesian grid (or, in the AUSM case, the mean value over the corresponding finite volume). For simplification of the next text, only the spatial index i in the $x-$ direction is preserved; the remaining two indexes are omitted. The high-order weighted ENO scheme [Jiang96] is chosen as the interpolation method. The original WENO interpolation uses an upwind bias and it can be formally written in the following form (function `weno5` is described in [Jiang96]):

$$\phi_{i+1/2} = \begin{cases} \phi^+_{i+1/2} = \texttt{weno5}(\phi_{i-2}, \phi_{i-1}, \phi_i, \phi_{i+1}, \phi_{i+2}) & \text{if } u_{i+1/2} > 0, \\ \phi^-_{i+1/2} = \texttt{weno5}(\phi_{i+3}, \phi_{i+2}, \phi_{i+1}, \phi_i, \phi_{i-1}) & \text{if } u_{i+1/2} \le 0. \end{cases} \tag{4}$$

It is still necessary to determine the velocity $u_{i+1/2}$.

This interpolation is applied to the incompressible case separately for the convective and pressure terms. In agreement with mathematical analysis the convective part is discretized by simple upwinding, the third component of the pressure is approximated by backward differencing and the fourth component by a forward difference. The final scheme takes the form

$$u_{i+1/2} := (u^+_{i+1/2} + u^-_{i+1/2})/2,\ p_{i+1/2} := (p^+_{i+1/2} + p^-_{i+1/2})/2, \tag{5}$$

$$F^c(W)_{i+1/2} := \begin{bmatrix} (\rho u)^{\pm}_{i+1/2} \\ (u^2)^{\pm}_{i+1/2} \\ (uv)^{\pm}_{i+1/2} \\ (uw)^{\pm}_{i+1/2} \\ 0 \end{bmatrix}, \quad F^p(W) := \begin{bmatrix} 0 \\ p_{i+1/2} + \beta \frac{u^+_{i+1/2} - u^-_{i+1/2}}{2} \\ 0 \\ 0 \\ u_{i+1/2} + \frac{p^+_{i+1/2} - p^-_{i+1/2}}{2\beta} \end{bmatrix}, \tag{6}$$

where + or − is taken in the convective flux according to the sign of $u_{i+1/2}$.

A similar algorithm is applied in other directions for the fluxes G, H. The resulting scheme has high-order accuracy in space. It was validated for the case of compressible inviscid flows by a computation of shock-vortex interaction; see [Furst96].

3.1.2 AUSM Scheme

The finite volume AUSM scheme was used for spatial discretization of the inviscid fluxes in our second scheme. Until now we have applied it only in the 2D case but an extension to 3D is being prepared.

$$\int_{\Omega} (F^{in}_x + G^{in}_y) dS = \oint_{\partial\Omega} (F^{in} n_x + G^{in} n_y) dl$$
$$\approx \sum_{k=1}^{4} \left[u_n \begin{pmatrix} \varrho \\ u \\ v \\ \beta^2 \end{pmatrix}_{L/R} + p \begin{pmatrix} 0 \\ n_x \\ n_y \\ 0 \end{pmatrix} \right] \Delta l_k \tag{7}$$

where n is the normal vector, u_n the normal velocity vector, and $(q)_{L/R}$ are quantities on the left/right hand side of the face. These quantities are computed using MUSCL reconstruction with the Hemker–Koren limiter:

$$q_R = q_{i+1} - \frac{1}{2}\delta_R, \quad q_L = q_i + \frac{1}{2}\delta_L,$$

$$\delta_{L/R} = \frac{a_{L/R}(b^2_{L/R} + 2) + b_{L/R}(2a^2_{L/R} + 1)}{2a^2_{L/R} + 2b^2_{L/R} - a_{L/R} b_{L/R} + 3},$$

$$a_R = q_{i+2} - q_{i+1}, \quad a_L = q_{i+1} - q_i, \quad b_R = q_{i+1} - q_i, \quad b_L = q_i - q_{i-1}.$$

Since the pressure is discretized using central differences, the scheme is stabilized following [Vier99] by a pressure diffusion of the form

$$F_{\mathbf{d}i+1/2,j} = \left(0,\ 0,\ 0,\ \eta \frac{p_{i+1,j} - p_{i,j}}{\beta_x}\right)^T, \qquad \beta_x = w_r + \frac{2\nu}{\Delta x}$$

where T denotes transpose and w_r is a reference velocity (in our case the maximum velocity in the flow field). Viscous fluxes are discretized using central differences on the dual mesh. This scheme is second-order accurate in space.

3.2 Time Integration

The spatial discretization yields a system of ODE in the physical time t variable, which is solved by the second-order BDF formula

$$P\frac{3W^{n+1} - 4W^n + W^{n-1}}{2\Delta t} + \tilde{F}_x(W^{n+1}) + \tilde{G}_y(W^{n+1}) + \tilde{H}_z(W^{n+1}) = \tilde{S}^{n+1}. \tag{8}$$

Here each tilde denotes a discrete approximation of F_x, G_y, H_z, S. Set

$$Res(W^{n+1}, W^n, W^{n-1}) = P(\frac{3}{2\Delta t}W^{n+1} - \frac{2}{\Delta t}W^n + \frac{1}{2\Delta t}W^{n-1}) + \\ + \tilde{F}_x(W^{n+1}) + \tilde{G}_y(W^{n+1}) + \tilde{H}_z(W^{n+1}) - \tilde{S}^{n+1}.$$

The above formula (8) is $Res(W, W^n, W^{n-1}) = 0$. It is solved by an artificial compressibility method in the dual time τ. The system of equations

$$\tilde{P}W_\tau + Res(W, W^n, W^{n-1}) = 0$$

where $\tilde{P} = diag(1, 1, 1, 1, \frac{1}{\beta^2})$, is solved by an explicit 3-stage second-order Runge–Kutta method.

4 Obstacle Modelling

We are interested in the solution of the stratified flows past a moving body. The obstacle is modelled very simply as a source term emulating a porous media with small permeability. This volume penalization technique was originally proposed by Arquis and Caltagirone [Cal84]. The source term $S(W)$ in this case is given by

$$\left[-v\frac{d\rho_0}{dz}, 0, 0, -\rho g, 0\right]^T + \frac{\chi(x, y, z, t)}{K}\left[0, U^{ob} - u, V^{ob} - v, W^{ob} - w, 0\right]^T, \tag{9}$$

where K corresponds to small permeability and $\chi(x, y, z, t)$ is the characteristic function of the obstacle, which moves with velocity (U^{ob}, V^{ob}, W^{ob}).

To estimate the influence of the permeability K, a very simple analytical model was developed. We suppose a 1D case, with the obstacle at rest and U_0 the velocity of the incoming flow. The flow at the obstacle is decelerated only by the resistance of the body; other terms are omitted. This situation leads to

$$\dot{u} = -u/K, \qquad u(0) = U_0.$$

Integrating the velocity as $t \to \infty$, we obtain an estimate of the depth of penetration of fluid into the body:

$$u(t) = U_0 e^{-\frac{t}{K}} \quad \Rightarrow \quad s = \int_0^\infty u(t)dt = U_0 K.$$

If we prescribe the depth of penetration (this may be interpreted as the effective diameter of the obstacle), we can estimate the permeability K. For instance, in the case of a sphere of radius $r = 0.1$ m, a velocity $U_0 = 1\,\mathrm{m\,s^{-1}}$ and a penetration depth of 10% of r lead to $K = s/U_0 = 1/100$.

5 Numerical Results

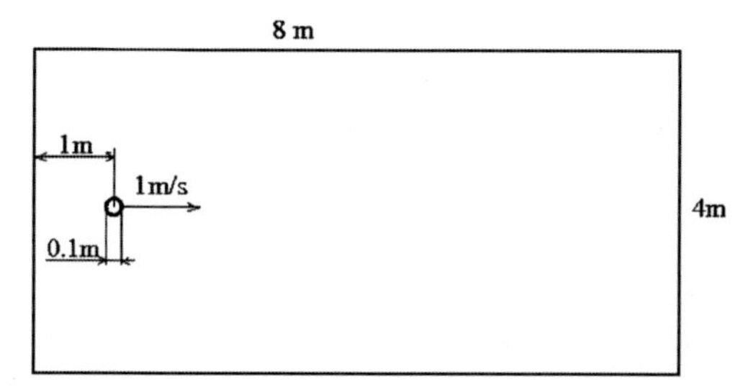

Towing tank

The obstacle is a sphere of radius 0.1 m, located 1 m from the left wall and at the midpoints of height and width see [Benes08]. At time $t = 0$ the obstacle starts moving to the right (in the positive x direction) with constant velocity $U^{ob} = 1\,\mathrm{m\,s^{-1}}$. The flow field is initially at rest with stable density gradient $d\rho_0/dz = -0.1\,\mathrm{kg\,m^{-4}}$. The average density is $\rho_* = 1\,\mathrm{kg\,m^{-3}}$ and the kinematic viscosity is $\nu = 10^{-4}\,\mathrm{m^2\,s^{-1}}$. Homogeneous Dirichlet boundary conditions for the velocity and Neumann conditions for the density and pressure disturbances were used in 2D. In 3D, these boundary conditions were extended by periodic boundary conditions in the y-direction.

The problem was solved on Cartesian grids. In 3D, a mesh with $320 \times 40 \times 160$ cells was used. In 2D, a mesh with 320×160 nodes and, for testing of the mesh independence, a fine grid with 640×320 nodes were used.

Various stratification levels were modelled. To describe the stratification, the following bulk Richardson number is used:

$$Ri = \frac{g \frac{d\rho_0}{dz}}{\varrho_* U^{ob}}$$

For the numerical tests, the towing tank problem was used. The towing tank is a channel with the obstacle inside. Motion of this obstacle generates disturbances in the flow field. In the cases we solved, the towing tank has dimensions 8 m × 4 m in 2D or 8 m × 4 m × 1 m in 3D.

The degree of stratification is unaffected by changes in the density gradient, but by modifying the gravity constant in the range $g \in< 0, 1000 >$. The corresponding Richardson numbers satisfy $Ri \in< 0, 100 >$. The influence of permeability was also tested for selected values in range $K^{-1} \in< 0, 1000 > \mathrm{s}^{-1}$. The two numerical methods were compared.

Figures 1 and 2 compare the schemes in 2D. In the first figure we can see the comparison of density isolines at the time $t = 5\,\mathrm{s}$. The second figure displays the distribution of selected quantities in the transversal direction. These figures exhibit good agreement between both methods, especially further from the obstacle, while small differences occur behind the sphere. The maximal values predicted by WENO 5 scheme at the height midpoint are somewhat lower. Next, Fig. 3 examines

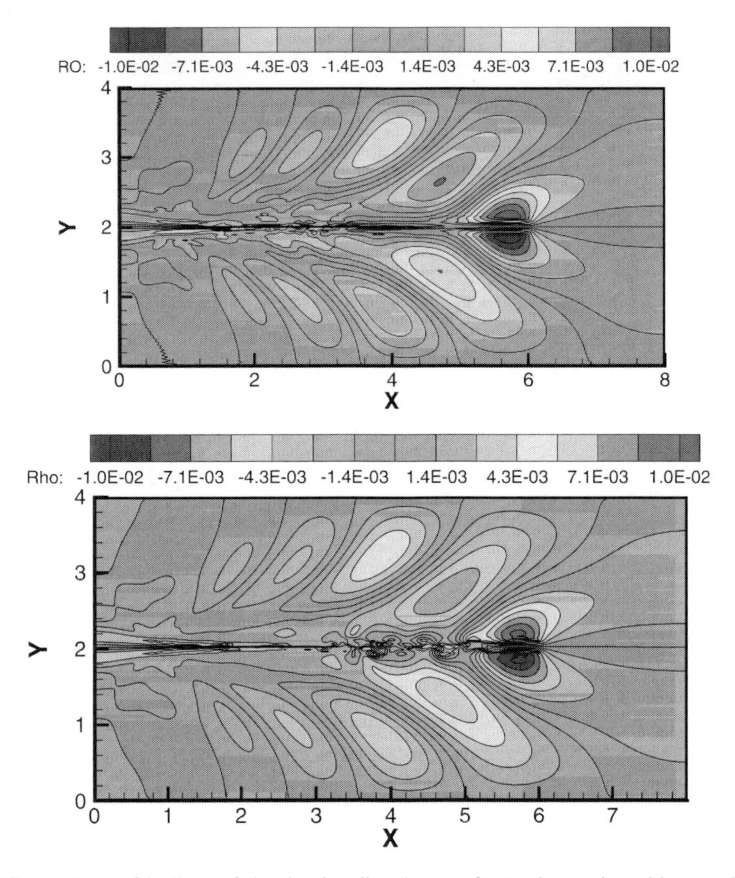

Fig. 1 Comparison of isolines of the density disturbances for towing tank problem at the time $t = 5\,\mathrm{s}$, $g = 100$, $Ri = 10$. AUSM MUSCL scheme (*top*) and WENO5 (*bottom*)

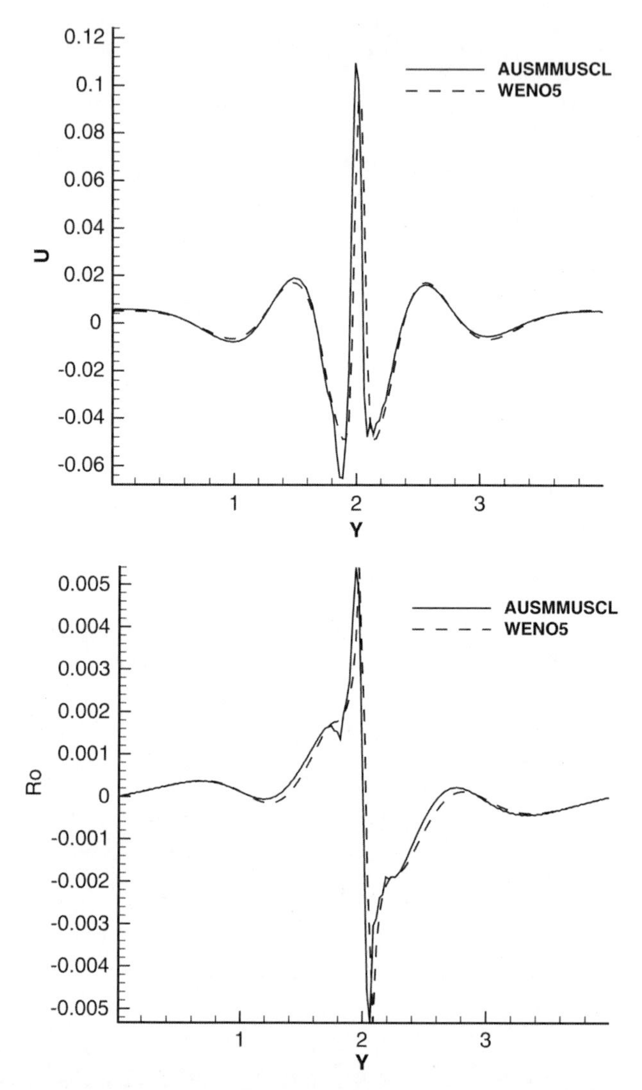

Fig. 2 Comparison of both schemes, $Ri = 10$, time $t = 5$ s. Transversal distribution of the u-velocity component (*top*) and density disturbances (*bottom*), $y = 2.25$

the dependence of the solution on the mesh and shows that the solution is relatively mesh independent. Only the maxima of quantities at the height midpoint behind the obstacle are lower and they are probably not resolved correctly on this coarse mesh.

Figure 4 shows the dependence of the solution on the permeability K for the three different values $1/K = 10, 100, 1000$. For the values 100 and 1,000 the solutions are very similar and the dependence on K is low. The obstacle can be considered as impermeable for $1/K \geq 100$. The results are also in good agreement with the predictions given by our simple analytical model.

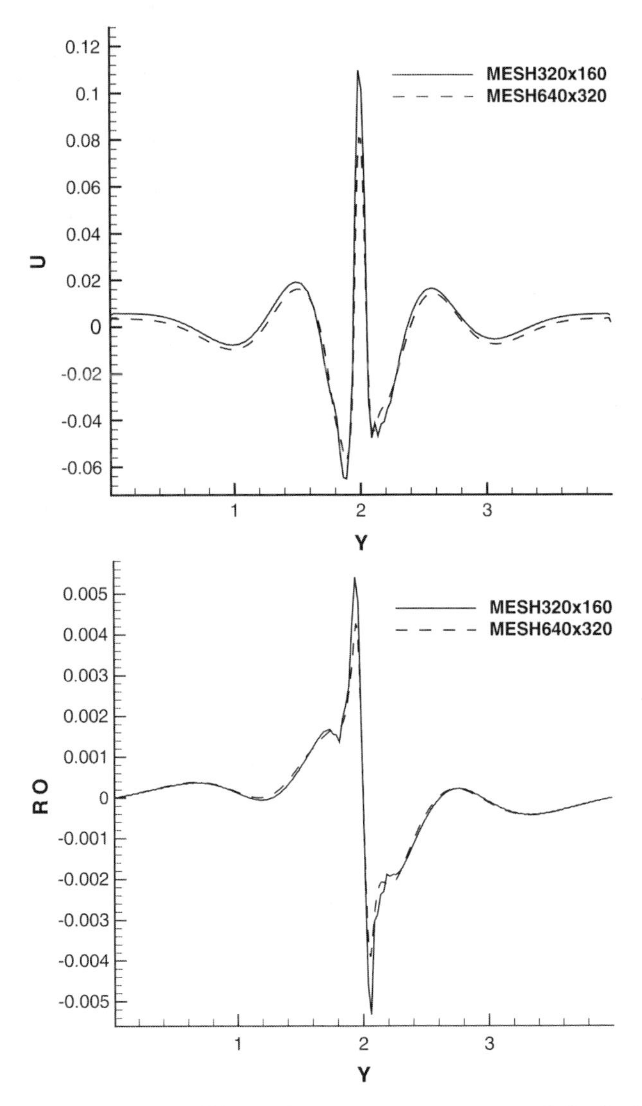

Fig. 3 Dependence on the mesh, $Ri = 10$, time $t = 5$ s. Transversal distribution of the u-velocity component (*top*) and density disturbances (*bottom*), $x = 1$

Figure 5 displays the dependence of the flow on the Richardson number. A comparison of the isolines of density perturbation for four different Richardson numbers ($Ri = 0.1, 1, 10, 100$) is presented at the time $t = 6$ s. At a lower level of stratification behind the obstacle, a Karman vortex street forms. When the level of stratification increases, the character of the flow changes; wake instabilities are damped by stratification and internal gravity waves are clearly visible. Beyond this level, the obstacle generates a strip with constant density. The changes in the character of the flow are clearly visible in Fig. 6, where transversal distribution of

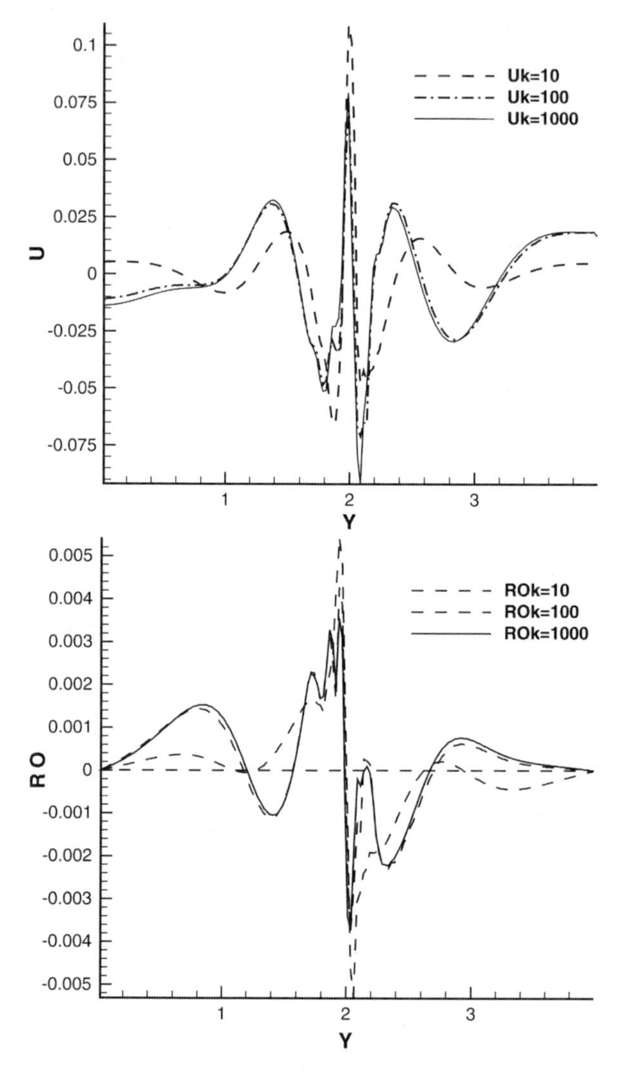

Fig. 4 Dependence on the permeability parameter $\boldsymbol{K}$, $\boldsymbol{Ri} = \mathbf{10}$, time $t = 5$ s. Transversal distribution of the $\boldsymbol{u}$-velocity component (*top*) and density disturbances (*bottom*), $x = 1$

computed quantities for different Richardson numbers are shown. For comparison see [Ber01].

The isosurfaces of the vorticity in 3D for the Richardson numbers $Ri = 1$ and $Ri = 10$ are shown in Fig. 7. The marked influence of stratification can be seen at the x–z cross-section. In the case $Ri = 1$, the influence of stratification is small and the shape of vorticity in the cross-section is close to a circle. On the other hand, for the higher level of stratification $Ri = 10$ the vortices are damped differently in different directions, which leads to an asymmetry in the vorticity isosurface.

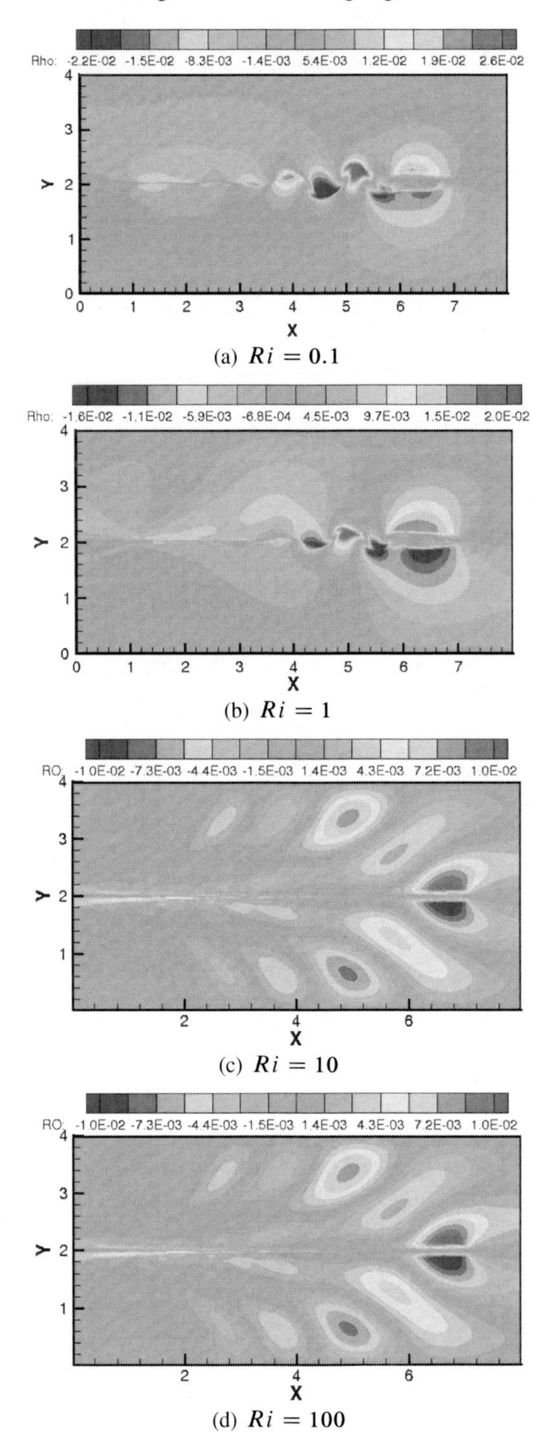

Fig. 5 Isolines of density perturbations for different values of Ri. Time $t = 6$ s

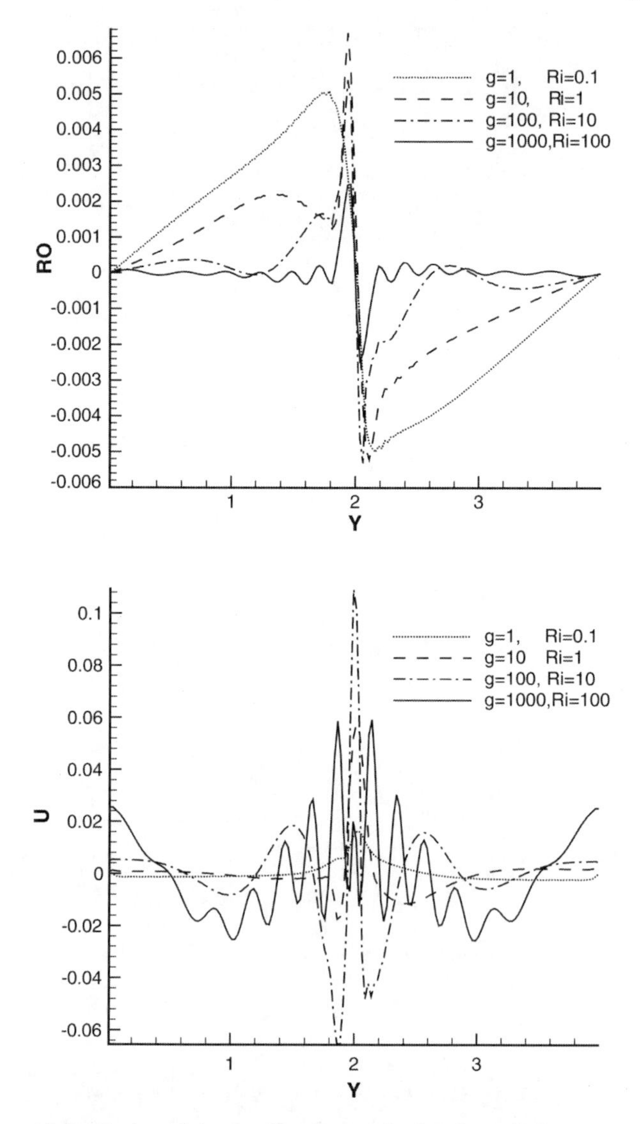

Fig. 6 Transversal distribution of density disturbances (*top*) and u-velocity component (*bottom*) for different Richardson numbers, $x = 1$, time $t = 6$ s

The isosurfaces of the density perturbations in 3D for the same Richardson numbers are shown in Fig. 8. The internal gravity waves with Brunt–Väisälä frequency are clearly visible.

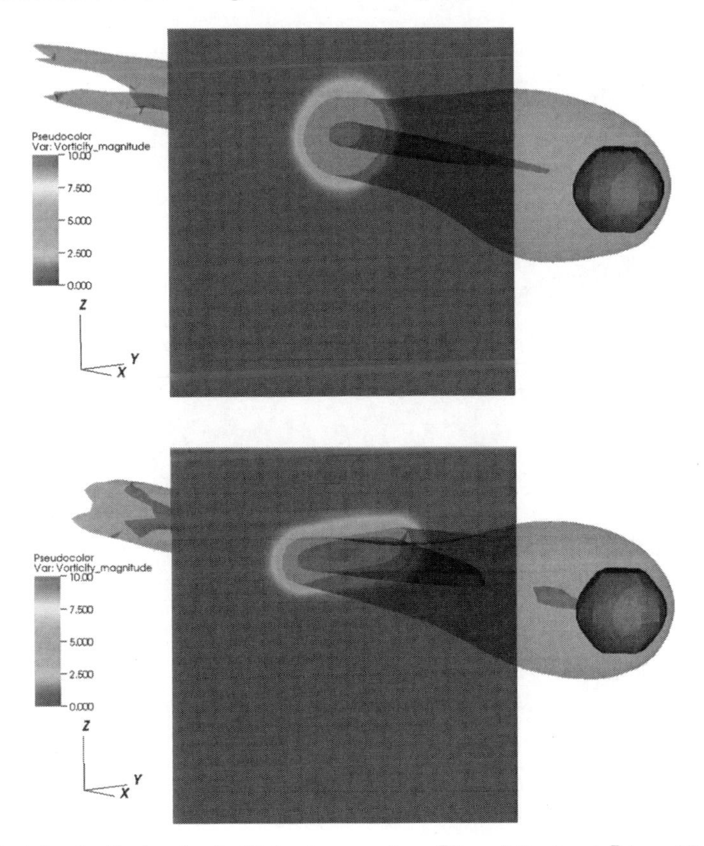

Fig. 7 Vorticity distribution for the Richardson numbers $Ri = 1$ (*top*) and $Ri = 10$ (*bottom*), time $t = 5$ s

6 Conclusion

Two numerical methods for simulation of 2D and 3D stratified flows have been developed. Such simulations are necessary for more complicated situations, where experimental data or other information about solution is no longer available. Since the solution can depend on the numerical scheme, a comparison of solutions obtained using different methods eliminates this dependence. Both methods have been used successfully for the towing tank problem. The numerical results obtained are in good mutual agreement and also match physical expectations.

Numerical results were obtained for Richardson numbers $Ri \in <0, 100>$ and permeability $K \in <1, 1000>$. From this, according to our simple analytical model, it follows that the minimal value of permeability is $K \geq 100$. The dependence of the solution on the mesh was also tested. The computations performed demonstrate the applicability of our methods to the simulations of stratified flows.

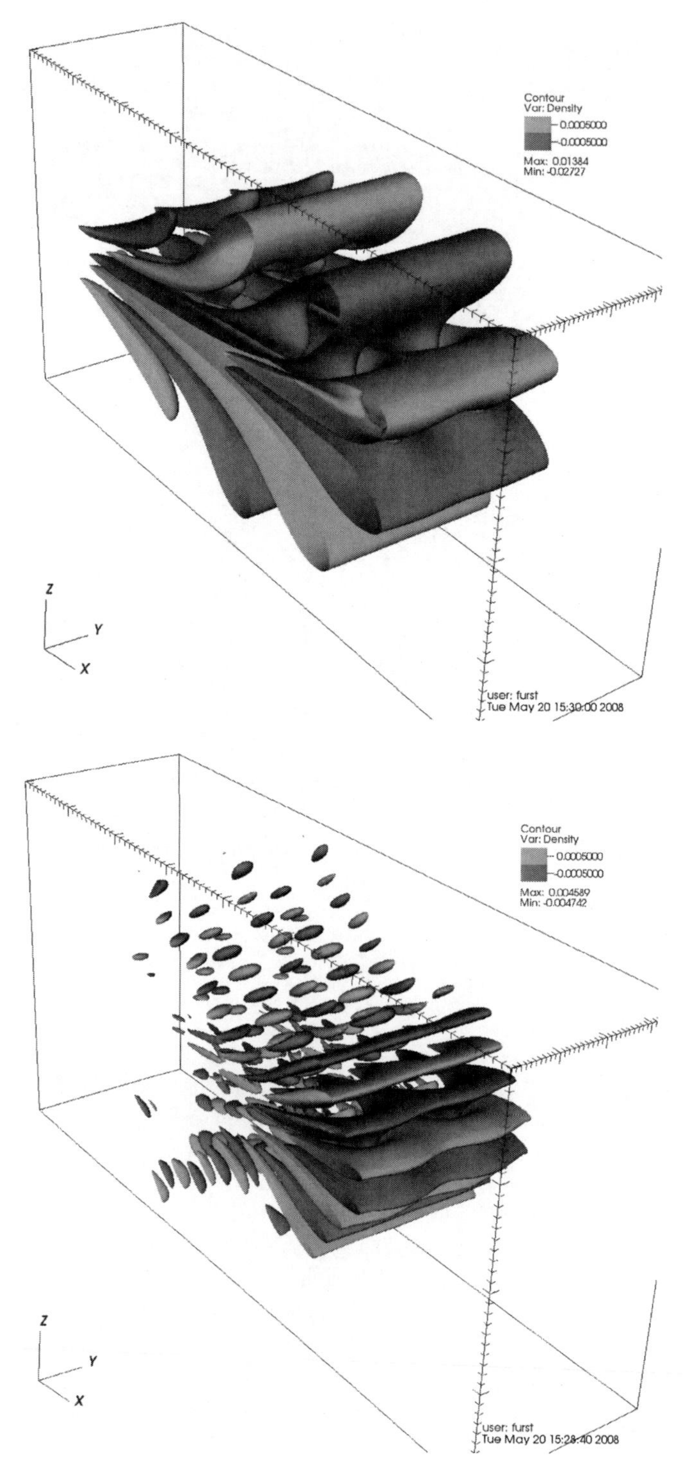

Fig. 8 Isosurfaces of the density perturbations at the time $t = 5$ s; $Ri = 10$ on *top*, $Ri = 100$ on *bottom*

An open question is the choice of appropriate boundary conditions; those used in the current approach are suitable for the simulation of flows in a domain bounded with walls. Alternative conditions should be considered for free atmosphere flows.

Acknowledgements This work was supported by Research Plan MSM 6840770003, by GACR Project No.205/06/0727 and by grant GA 201/08/0012.

References

[Cal84] Arquis E., Caltagirone J.P.: Sur les conditions hydrodynamiques au voisinage dune interface milieu fluide milieu poreux: application a la convection naturelle. C.R. Acad. Sci. Paris II **299**, 1–4 (1984)

[Bla01] Blazek J.: Computational Fluid Dynamics: Principles and Applications. Elsevier Science, Amsterdam, 2001, ISBN 0080430090

[Ber01] Berrabaa S., Fraunie P.H., Crochet M.: 2D large eddy simulation of highly stratified flows: the stepwise structure effect. Advances in Computation. Scientific Computing and Applications, volume 7. Nova, Hauppauge, NY, 2001, pp. 179–185

[Jiang96] Chi-Wang Shu, Guang-Shan Jiang: Efficient implementation of weighted eno schemes. J. Comput. Phys. **126**, 202–228 (1996)

[Furst96] Kozel K., Angot Ph., Fürst J.: TVD and ENO schemes for multidimensional steady and unsteady flows. In: Benkhaldoun F., Vilsmeier R. (eds) Finite Volumes for Complex Applications. Hermes, Paris, 1996, pp. 283–290

[Vier99] Dick E., Vierendeels J., Riemslagh K.: A multigrid semi-implicit line-method for viscous incompressible and low-mach-number flows on high aspects ratio grids. J. Comput. Phys. **154**, 310–341 (1999)

[Issa85] Issa R.I.: Solution of the implicitly discretized fluid flow equations by operator-splitting. J. Comput. Phys. **62**, 40–65 (1985)

[Benes08] Fraunie Ph., Beneš L., Fürst J.: Numerical simulation of the stratified flow. In: Proceeding of conference Topical Problems of Fluid Mechanics 2008, 5–8 (2008)

Nonlinear Singular Kelvin Modes in a Columnar Vortex

Philippe Caillol and Sherwin A. Maslowe

Abstract This paper considers the propagation of helical neutral modes within a cylindrical vortex and the subsequent formation of nonlinear critical layers around the radius where the mean-flow angular velocity and the mode frequency are comparable. Analogy can be done with the stratified critical layers. We formulate a steady-state theory valid when the analogous Richardson number is small at the critical radius. The apparent singularity is removed by retaining nonlinear terms in the critical-layer equations of motion. The result from the interaction is the emergence of multipolar vortices whose poles are located around the critical radius, spiral along the basic vortex axis and are embedded in a distorted mean flow caused by a slow diffusion of the three-dimensional vorticity field from the critical layer.

1 Introduction

The propagation of helical perturbations to a columnar and bounded vortex has been studied first by Lord Kelvin. In cylindrical coordinates (r, θ, z), the problem involves the investigation of infinitesimal perturbations (u_r, u_θ, u_z) superimposed on a flow with azimuthal velocity profile $\overline{V}(r)$. In this paper, we are interested in waves propagating in an unbounded vortex. A model that has often been employed is the discontinuous Rankine vortex, a constant-vorticity cylinder embedded in a zero-vorticity space. The related modes are called Kelvin modes. The motivation stems with the study of the stability of interacting vortices. For instance, we can cite the aircraft trailing vortices: a pair of counter-rotating vortex filaments shed from the wingtips of aircraft. A prevailing instability in such problems is the elliptic instability that involves resonantly interacting Kelvin waves. Tsai and Widnall [11] found that the most unstable perturbations of the Rankine vortex corresponded to a pair of

P. Caillol (✉)
Department of Applied Mathematics, Sheffield University, Sheffield, S3 7RH, UK,
E-mail: P.L.Caillol@shef.ac.uk

A.F. Hegarty et al. (eds.), *BAIL 2008 - Boundary and Interior Layers*.
Lecture Notes in Computational Science and Engineering,
DOI: 10.1007/978-3-642-00605-0,

Kelvin modes having zero frequency and azimuthal wavenumbers $m = \pm 1$. Real vortices, however, have continuous profiles and it is important to ask what effect the use of a continuous-vorticity profile might have on this instability mechanism. Sipp and Jacquin [9] have recently done so in a linear study and they concluded that the "Widnall instabilities" would not occur because of the presence of a critical layer. The neutral Kelvin modes required for the resonant interaction would be damped in the continuous case. In this paper, we reexamine the question by emphasizing the effect of nonlinearity rather than viscosity in the critical layer.

Due to the similarity between both critical-layer singularities, it is possible to anticipate certain results based on those that have been demonstrated for stratified shear flows in [5] which is a companion paper. For Kelvin modes on vortices, we will extract an equivalent Richardson number and will show that when the latter is small at the critical level, inviscid nonlinear modes exist while they would be damped if viscosity were used to deal with the critical layer.

The reason of the nonlinear neutral mode existence is the absence of any phase change across the critical point. We will show in Sect. 4 by means of an inviscid analysis valid when the vorticity is small at the critical level that the only solution compatible with a nonlinear critical layer has no phase jump. Section 5 yields the same result when the axial wavelength is large. This result was found by Caillol and Grimshaw (2004) in the two-dimensional-motion assumption with the same small-vorticity approximation and for a Bessel function J_1 azimuthal-velocity basic profile [3]. In that particular case, neutral modes have an analytical expression.

2 Outer Flow

We consider small-amplitude helical perturbations to a swirling flow $\overline{V}(r)$ corresponding to a pressure distribution $\overline{p}(r)$, of phase $\xi = kz + m\theta - \omega t$, k and m being respectively the axial and azimuthal wavenumbers, and ω the frequency. Dealing with neutral modes, ξ can be used as an independent variable. The momentum and continuity equations can then be written

$$\frac{Du_r}{Dt} = \frac{u_\theta^2}{r} - \frac{\partial p}{\partial r} + \frac{1}{Re}\frac{1}{r}\frac{\partial}{\partial r}\Big(r\frac{\partial u_r}{\partial r}\Big), \tag{1}$$

$$\frac{Du_\theta}{Dt} = -\frac{u_r\,u_\theta}{r} - \frac{m}{r}\frac{\partial p}{\partial \xi} + \frac{1}{Re}\frac{1}{r}\frac{\partial}{\partial r}\Big(r\frac{\partial u_\theta}{\partial r}\Big) + F_b,$$

$$\frac{Du_z}{Dt} = -k\frac{\partial p}{\partial \xi} + \frac{1}{Re}\frac{1}{r}\frac{\partial}{\partial r}\Big(r\frac{\partial u_z}{\partial r}\Big),$$

where $$\frac{D}{Dt} = \Big(\frac{m}{r}u_\theta + ku_z - \omega\Big)\frac{\partial}{\partial \xi} + u_r\frac{\partial}{\partial r}$$

and $$\frac{\partial(r\,u_r)}{\partial r} + m\frac{\partial u_\theta}{\partial \xi} + k\,r\frac{\partial u_z}{\partial \xi} = 0\,.$$

Our analysis being primarily inviscid, we have retained in the momentum equations only those viscous terms that will be the largest in the critical layer and, consequently, required in the analysis to follow. The basic-vortex small viscous damping is balanced by a body force F_b whose expression will be given later on. The equations have been nondimensionalized by using the angular velocity of the vortex at its center and a vortex characteristic radius.

2.1 The Singular Mode

$O(\varepsilon)$ disturbances are superimposed to a mean flow $\overline{V}$ and $\overline{W}$:

$$u_r = \varepsilon\, U_r\,, \quad u_\theta = \overline{V} + \varepsilon\, U_\theta\,, \quad u_z = \overline{W} + \varepsilon\, U_z\,, \quad p = \overline{p} + \varepsilon\, P_r. \qquad (2)$$

ε is a small dimensionless parameter. The mean shear flow induces axial and azimuthal mean vorticities $Q_z = D_*[\overline{V}]$ and $Q_\theta = -D[\overline{W}]$ where $D = d/dr$ and $D_* = D + 1/r$. The angular rotation of the vortex is denoted $\Omega = \overline{V}/r$. We study the asymptotic steady régime following the critical layer formation induced by the wave/vortex interaction. Mean axial and azimuthal motions are generated while the critical layer is forming as results from this interaction. To have an analytically tractable problem, $\overline{W}$ will be of smaller amplitude than the basic-vortex azimuthal velocity $\overline{V}_0$. In the same way, $\overline{V}$ contains additional smaller contributions to $\overline{V}_0$. Such a mean flow is produced by viscous diffusion of momentum through the critical layer over a very long time due to small viscosity [4]. Away from the critical layer, the perturbations are taken sinusoidal: $U_r = u \sin\xi$, $U_\theta = v \cos\xi$, $U_z = w \sin\xi$ and $P_r = p_r \cos\xi$. Introducing these into (1), the linearized system can be reduced to the Howard–Gupta equation [7]

$$D[S(r)D_*u] + \Big[\frac{m}{\gamma(r)r^2}\Big(2S(r)Q_z(r) - rD[S(r)Q_z(r)]\Big) + 2k^2S(r)\Omega(r)\frac{Q_z(r)}{\gamma^2(r)}$$
$$+ \frac{k}{r\gamma(r)}\Big(rD[S(r)Q_\theta(r)] - S(r)Q_\theta(r)\Big) + 2mk\frac{\Omega(r)}{r\gamma^2(r)}S(r)Q_\theta(r) - 1\Big]u = 0, \qquad (3)$$

where $\gamma = m\Omega + k\overline{W} - \omega$ and $S = r^2/(m^2 + k^2r^2)$. This equation admits a singularity at the critical radius r_c where $\gamma(r_c) = 0$. Following [8], we expand all terms in (3) around r_c to obtain a solution valid locally having the form

$$u(\eta) = A\, u_+(\eta) + B\, u_-(\eta)\,, \;\; u_\pm(\eta) = \eta^{\frac{1}{2}(1\pm\nu)}\, \hat{u}_\pm(\eta), \text{ and } \eta = r - r_c. \qquad (4)$$

The functions $\hat{u}_\pm(\eta)$ are regular in 0. We define ν as $\nu = (1 - 4\,J_c)^{1/2}$, and the equivalent local Richardson number as

$$J_c = \frac{2k^2\,\Omega_c}{(m\Omega_c' + k\overline{W}_c')^2}\Big(Q_{z,c} + \frac{m}{kr_c}Q_{\theta,c}\Big). \qquad (5)$$

3 Critical Layer Analysis

The critical-layer scaling is determined by balancing the perturbation with the swirling flow in a frame moving with the wave angular speed $\omega/m = \Omega_c$. Let us concentrate on the case $J_c < 1/4$ corresponding to (4). The most singular Frobenius solution is characterized by the exponent $\delta = (1-\nu)/2$. Consideration of the system (1) leads to the conclusion that the inner cross-stream coordinate is $R = (r-r_c)/\varepsilon^\beta$ where $\beta = (2-\delta)^{-1}$. The azimuthal velocity V in the new frame is defined by

$$u_\theta - \overline{V}_c \sim \overline{V}_c{}'(r-r_c) + \varepsilon\, v(r)\cos\xi = \varepsilon^\beta [V(R,\xi) + \Omega_c\, R] .$$

The remaining dependent variables are scaled as

$$u_r = \varepsilon^{2\beta} U(R,\xi),\ u_z = \varepsilon^\beta W(R,\xi) \quad \text{and} \quad p - \frac{1}{2}\Omega_c^2 r^2 = \varepsilon^{2\beta} P(R,\xi) .$$

The Reynolds number scales as $1/Re = \lambda\,\varepsilon^{3\beta}$. Substituting these new variables into (1) leads to the critical-layer equations. The latter are highly nonlinear and the solution even at lowest order involves all the harmonics. For that reason, we consider the case of small Richardson number because a simple closed form solution is possible. This number is small for different régimes according to the value of the shear, ratio of the vorticity over the inertial frequency at the critical radius: (1) $S_z,\ S_\theta \ll 1$, (2) $kr_c,\ S_\theta \ll 1$ where $S_{z/\theta} = Q_{z/\theta,c}/(2\Omega_c)$. (1) may apply to a rapidly rotating vortex or a critical layer occurring far away from the core of the vortex. Numerical solutions of the eigenvalue problem reveal that it is indeed often the case. (2) corresponds to a long axial wavelength mode. Assuming J_c small, the expansions (4) become

$$\phi_a = \eta + \sum_{n=2}^{\infty} a_{0,n}\eta^n,\ \phi_b = 1 + \sum_{n=2}^{\infty} b_{0,n}\eta^n + b_0\phi_a(\eta)\ln\eta^* . \tag{6}$$

η^* is a normalized cross-stream coordinate: $\eta^* = \eta/\eta_0$ where η_0 is determined while matching the outer flow with the critical-layer flow. So, u becomes

$$u(\eta) = (\Theta\, r_c\,\phi_a + \phi_b)\ \sin\xi + b_0 r_c\,\Phi\,\phi_a\ \cos\xi. \tag{7}$$

The logarithmic term in (6) is expressed by writing $\ln|r-r_c|$ for $r > r_c$ and $\ln|r-r_c| + i\,\Phi$ when $r < r_c$; $\Phi(\lambda)$ is defined as the phase change. On either side of the critical level, Θ takes different values denoted $\Theta^\pm$. u must be continuous at $r = r_c$, so the integration constant in front of ϕ_b is unique and chosen without loss of generality equal to 1.

4 The Small-Vorticity Limit

Mean fields are expanded in the outer flow in this way

$$\overline{V} = \overline{V}_0 + \varepsilon^{\frac{1}{2}}\,\overline{V}_1 + \varepsilon\,\overline{V}_2 + \ldots,\ \overline{W} = \varepsilon^{\frac{1}{2}}\overline{W}_1 + \varepsilon\,\overline{W}_2 + \ldots \tag{8}$$

with the related vorticities

$$Q_z = Q_0 + \varepsilon^{\frac{1}{2}}\,Q_{z,1} + \ldots,\ Q_\theta = \varepsilon^{\frac{1}{2}}\,Q_{\theta,1} + \varepsilon Q_{\theta,2} + \ldots$$

and similar expressions for U_θ, U_z and P_r. We omit the subscript z to the zero-order axial mean vorticity. The first-order Richardson number is then since $Q_{0,c} = 0$

$$J_{1,c} = \frac{k^2 r_c^2}{2m^2\,\Omega_{0,c}}\left(Q_{z,1,c} + \frac{m}{k r_c} Q_{\theta,1,c}\right). \tag{9}$$

The additional mean flow is likely to be distorted, that is the velocity and its successive derivatives may be different at either side of the critical radius. Similar velocity and temperature distortions were shown by [10] to be necessary for the stratified critical layer. Distortions enable one to match the three components of vorticity and the normal velocity on the separatrices bounding open and closed-streamline flows within the critical layer.

4.1 Critical-Layer Flow Outside of the Separatrices

The critical-layer equations are in the small-vorticity limit ($\beta = 1/2$)

$$\frac{\partial P}{\partial R} - 2\,\Omega_c V = \varepsilon^{\frac{1}{2}}\,\frac{V^2}{r_c}$$

$$\frac{m}{r}\frac{\partial P}{\partial \xi} + U\left(2\Omega_c + \frac{\partial V}{\partial R}\right) + \left(\frac{m}{r}V + kW\right)\frac{\partial V}{\partial \xi} + \varepsilon^{\frac{1}{2}}\frac{UV}{r_c} = \lambda\left(\frac{\partial^2 V}{\partial R^2} + \frac{\varepsilon^{\frac{1}{2}}}{r_c}\frac{\partial V}{\partial R}\right) + \frac{F_b}{\varepsilon}$$

$$k\frac{\partial P}{\partial \xi} + U\frac{\partial W}{\partial R} + \left(\frac{m}{r}V + kW\right)\frac{\partial W}{\partial \xi} = \lambda\left(\frac{\partial^2 W}{\partial R^2} + \frac{\varepsilon^{\frac{1}{2}}}{r_c}\frac{\partial W}{\partial R}\right)$$

$$\text{and}\qquad \frac{\partial U}{\partial R} + \frac{m}{r}\frac{\partial V}{\partial \xi} + k\frac{\partial W}{\partial \xi} = 0\,. \tag{10}$$

The body force is $F_b = -\lambda\varepsilon^{\frac{3}{2}}\,\Delta\overline{V}_0 \simeq -\lambda\varepsilon^{\frac{3}{2}}\,(\overline{V}_0'' + \overline{V}_0'/r_c)$. Writing the outer expansion in inner coordinate determines how the inner variables are to be expanded in the critical layer. For example, for U,

$$U \sim U^{(0)} + \varepsilon^{1/2}\log\varepsilon\, U^{(1)} + \varepsilon^{1/2} U^{(2)} + \cdots,\ U^{(j)} = U_i^{(j)} + \lambda U_v^{(j)} + O(\lambda^2)\,, \tag{11}$$

with similar expansions for V, W and P. Each field at each order j, as $\lambda \to 0$, can be expanded in an inviscid and a viscous parts. Injecting such a decomposition in (10) leads to secularity conditions on the viscous velocity and pressure components. Requiring them to match with the secular terms in the outer flow will fix the arbitrary functions that arise from integrating the governing PDEs [2]. In contrast with the $O(1)$ vorticity case, it is straightforward to find a leading-order solution of the system (10). It consists simply of a radial oscillation superimposed on the mean flow, specifically,

$$\begin{aligned} U^{(0)} &= \sin\xi, V^{(0)} = \overline{V}_{1,c} - 2\,\Omega_{0,c}R, W^{(0)} = \overline{W}_{1,c} \;\; P^{(0)} \\ &= 2\,\Omega_{0,c}R\,(\overline{V}_{1,c} - \Omega_{0,c}\,R). \end{aligned}$$

The $O(\varepsilon^{1/2}\log\varepsilon)$ solution is also obtained from the outer flow. The non trivial equations are obtained at the order $O(\varepsilon^{\frac{1}{2}}.)$ Eliminating the pressure and replacing $V^{(2)}$ by a streamfunction-like variable $\psi^{(2)}$ lead to two coupled PDEs that can be integrated once with respect to R. Before displaying these, we accomplish a transformation in order to have the streamwise-motion equation written in a standard way [6].

$$\xi = X - \frac{\pi}{2}\,[1 + s_i\,],\;\; R = s_i\,R_0R^*,\;\; \psi^{(2)} = s_i\,\frac{\Omega_{0,c}}{r_c}\,R_0^3\hat{\psi}^{(2)},\;\; U^{(2)} = \frac{R_0}{r_c}\,\hat{U}^{(2)},$$

$$V^{(2)} = \frac{s_i}{2\,m}\,\hat{V}^{(2)},\;\; W^{(2)} = s_i\,\frac{k\,r_c}{2\,m^2}\,\hat{W}^{(2)},\;\; P^{(2)} = 2\,s_i\,\Omega_{0,c}^2\,\frac{R_0^3}{r_c}\,\hat{P}^{(2)},$$

$$V^{(2)} = \psi_R^{(2)} + R^2,\quad s = \mathrm{sgn}(R)\,,\quad s_i = \mathrm{sgn}(m\,\Omega_{0,c})\,,\text{ and } R_0 = \left|\frac{r_c}{2\,m\,\Omega_{0,c}}\right|^{1/2}.$$

We then get

$$\sin X\,\hat{\psi}^{(2)}_{R^*R^*} + R^*\,\hat{\psi}^{(2)}_{XR^*} = \hat{\psi}^{(2)}_X - \overline{\lambda}\left(\hat{\psi}^{(2)}_{R^*R^*R^*} - \frac{r_c\,Q'_{0,c}}{\Omega_{0,c}}\right) - \frac{r_c\,F(X)}{R_0\Omega_{0,c}}, \tag{12}$$

$$\text{and}\qquad \sin X\,\hat{W}^{(2)}_{R^*} + R^*\,\hat{W}^{(2)}_X = \hat{\psi}^{(2)}_X - \overline{\lambda}\,\hat{W}^{(2)}_{R^*R^*} - \frac{r_c\,F(X)}{R_0\Omega_{0,c}}, \tag{13}$$

where $\overline{\lambda} = \lambda/R_0$. In the following analysis, we drop the hats with the understanding that it is the new variables with which we are dealing. The radial-momentum equation integrated with respect to R determines $P^{(2)}$. Finally, the continuity equation provides an expression for $U^{(2)}$. $F(X)$ is an arbitrary function arising from the integration. Matching to the outer solution leads to the expression of $F(X)$ and then to

$$Q_{\theta,1,c} = -\frac{kr_c}{m}\,Q_{z,1,c} \quad\text{and}\quad J_{1,c} = 0. \tag{14}$$

In order to relate the mean-vorticity jumps to the phase change, we integrate (12) and (13) over one wavelength in X and then over R. The obtained relations are valid as $R \to \infty$, so $\psi^{(2)}$ and $W^{(2)}$ are replaced by their asymptotic expansions. Clearly, when there is a phase change $\Phi \neq 0$, the only way is if $Q_{z,1,c}$ and $Q_{\theta,1,c}$ are discontinuous across the critical layer. Finally, the relations of the vorticity jumps to the phase change are given by

$$[Q_{z,1,c}]_-^+ = -\frac{m}{kr_c}[Q_{\theta,1,c}]_-^+ = -s_i\,\frac{1}{2}\,Q'_{0,c}R_0\,\frac{\Phi(\overline{\lambda})}{\overline{\lambda}}\,, \tag{15}$$

where $\Phi(\overline{\lambda})$ must be determined by solving (12) and (13) numerically. When we consider the limit $\overline{\lambda} \ll 1$, as in other critical-layer problems, there are regions of closed flow in the cartesian frame (X, R) and the solutions within such regions must be matched to those outside across separatrices. $\psi^{(2)}$ and $W^{(2)}$ can be then determined by solving (12) and (13). Each streamline can be defined univoquely by the variable: $Z = 1/2R^2 + \cos X$, Fig. 1 for a picture of the current lines projected on the plane $z = Cst$ in the case of the nonlinear neutral Kelvin mode $m = 2$.

4.2 Flow Within the Separatrices

First, the three vorticity components are matched across the separatrices. In Appendix, we have extended the Prandtl–Batchelor theorem and shown that $\psi_{RR}^{(2,\odot)} = const. = Q^{(2,\odot)}$ within a region of closed flow according to (A4). $\odot$ defines the flow within the separatrices. Matching the axial vorticities along the upper and lower separatrices $Z = 1$ give

$$\begin{aligned} Q^{(2,\odot)} &= \frac{1}{2}\,\frac{s_i\,r_c}{\Omega_{0,c}\,R_0}\,(Q_{z,1,c}^+ + Q_{z,1,c}^-) \quad \text{and} \\ [Q_{z,1,c}]_-^+ &= -2\,s_i\,Q'_{0,c}\,R_0\,[K(1) + \sqrt{2}]\,. \end{aligned} \tag{16}$$

The second equation in (16) shows that a jump in axial vorticity takes place across the critical layer even in the inviscid limit. Equating now the two expressions for $[Q_{z,1,c}]_-^+$ derived in (15) and in (16), we obtain

$$\frac{\Phi(\overline{\lambda})}{\overline{\lambda}} \sim 4\,[K(1) + \sqrt{2}] \quad \text{as} \quad \overline{\lambda} \to 0\,. \tag{17}$$

This is exactly the result obtained in [6] but with the opposite sign. Numerical evaluation of the integral below defining K yields $K(1) + \sqrt{2} \simeq 1.3788$,

$$K(Z) = \frac{1}{\sqrt{2}} \int_\infty^Z \Big[\frac{2\pi}{\int_0^{2\pi} (Z_1 - \cos X)^{\frac{1}{2}}\,dX} - \frac{1}{Z_1^{\frac{1}{2}}}\Big]\,dZ_1\,.$$

There are two conditions that should be satisfied along the separatrix, namely, a kinematic condition and, secondly, continuity of pressure. The kinematic condition

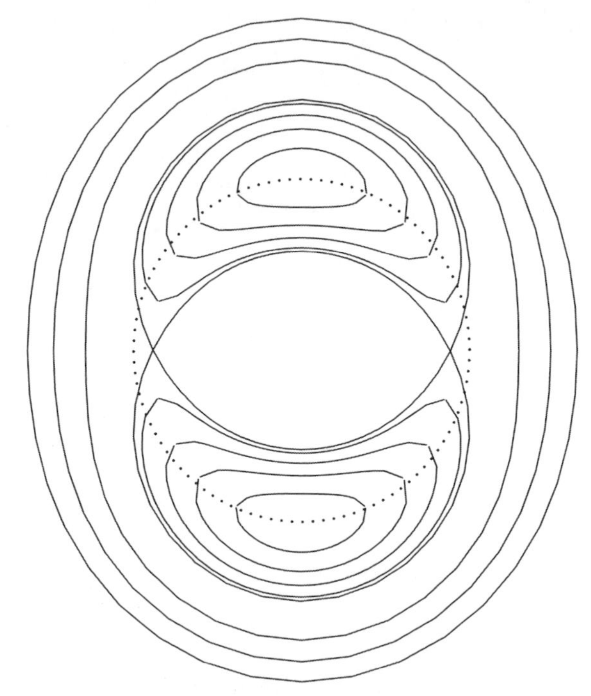

Fig. 1 Nonlinear neutral mode ($\varepsilon = 0.2$) $m = 2$, $k = 2$, $r_c = 1$ and $r_c Q'_{z,0,c} = Q^+_{z,1,c} = \Omega_{0,c}$; view taken at a height $z = Cst$. The *dotted circle* is $r = r_c$.

requires the normal velocity to the separatrix $Z = 1$ to be continuous. At the order $\varepsilon^{\frac{1}{2}}$, the kinematic condition plus the azimuthal-vorticity matching yield jointly with the PB theorem the determination of $W^{(2,\odot)}$. Radial-vorticity matching is equivalent to azimuthal-vorticity matching. Moreover, if we require continuities of $V^{(2,\odot)}$ across $R = 0$ and $V^{(2)}$ across the separatrix, then $\Theta^+ = \Theta^-$; as a result, there is no phase change across the nonlinear critical layer in the inviscid limit.

5 The Long-Wave Nonlinear Critical Layer

In this section, the Richardson number J_c is still taken to be $O(\varepsilon^{1/2})$. However, J_c is small because $k \ll 1$. Specifically, (5) shows that we must scale $kr_c = \kappa\,\varepsilon^{1/4}$. u_z then possesses a new scaling; $u_z = \varepsilon^{1/4}\kappa\, W(R, \xi)$. As in Sect. 4, a simple leading-order solution of the system (10) can readily be found and consists of a radial oscillation superimposed on the mean flow with an additional oscillatory component to the pressure. The axial-vorticity jump and the interior axial vorticity turn out to be the same as in (15) and (16), where R_0 is now defined as

$R_0 = |r_c/[m\,(2\,\Omega_{0,c} - Q_{0,c})]|^{1/2}$ and $J_{1,c}$ as

$$J_{1,c} = \frac{2\,\kappa^2}{m^2}\,\Omega_{0,c}\,\frac{Q_{0,c} + mQ_{\theta,1,c}}{(Q_{0,c} - 2\,\Omega_{0,c})^2}. \tag{18}$$

Matching of the azimuthal vorticity across the separatrix $Z = 1$ leads to

$$m\,Q_{\theta,1,c} = -Q_{0,c}\,, \quad \text{or} \quad J_{1,c} = 0\,, \quad \text{and} \quad [Q_{\theta,1,c}]_-^+ = 0 \quad \text{or} \quad [\overline{W}_1^{\prime}]_-^+ = 0. \tag{19}$$

Matching of the normal velocity across the separatrix determines $W^{(2,\odot)}$. Continuities of $V^{(2,\odot)}$ at $R = 0$ and $Z = 1$ respectively give $\Theta^+ = \Theta^-$ and $[\overline{W}_{1,c}^{\prime\prime}]_-^+ = 0$. To conclude this section, we say a few words about the mean-flow distortions that are present even in the limit $\lambda \to 0$. Caillol and Grimshaw [4] have used the method of strained coordinates to parametrize the streamlines in the critical layer in order to have the velocity at the cat's eye core and at the stagnation points obeying certain topological conditions. We have done this here as well, but omit the details; these points belong to helices, so all the velocity components are linked.

6 Concluding Remarks

We have analytically investigated the critical-layer like interaction of a neutral Kelvin mode with a swirling shear flow. The linear theory is similar to this of a stratified shear flow. The nonlinear study is made possible due to the small-Richardson-number assumption relevant for instance, to rapidly rotating vortices and yields a classic Kelvin cat's eye pattern within the critical layer. The result of this interaction is an additional and distorted mean flow of higher magnitude than the mode amplitude. Axial and azimuthal mean vorticities may be distorted. The vorticity jump is then proportional to the gradient of the basic axial vorticity. The equivalent Richardson number J_c reveals to be smaller than expected, of order the mode amplitude, which implies that the streamlines within the critical layer need to be even more distorted, in order to describe an $O(\varepsilon^{\frac{1}{2}})$ J_c critical-layer flow.

Appendix: Generalized Prandtl–Batchelor Theorem

Batchelor proved that for a steady, inviscid and plane flow the vorticity inside a bounded region is constant. We follow that basic procedure, but the three dimensionality of the present problem naturally adds complications. Our starting point is the momentum equation

$$\partial_t \mathbf{u} + \mathbf{Q} \times \mathbf{u} + \nabla H = \frac{1}{Re}\,\triangle\,\mathbf{u} + \mathbf{F}_b, \tag{1}$$

$H = p/\rho + |\mathbf{u}|^2/2$, $\mathbf{Q}$ is the vorticity and $\mathbf{F}$ is a body force to enable a viscous, parallel flow. We decompose the motion into an inviscid and a viscous components. The development by Batchelor, at this point, involved a curvilinear integration along a streamline, whereas in our case the integration is on a surface $Z = const$ (recall that $Z = R^2/2 + \cos X$). This surface is a cylinder that spirals with respect to the axis of the vortex. The integration will be done either in a plane $z = const$, in which case θ varies over a $2\pi/m$ range or in a plane $\theta = const$, in which case z varies over a $2\pi/k$ range. In that way, we obtain two conditions that are sufficient to determine the flow inside the separatrices. Performing the integration now, we obtain at the leading orders

$$\oint(\mathbf{Q}_i \times \mathbf{u}_v) \cdot d\mathbf{l} + \oint(\mathbf{Q}_v \times \mathbf{u}_i) \cdot d\mathbf{l} + \varepsilon^{3/2} \oint[\nabla \times \mathbf{Q}_i + \triangle \overline{V}_0\, \mathbf{e}_\theta] \cdot d\mathbf{l} = \mathbf{0}. \tag{2}$$

In the two dimensional case, the first two integrals vanish, but that does not happen here because $\mathbf{u}$ is three dimensional. First, we carry out an integration with respect to θ in a plane $z = const$. The integration is along a "streamline" $Z = Z_0$, say, where Z_0 is constant. At the lowest order, we obtain

$$\oint s[R\,\psi_{RRZ}^{(2,\odot)} - 1 - r_c \frac{Q'_{0,c}}{\Omega_{0,c}}]\, d\theta = 0,$$

where the integration is in the clockwise sense. The body force vanishes because of symmetry which permits us to write the integral as

$$\int_0^{2\pi/m} [\psi_{RRZ}^{(2,\odot)+} + \psi_{RRZ}^{(2,\odot)-}]\sqrt{Z_0 - \cos[m\theta]}\, d\theta = 0\,. \tag{3}$$

Given that $\psi_{RR}^{(2,\odot)}$ depends only on Z, (A3) leads us to conclude that

$$\psi_{RR}^{(2,\odot)} = const. \tag{4}$$

A second condition is determined by integrating in a plane $\theta = const$ with z traversing a $2\pi/k$ path. At leading order, this leads to a condition that helps determine the axial velocity, namely,

$$\oint sR[(RW_Z^{(2,\odot)})_Z + W_{v,X}^{(2,\odot)}]\, dz = 0. \tag{5}$$

The governing equations for $\psi^{(2,\odot)}$ and $W^{(2,\odot)}$ used in the foregoing development were those appropriate to the small-vorticity case $Q_{z,c} \ll 1$. A similar analysis can be carried out in the long-wavelength problem. Again, we begin with an integration with respect to θ in a plane $z = const$. At the lowest order, we obtain (A3). A second condition will now be determined by integrating in a plane $\theta = const$. The leading-order equation is the following:

$$\oint sR\, W_{RZ}^{(2,\odot)}\, dz = 0. \tag{6}$$

Writing $W^{(2,\odot)} = \kappa Q_{0,c}/m\, R + \Psi_R^{\odot}$, further differentiations with respect to R and Z followed by a substitution into (6) leads to

$$\int_0^{2\pi/k} [\Psi_{RRZ}^{(\odot)+} + \Psi_{RRZ}^{(\odot)-}]\sqrt{Z_0 - \cos[kz]}\, dz = 0\,. \tag{7}$$

That leads to the conclusion that $\Psi_R^{(\odot)} = 0$, the reason being that the general solution involves $\sqrt{2\,Z}$ and when substituted into (7), a singularity at $Z = 0$ would result from differentiating with respect to Z.

References

1. G.K. Batchelor. On steady laminar flow with closed streamlines at large Reynolds number. *J. Fluid Mech.*, 1:177–190, 1956.
2. D.J. Benney and R.F. Bergeron. A new class of nonlinear waves in parallel flows. *Stud. Appl. Maths.*, 48:181–204, 1969.
3. P. Caillol and R.H. Grimshaw. Steady multi-polar planar vortices with nonlinear critical layers. *Geophys. Astrophys. Fluid Dyn.*, 98:473–506, 2004.
4. P. Caillol and R.H. Grimshaw. Rossby solitary waves in the presence of a critical layer. *Stud. Appl. Maths.*, 118:313–364, 2007.
5. P. Caillol and S. Maslowe. The small-vorticity nonlinear critical layer for Kelvin modes on a vortex. *Stud. Appl. Maths.*, 118:221–254, 2007.
6. R. Haberman. Critical layers in parallel flows. *Stud. Appl. Maths.*, L1(2):139–161, 1972.
7. L. Howard and A. Gupta. On the hydrodynamic and hydromagnetic stability of swirling flows. *J. Fluid Mech.*, 14(3):463–476, 1962.
8. J.W. Miles. On the stability of heterogeneous shear flows. *J. Fluid Mech.*, 10:496–508, 1961.
9. D. Sipp and A.L. Jacquin. Widnall instabilities in vortex pairs. *Phys. Fluids*, 15:1861–1874, 2003.
10. Y. Troitskaya. The viscous-diffusion nonlinear critical layer in a stratified shear flow. *J. Fluid Mech.*, 233:25–48, 1991.
11. C.Y. Tsai and S.E. Widnall. The stability of short waves on a straight vortex filament in a weak externally imposed strain field. *J. Fluid Mech.*, 73:721–733, 1976.

High Order Schemes for Reaction–Diffusion Singularly Perturbed Systems

C. Clavero, J.L. Gracia, and F.J. Lisbona

Abstract In this paper we are interested in solving efficiently a singularly perturbed linear system of differential equations of reaction–diffusion type. Firstly, a non-monotone finite difference scheme of HODIE type is constructed on a Shishkin mesh. The previous method is modified at the transition points such that an inverse monotone scheme is obtained. We prove that if the diffusion parameters are equal it is a third order uniformly convergent method. If the diffusion parameters are different some numerical evidence is presented to suggest that an uniformly convergent scheme of order greater than two is obtained. Nevertheless, the uniform errors are bigger and the orders of uniform convergence are less than in the case corresponding to equal diffusion parameters.

1 Introduction

In this work we consider the singularly perturbed boundary value problem given by the linear reaction–diffusion system

$$L_\varepsilon \mathbf{u} = \mathbf{f}, \qquad x \in \Omega = (0,1), \quad \mathbf{u}(0) = \mathbf{u}_0,\ \mathbf{u}(1) = \mathbf{u}_1, \tag{1}$$

where the differential operator L_ε is defined by

$$L_\varepsilon \equiv -\text{diag}\left\{\varepsilon_1 \frac{d^2}{dx^2}, \varepsilon_2 \frac{d^2}{dx^2}\right\} + A, \quad A = \begin{pmatrix} a_{11}(x) & a_{12}(x) \\ a_{21}(x) & a_{22}(x) \end{pmatrix}.$$

We will assume that the diffusion parameters $0 < \varepsilon_1 \le \varepsilon_2 \le 1$, can take arbitrary small values having, in general, different order of magnitude, that the data of

C. Clavero (✉)
Department of Applied Mathematics, University of Zaragoza, Saragossa, Spain,
E-mail: clavero@unizar.es

A.F. Hegarty et al. (eds.), *BAIL 2008 - Boundary and Interior Layers.*
Lecture Notes in Computational Science and Engineering,
DOI: 10.1007/978-3-642-00605-0,

problem (1) are sufficiently smooth functions and also that the coefficients of the coupling reaction term satisfy

$$\sum_{j=1}^{2} a_{ij} \geq \alpha > 0,\ a_{ii} > 0,\ i = 1,2,\ a_{ij} \leq 0 \text{ if } i \neq j, \tag{2}$$

i.e., the reaction matrix is an M-matrix.

First order uniform convergence of the central finite difference scheme constructed on a Shishkin mesh was proved in [7] and in [4] this was improved to almost second order. Linß and Madden [6] extended this result to the case of an arbitrary number of equations, when the reaction coefficient matrix A satisfy another type of conditions, which include these ones given in (2) for the case that the coupled system has only two equations as problem (1) here considered. Also, in [3] precise information of the asymptotic nature of the solution and its derivatives, for a problem having n equations with n diffusion parameters, has been recently established by means of an appropriate decomposition of the solution, revealing that the solution exhibits overlapping boundary layers with a width $\mathcal{O}(\varepsilon_i^{-1/2})$, $i = 1, 2, \ldots, n$ at both endpoints $x = 0$ and $x = 1$. It was also proved that the central finite difference scheme constructed on a piecewise uniform mesh of Shishkin type, is first order uniformly convergent in the maximum norm.

High order convergence schemes are very interesting in practice because they provide accurate numerical approximations with a low computational cost. Nevertheless, at the moment, in the literature there are not numerical methods for problem (1) with this desirable property. The aim of this work is to see how the HODIE technique permits to obtain a uniformly convergent method having order bigger than two. In some cases the proof of the uniform convergence of the method is fulfilled, but in general we only dispose of numerical evidences showing the efficiency of the HODIE method.

Henceforth, C denotes a generic positive constant independent of the diffusion parameters, and also of the discretization parameter.

2 The Numerical Method

To construct the numerical method we first define a piecewise uniform Shishkin mesh. Following [7] the mesh points are

$$x_j = \begin{cases} j h_{\varepsilon_1}, & j = 0, \ldots, N/8, \\ x_{N/8} + (j - N/8) h_{\varepsilon_2}, & j = N/8 + 1, \ldots, N/4, \\ x_{N/4} + (j - N/4) H, & j = N/4 + 1, \ldots, 3N/4, \\ x_{3N/4} + (j - 3N/4) h_{\varepsilon_2}, & j = 3N/4 + 1, \ldots, 7N/8, \\ x_{7N/8} + (j - 7N/8) h_{\varepsilon_1}, & j = 7N/8 + 1, \ldots, N, \end{cases}$$

where $h_{\varepsilon_1} = 8\tau_{\varepsilon_1}/N,\ h_{\varepsilon_2} = 8(\tau_{\varepsilon_2} - \tau_{\varepsilon_1})/N,\ H = 2(1 - 2\tau_{\varepsilon_2})/N$, and the transition parameters are given by

$$\tau_{\varepsilon_2} = \min\left\{1/4, \sigma_0\sqrt{\varepsilon_2}\ln N\right\},\ \tau_{\varepsilon_1} = \min\left\{\tau_{\varepsilon_2}/2, \sigma_0\sqrt{\varepsilon_1}\ln N\right\},$$

and $\sigma_0 \geq 4$. If $\tau_{\varepsilon_1} \neq 1/8$ and $\tau_{\varepsilon_2} = 1/4$, we modify slightly the mesh points; now they are given by

$$x_j = \begin{cases} jh_{\varepsilon_1}, & j = 0, \dots, N/8, \\ x_{N/8} + (j - N/8)\hat{H}, & j = N/8 + 1, \dots, 7N/8, \\ x_{7N/8} + (j - 7N/8)h_{\varepsilon_1}, & j = 7N/8 + 1, \dots, N, \end{cases}$$

where $\hat{H} = 4(1 - 2\tau_{\varepsilon_1})/3N$. Below we denote the local step sizes by $h_j = x_j - x_{j-1},\ j = 1, \dots, N$. On this mesh we impose that the local error be zero on the set of vector polynomials of the form

$$\alpha_0 \begin{pmatrix} 1 \\ 1 \end{pmatrix} + \alpha_1 \begin{pmatrix} x \\ x \end{pmatrix} + \alpha_2 \begin{pmatrix} x^2 \\ x^2 \end{pmatrix} + \alpha_3 \begin{pmatrix} x^3 \\ x^3 \end{pmatrix},\ \alpha_i \in \mathbf{R},\ i = 0, \dots, 3.$$

Following the construction made in [1] for the scalar case, we write the finite difference scheme $\mathbf{L}^N = (L_1^N, L_2^N)^T$ in the form

$$\begin{aligned} L_i^N \mathbf{U}(x_j) \equiv\ & r_{i,j}^- U_i(x_{j-1}) + r_{i,j}^c U_i(x_j) + r_{i,j}^+ U_i(x_{j+1}) + \\ & + q_{i,j}^1 a_{i,k}(x_{j-1}) U_k(x_{j-1}) + q_{i,j}^2 a_{i,k}(x_j) U_k(x_j) + \\ & + q_{i,j}^3 a_{i,k}(x_{j+1}) U_k(x_{j+1}) = \\ & = q_{i,j}^1 f_i(x_{j-1}) + q_{i,j}^2 f_i(x_j) + q_{i,j}^3 f_i(x_{j+1}), \end{aligned} \tag{3}$$

for $j = 1, \cdots, N-1$, $i = 1, 2$, where $k = 2$ if $i = 1$, $k = 1$ if $i = 2$ and the coefficients $q's$ satisfy the normalization condition $q_{i,j}^1 + q_{i,j}^2 + q_{i,j}^3 = 1,\ i = 1, 2$. Then, it is not difficult to prove that for $j = 1, \cdots, N-1, i = 1, 2$ the coefficients $r's$ of the scheme are given, in function of the coefficients $q's$, by

$$\begin{aligned} r_{i,j}^c &= q_{i,j}^1 a_{i,i}(x_{j-1}) + q_{i,j}^2 a_{i,i}(x_j) + q_{i,j}^3 a_{i,i}(x_{j+1}) - r_{i,j}^- - r_{i,j}^+, \\ r_{i,j}^+ &= -2\varepsilon_i/(h_{j+1}(h_j + h_{j+1})) + q_{i,j}^3 a_{i,i}(x_{j+1}), \\ r_{i,j}^- &= -2\varepsilon_i/(h_j(h_j + h_{j+1})) + q_{i,j}^1 a_{i,i}(x_{j-1}), \end{aligned} \tag{4}$$

and also that it holds

$$q_{i,j}^1 = (h_j - h_{j+1})/(3h_j) + q_{i,j}^3 h_{j+1}/h_j. \tag{5}$$

The value of the free parameter $q_{i,j}^3$ is taken equal to the one obtained for the scalar case in [1] and we will see that this choice is also appropriate for the case of systems. This value depends on the location of the mesh points and also on the ratio between the step sizes of the Shishkin mesh and the diffusion parameters. Concretely, for

$j=1,\cdots,N/8-1,7N/8+1,\cdots,N-1$, i.e., $x_j \in (0,\tau_{\varepsilon_1})\bigcup(1-\tau_{\varepsilon_1},1)$ and $i=1,2$, the coefficients $q's$ are

$$q^1_{i,j}=\frac{1}{6}\left(1-\frac{h^2_{j+1}}{h_j(h_j+h_{j+1})}\right),\ q^3_{i,j}=\frac{1}{6}\left(1-\frac{h^2_j}{h_{j+1}(h_j+h_{j+1})}\right), \tag{6}$$
$$q^2_{i,j}=1-q^1_{i,j}-q^3_{i,j}.$$

For $j=N/4,\cdots,3N/4$, i.e., $x_j\in[\tau_{\varepsilon_2},1-\tau_{\varepsilon_2}]$, and $i=1,2$ we distinguish two cases: first, if $2H^2\|a_{ii}\|_\infty/3\le\varepsilon_i$, then the coefficients are defined again by (6); in the other case, when $2H^2\|a_{ii}\|_\infty/3>\varepsilon_i$, they are given by

$$q^1_{i,j}=q^3_{i,j}=0,\ q^2_{i,j}=1, \tag{7}$$

corresponding to the classical discretization of central differences. Note that in this case (5) does not hold.

Last case is when $j=N/8,\cdots,N/4-1,3N/4+1,\cdots,7N/8$, i.e., $x_j\in[\tau_{\varepsilon_1},\tau_{\varepsilon_2})\bigcup(1-\tau_{\varepsilon_2},1-\tau_{\varepsilon_1}]$. Now, for the second equation, $i=2$, the coefficients are again given by (6). Nevertheless for the first equation, $i=1$, again we must distinguish two cases; first, when $2h^2_{\varepsilon_2}\|a_{11}\|_\infty/3\le\varepsilon_1$, the coefficients are given by (6); in the other case, $2h^2_{\varepsilon_2}\|a_{11}\|_\infty/3>\varepsilon_1$, they are given by (7).

Note that, in general, the coefficients defined in (6) are not positive and then the associated matrix to the numerical scheme is not an M-matrix. Nevertheless, we will see the efficiency of this scheme. As an example, we solve the particular problem (see [7]) setting by

$$A=\begin{pmatrix}2(x+1)^2 & -(x^3+1)\\ -2\cos(\pi x/4) & 2.2e^{-x+1}\end{pmatrix},\ f=\begin{pmatrix}2e^x\\ 10x+1\end{pmatrix}, \tag{8}$$

with $\mathbf{u}_0=\mathbf{u}_1=\mathbf{0}$. For this problem the exact solution is unknown and therefore to approximate the pointwise errors $|(\mathbf{U}-\mathbf{u})(x_j)|,\ j=0,\cdots,N$, we use a variant of the double mesh principle. So, we calculate a numerical approximation $\widehat{\mathbf{U}}$ to $\mathbf{u}$ given by the scheme (3) on the mesh $\{\hat{x}_j\}$ that contains the mesh points of the original piecewise Shishkin mesh and their midpoints, i.e., the mesh points are defined by $\hat{x}_{2j}=x_j,\ j=0,\ldots,N,\ \hat{x}_{2j+1}=(x_j+x_{j+1})/2,\ j=0,\ldots,N-1$. Then, at the original mesh points $x_j,\ j=0,1,\cdots,N$, the maximum errors and the uniform errors are approximated by

$$\mathbf{d}_{\varepsilon,N}=\max_{0\le j\le N}|\mathbf{U}(x_j)-\widehat{\mathbf{U}}(\hat{x}_{2j})|,\ \mathbf{d}_N=\max_S d_{\varepsilon,N},$$

where, in order to permit the stabilization of the errors, we take S as the set

$$S=\{(\varepsilon_1,\varepsilon_2)\mid\varepsilon_2=2^0,2^{-2},\ldots,2^{-30},\ \varepsilon_1=\varepsilon_2,2^{-2}\varepsilon_2,\ldots,2^{-50}\}. \tag{9}$$

From these estimates of the pointwise errors we obtain the corresponding orders of convergence and the uniform orders of convergence in a standard way, by using

$$\mathbf{p}=\log_2(\mathbf{d}_{\varepsilon,N}/\mathbf{d}_{\varepsilon,2N}),\ \mathbf{p}_{uni}=\log_2(\mathbf{d}_N/\mathbf{d}_{2N}).$$

Table 1 Uniform errors and orders of convergence for the scheme given by (3)–(7)

	$N=32$	$N=64$	$N=128$	$N=256$	$N=512$	$N=1{,}024$	$N=2{,}048$	$N=4{,}096$
$d_{1,N}$	9.018E-02	4.755E-02	2.452E-02	7.073E-03	1.091E-03	1.037E-04	1.007E-05	1.096E-06
$p_{1,uni}$	0.923	0.956	1.793	2.696	3.395	3.364	3.200	
$d_{2,N}$	3.609E-01	1.613E-01	4.026E-02	5.665E-03	5.723E-04	5.593E-05	5.111E-06	4.545E-07
$p_{2,uni}$	1.162	2.002	2.829	3.307	3.355	3.452	3.491	

In all cases we take the constant $\sigma_0 = 4$; in practice if this constant is smaller, the desired order of uniform convergence is not achieved. On the other hand, if it is greater than 4 the numerical errors are bigger but the order of uniform convergence is preserved.

Table 1 displays the results obtained with the HODIE scheme; from these results we observe that the order of uniform convergence is four except by a logarithmic factor, as it is usual on Shishkin meshes. Nevertheless, the discrete operator of this scheme is not of positive type and we do not have the proof of the uniform (l_∞, l_∞)-stability. In [6] this uniform stability was proved without using the inverse monotonicity of the discrete operator, but unfortunately so far we have not been able to apply this technique to the HODIE operator. In [4] a non-monotone FEM scheme was used to solve a scalar reaction–diffusion problem, proving also its uniform stability in the maximum norm.

Therefore we propose a slight modification of scheme (3) to have a new scheme satisfying the discrete maximum principle. We clearly see that only the discretization associated with the transition points does not give the correct coefficients sign pattern to have an M-matrix. Then, we change the discretization corresponding to the indexes $j = N/8, N/4, 3N/4, 7N/8$, such that $r^-_{i,j} < 0, r^+_{i,j} < 0, r^-_{i,j} + r^+_{i,j} + r^c_{i,j} < 0$ and $q^1_{i,j}, q^2_{i,j}, q^3_{i,j}$ be positive. It is straightforward to obtain that the coefficients $q's$ are given by

$$q^1_{i,j} = q^3_{i,j} = q^2_{i,j} = 1/3, \quad j = N/8, N/4, 3N/4, 7N/8, i = 1, 2. \tag{10}$$

It is easy to proof that the discrete operator is of positive type and therefore it satisfies the discrete maximum principle.

3 The Case of Equal Diffusion Parameters

To find a theoretical proof of the uniform convergence of the method, we begin with the case where both diffusion parameters take the same value, $\varepsilon_1 = \varepsilon_1 = \varepsilon$. Note that in this case really there are only two transition points in the Shishkin mesh and the transition parameter is defined by $\tau = \min\left\{1/4, \sigma_0\sqrt{\varepsilon_2}\ln N\right\}$. Following the idea of extending the domain introduced by Shishkin in [8], which was also used in [2] to find a decomposition of the exact solution of a two dimensional scalar equation of reaction–diffusion type, it can be proved the following result showing the asymptotic behavior of the exact solution.

Lemma 1. *Let assume* $a_{ij}, f \in C^4(\overline{\Omega}),\ i,j=1,2$. *Then, for* $\varepsilon_1=\varepsilon_2=\varepsilon$, *the exact solution of (1) can be decomposed as* $\mathbf{u}=\mathbf{v}+\mathbf{w}$, *where for* $x\in[0,1]$, $0\le k\le 6$ *and* $i=1,2$ *it holds*

$$\|v_i^{(k)}(x)\|_\infty \le C\left(1+\varepsilon^{(4-k)/2}\right), \tag{11}$$

and

$$\|w_i^{(k)}(x)\|_\infty \le C\varepsilon^{-k/2}\left(e^{-x\sqrt{\alpha/\varepsilon}}+e^{-(1-x)\sqrt{\alpha/\varepsilon}}\right). \tag{12}$$

Note that we have appropriate bounds of the regular and singular components and their derivatives up to sixth order, which we will need in the analysis of the truncation error.

Theorem 1. *Let* $\mathbf{u}$ *be the solution of continuous problems (1) and* $\mathbf{U}$ *the solution of the discrete operator (3)–(7) and (10) defined on the previous Shishkin mesh, when* $\varepsilon_1=\varepsilon_2=\varepsilon$. *Then, the error satisfies*

$$\|\mathbf{U}-\mathbf{u}\|_\infty \le C(N^{-3}+N^{-4}\ln^4 N).$$

Proof. In the case $\tau=1/4$, using that $\varepsilon^{-1/2}\le C\ln N$ and the crude bounds $\|\mathbf{u}^{(i)}\|_\infty \le C\varepsilon^{-i/2}$, $i=0,\cdots,6$, a classical analysis proves that $\|\mathbf{U}-\mathbf{u}\|_\infty \le CN^{-4}\ln^4 N$. When $\tau<1/4$, first we study the error for the regular component. Then, if $\varepsilon\le 2H^2 \min_i \|a_{ii}\|_\infty/3$, taking Taylor expansions the local error can be bounded as in [1] for a single equation, and therefore we have

$$|\mathbf{L}^N(\mathbf{V}-\mathbf{v})(x_j)| \le \begin{cases} \mathbf{C}N^{-2}\varepsilon\|\mathbf{v}^{(4)}\|_\infty \le \mathbf{C}N^{-2}\varepsilon \le \mathbf{C}N^{-4}, & x_j\in(\tau,1-\tau),\\ \mathbf{C}N^{-1}\varepsilon\|\mathbf{v}^{(3)}\|_\infty \le \mathbf{C}N^{-1}\varepsilon \le \mathbf{C}N^{-3}, & x_j\in\{\tau,1-\tau\},\\ \mathbf{C}\varepsilon(N^{-1}\varepsilon^{1/2}\ln N)^4\|\mathbf{v}^{(6)}\|_\infty \le & \\ \qquad \le \mathbf{C}N^{-4}, & x_j\in(0,\tau)\cup(1-\tau,1). \end{cases}$$

Then, the discrete maximum principle proves that

$$\|\mathbf{V}-\mathbf{v}\|_\infty \le CN^{-3}. \tag{13}$$

On the other hand, if $\varepsilon > 2H^2\min_i \|a_{ii}\|_\infty/3$, we can obtain

$$|\mathbf{L}^N(\mathbf{V}-\mathbf{v})(x_j)| \le \begin{cases} \mathbf{C}N^{-4}\varepsilon\|\mathbf{v}^{(6)}\|_\infty \le \mathbf{C}N^{-4}, & x_j\in(\tau,1-\tau),\\ \mathbf{C}\varepsilon(N^{-1}\varepsilon^{1/2}\ln N)^4\|\mathbf{v}^{(6)}\|_\infty \le & \\ \qquad \le \mathbf{C}N^{-4}, & x_j\in(0,\tau)\cup(1-\tau,1). \end{cases}$$

At the transition points, using that for any $z\in C^4(\overline{\Omega})$ it holds

$$\left|-\frac{2(z_{j+1}-z_j)}{h_{j+1}(h_j+h_{j+1})}-\frac{2(z_{j-1}-z_j)}{h_j(h_j+h_{j+1})}+\frac{1}{3}(z''_{j+1}+z''_j+z''_{j-1})\right| \le$$

$$\le C\max\{h_j^2,h_{j+1}^2\}\|z^{(4)}\|_\infty,$$

we deduce

$$|\mathbf{L}^N(\mathbf{V}-\mathbf{v})(x_j)| \le \mathbf{C}N^{-2}\varepsilon\|\mathbf{v}^{(4)}\|_\infty \le \mathbf{C}N^{-2}\varepsilon,\ x_j \in \{\tau, 1-\tau\}.$$

Defining the barrier function $\mathbf{Z}(x_j) = \mathbf{C}(N^{-3}\varepsilon^{1/2}\ln N + N^{-4})(\theta(x_j)+1)$, where θ is the piecewise linear function

$$\theta(x) = \begin{cases} \dfrac{x}{\tau}, & \text{if } x \in [0,\tau], \\ 1, & \text{if } x \in [\tau, 1-\tau], \\ \dfrac{1-x}{\tau}, & \text{if } x \in [1-\tau, 1], \end{cases}$$

using the maximum principle, it can be proven that

$$\|\mathbf{V}-\mathbf{v}\|_\infty \le C(N^{-3}\varepsilon^{1/2}\ln N + N^{-4}),$$

and taking into account that $\varepsilon^{1/2}\ln N \le 1$, it follows

$$\|\mathbf{V}-\mathbf{v}\|_\infty \le CN^{-3}. \tag{14}$$

For the singular component we distinguish two cases depending on the location of the mesh point. For $x_j \in [\tau, 1-\tau]$, using the exponential character of this component, it is not difficult to deduce

$$|(\mathbf{W}-\mathbf{w})(x_j)| \le |\mathbf{W}(x_j)| + |\mathbf{w}(x_j)| \le \mathbf{C}N^{-4},\ x_j \in [\tau, 1-\tau].$$

In the second case, $x_j \in (0,\tau) \cup (1-\tau, 1)$, the local error is bounded by

$$|\mathbf{L}^N(\mathbf{W}-\mathbf{w})(x_j)| \le \mathbf{C}\varepsilon(N^{-1}\varepsilon^{1/2}\ln N)^4\|\mathbf{w}^{(6)}\|_\infty \le \mathbf{C}(N^{-1}\ln N)^4.$$

Applying again the maximum principle, now on $[0,\tau] \cup [1-\tau, 1]$, we deduce

$$\|(\mathbf{W}-\mathbf{w})(x_j)\|_\infty \le C(N^{-1}\ln N)^4,\ x_j \in [0,\tau] \cup [1-\tau, 1]. \tag{15}$$

From (13)–(15) the result follows.

For the same example as before, with $\varepsilon = 2^0, 2^{-2}, \ldots, 2^{-50}$, Table 2 displays the results obtained; from it we clearly observe that the order of uniform convergence is similar to that for the unmodified HODIE scheme. Note that the numerical results indicate an order of uniform convergence higher than this one proven in Theorem 1.

Table 2 Uniform errors and orders of convergence for the modified scheme given by (3)–(7) and (10)

	$N=32$	$N=64$	$N=128$	$N=256$	$N=512$	$N=1{,}024$	$N=2{,}048$	$N=4{,}096$
$d_{1,N}$	4.519E-02	1.330E-02	2.589E-03	3.307E-04	3.372E-05	3.241E-06	2.980E-07	2.641E-08
$p_{1,uni}$	1.765	2.361	2.969	3.294	3.379	3.443	3.497	
$d_{2,N}$	9.864E-02	2.132E-02	2.666E-03	2.960E-04	2.994E-05	2.856E-06	2.615E-07	2.315E-08
$p_{2,uni}$	2.210	2.999	3.171	3.305	3.390	3.449	3.498	

Table 3 Uniform errors and orders of convergence for the modified scheme given by (3)–(7) and (10)

	$N=32$	$N=64$	$N=128$	$N=256$	$N=512$	$N=1{,}024$	$N=2{,}048$	$N=4{,}096$
$d_{1,N}$	1.722E-01	8.145E-02	3.028E-02	8.314E-03	1.830E-03	3.500E-04	6.161E-05	1.049E-05
$p_{1,uni}$	1.080	1.428	1.865	2.184	2.386	2.506	2.554	
$d_{2,N}$	6.890E-01	3.258E-01	1.211E-01	3.324E-02	7.306E-03	1.392E-03	2.418E-04	4.087E-05
$p_{2,uni}$	1.081	1.428	1.865	2.186	2.392	2.525	2.565	

4 The General Case: $\varepsilon_1 \leq \varepsilon_2$

In the general case, when the diffusion parameters can be different, the theoretical question is more complicated. An important question is related with the decomposition of the exact solution. In this case it is possible to find a decomposition into a regular and singular part (see [5–7] for instance), but it is not clear how it is possible to obtain the bounds (11) for the regular component of the solution; note that we need the bounds of the derivatives up to sixth order, to find appropriate bounds for the local error associate to the scheme. Nevertheless, for us it is interesting to confirm in practice that this scheme gives an order of uniform convergence bigger than two.

Table 3 displays the results obtained with the new scheme when the diffusion parameters are not equal. From this table we observe that the method gives an almost third order uniformly convergent method, which is less than the order obtained in the case of equal diffusion parameters.

Nevertheless, this method improves both the maximum errors and the numerical order of uniform convergence with respect to central finite difference scheme. Because the modified finite difference scheme satisfies the maximum principle, having appropriate bounds for the derivatives of the regular and singular part of the solution, would allow us carry out the analysis of the local error, and therefore prove the desired uniform convergence.

Acknowledgements This research was partially supported by the project MEC/FEDER MTM-2007-63204 and the Diputación General de Aragón.

References

1. Clavero, C., Gracia, J.L.: High order methods for elliptic and time dependent reaction–diffusion singularly perturbed problems. Appl. Math. Comp., **168**, 1109–1127 (2005).
2. Clavero, C., Gracia, J.L., O'Riordan, E.: A parameter robust numerical method for a two dimensional reaction–diffusion problem. Math. Comp., **74**, 1743–1758 (2005).
3. Gracia, J.L, Lisbona, F.J., O'Riordan, E.: A coupled system of singularly perturbed parabolic reaction–diffusion equations. Adv. Comput. Math. Published online: 24 June 2008.
4. Linß, T.: Maximum-norm error analysis of a non-monotone FEM for a singularly perturbed reaction–diffusion problem . BIT Numer. Math., **47**, 379–391 (2007).

5. Linß, T., Madden, N.: Accurate solution of a system of coupled singularly perturbed reaction–diffusion equations. Computing, **73**, 121–133 (2004).
6. Linß, T., Madden, N.: Layer-adapted meshes for a system of coupled singularly perturbed reaction–diffusion problems. IMA J. Numer. Anal., **29**, 109–125 (2009). doi:10.1093/imanum/drm053.
7. Madden, N., Stynes, M.: A uniformly convergent numerical method for a coupled system of two singularly perturbed linear reaction–diffusion problems. IMA J. Numer. Anal., **23**, 627–644 (2003).
8. Shishkin, G.I.: Discrete aproximation of singularly perturbed elliptic and parabolic equations, Russian Academy of Sciences, Ural section, Ekaterinburg, 1992. (In Russian).

A Patched Mesh Method for Singularly Perturbed Reaction–Diffusion Equations

C. de Falco and E. O'Riordan

Abstract A singularly perturbed elliptic problem of reaction–diffusion type is examined. The solution is decomposed into a sum of a regular component, boundary layer components and corner layer components. Numerical approximations are generated separately for each of these components. These approximations are patched together to form a global approximation to the solution of the continuous problem. An asymptotic error bound in the pointwise maximum norm is established; whose dependence on the values of the singular perturbation parameter is explicitly identified. Numerical results are presented to illustrate the performance of the numerical method.

1 Introduction

Consider the singularly perturbed diffusion reaction problem

$$-\varepsilon\, \triangle u(\mathbf{x}) + b(\mathbf{x})\, u(\mathbf{x}) = f(\mathbf{x}),\ \mathbf{x} \in \Omega \subset \mathbb{R}^d,\ \ u|_{\partial\Omega} = g(\mathbf{x}), \tag{1}$$

with $0 < \varepsilon << 1$ and $b(\mathbf{x}) \geq \beta > 0$ for $\mathbf{x} \in \overline{\Omega}$. The solution displays boundary layers whose width depends on the parameter ε. For $d = 1$ a very simple yet effective strategy to construct parameter uniform numerical methods is the use of piecewise uniform Shishkin meshes [1], i.e. meshes with a refinement region near the boundary whose width is selected a priori to match the length-scale of the layer. In the case of $d = 2$ and when the domain Ω is a rectangle, it is well established [1,3] that the natural extension of this approach to a tensor product of two one dimensional piecewise uniform Shishkin meshes yields a parameter uniform [1] second order (ignoring logarithmic factors) rate of convergence. The extension of this approach to other geometries is non-trivial. Curvilinear tensor product meshes [8] can deal with

C. de Falco (✉)
School of Mathematical Sciences, Dublin City University, Dublin 9, Ireland,
E-mail: carlo.defalco@dcu.ie

A.F. Hegarty et al. (eds.), *BAIL 2008 - Boundary and Interior Layers.*

Lecture Notes in Computational Science and Engineering,
DOI: 10.1007/978-3-642-00605-0,

a limited set of geometries, while creating a single globally conforming unstructured triangulation with a uniform refinement in the layer region can produce inefficient or pathologically deformed meshes when ε is small. Although such inconveniences might be overcome by discretisation methods allowing for non conforming meshes (see, e.g. [9, Chap. 2, Sect. 2.5]) at the interface between the interior and boundary layer region, this would still involve producing a different triangulation for the whole domain Ω for each value of ε. This may require a significant computational effort which, for general domains, may outweigh that required for the solution of the discrete problem itself. To cope with these issues we investigate a method inspired by Chimera Overset Grid Methods [2] and by the method of Patches of Finite Elements of [6]. Note that one cannot expect this general approach to be parameter uniform without some modification that would resolve all layers within the solution. In contrast to the methods in [2, 6], which can be viewed as variants of the Schwartz iterative technique, our approach makes use of an a priori expansion to decompose the solution u of (1) into a sum of a *regular component* v, a set of *boundary layer components* w_q, $q=1,\dots n_w$ and a set of *corner layer components* z_p, $p=1,\dots n_z$. Each component is implicitly defined as the solution of a boundary value problem. In this paper, we consider the case of $\varepsilon \leq CN^{-1}$, where N^d is the dimension of the discrete problem. Hence, quantities of order ε are considered negligible compared to the discretisation errors. In Sect. 3, the pointwise bounds established on the layer components allow us identify subdomains or *patches* $\Omega_q, \Omega_p \subset \Omega$, $q=1,\dots n_w$, $p=n_w+1,\dots n_w+n_z$ outside of which a component is negligible. This decomposition also allows one to compute a discrete approximation to u by solving n_w+n_z+1 problems *once* without any further iteration. Furthermore, as the decomposition is performed at the continuous level, this approach does not pose restrictions on the method used to discretise each boundary value problem. For example, in the case of the regular component defined in (4), one could use the results in [10] to analyze the error (in the case of a sufficiently smooth regular component) if one employed a finite element method on an unstructured quasi-uniform mesh instead of the numerical method analyzed in Sects. 4 and 5, which is based on a standard finite difference operator on a tensor product mesh. We finally point out that, although in the sections below we present theoretical results for a problem posed on the simple geometry of the unit square, the encouraging numerical results presented in [4] and Sect. 6.2 indicate the practical viability of the same approach for singularly perturbed problems on more complicated geometries. Throughout the paper $\|\cdot\|$ denotes the global pointwise maximum norm over the domain $\overline{\Omega}$ and C is a constant independent of ε and N.

2 Continuous Problem

Consider the singularly perturbed elliptic problem

$$L_\varepsilon u := -\varepsilon\Delta u + b(x,y)u = f(x,y), (x,y) \in \Omega = (0,1)^2, \tag{2a}$$

$$u = g,\ (x,y) \in \partial\Omega, \tag{2b}$$

$$f, b \in C^{4,\alpha}(\overline{\Omega}),\ g \in C(\partial\Omega),\ b(x,y) \geq \beta > 0,\ (x,y) \in \bar{\Omega}, \tag{2c}$$

where $0 < \varepsilon \leq 1$ is a singular perturbation parameter. We adopt the following notation for the edges and the boundary conditions:

$$\partial\Omega_1 = \{(x,0)|0 \leq x \leq 1\}, \partial\Omega_2 = \{(1,y)|0 \leq y \leq 1\},$$
$$\partial\Omega_3 = \{(x,1)|0 \leq x \leq 1\}, \partial\Omega_4 = \{(0,y)|0 \leq y \leq 1\},$$
$$g(x,y) = g_i(x), (x,y) \in \partial\Omega_i, i = 1,3;\ g(x,y) = g_i(y), (x,y) \in \partial\Omega_i, i = 2,4.$$

Assume further that $g_s \in C^{4,\alpha}([0,1]),\ s = 1,2,3,4$. From Han and Kellogg [7] and Andreev [1] we note the following levels of compatibility conditions: for the corner $(0,0)$,

$$g_1(0) = g_4(0), \tag{3a}$$
$$-\varepsilon g_1''(0) - \varepsilon g_4''(0) + b(0,0)g_1(0) = f(0,0), \tag{3b}$$

and similarly for the other corners. If (3a) is assumed at all four corners then $u \in C^{1,\alpha}(\overline{\Omega})$ and if (3a) and (3b) are assumed at all four corners then $u \in C^{3,\alpha}(\overline{\Omega})$. The reduced solution u_0 is defined via the reduced problem

$$b(x,y)u_0(x,y) = f(x,y),\ (x,y) \in \overline{\Omega}.$$

The regular component v of u is the solution of the elliptic problem

$$L_\varepsilon v = f(x,y), (x,y) \in \Omega,\ v = u_0,\ (x,y) \in \partial\Omega. \tag{4}$$

Note that the regular component can be written as $v = u_0 + \varepsilon R$, where

$$L_\varepsilon R = \Delta u_0, (x,y) \in \Omega,\ R = 0,\ (x,y) \in \partial\Omega.$$

Hence $R \in C^{0,\alpha}(\overline{\Omega}) \cap C^{2,\alpha}(\Omega)$ and by the maximum principle $\|R\| \leq C$.

Remark 1. Note that at the corner $(0,0)$ the necessary compatibility condition for $u \in C^{3,\alpha}(\overline{\Omega})$ is that $b(0,0)u(0,0) = f(0,0) + \varepsilon(g_4''(0) + g_1''(0))$ which is that

$$u(0,0) - u_0(0,0) = O(\varepsilon). \tag{5}$$

3 Solution Decomposition

The solution is decomposed into a sum of a regular component v, boundary layer components $w_i(x,y), i = 1,2,3,4$ and corner layer components $z_i(x,y), i = 1,2,3,4$

$$u = v + \sum_{i=1}^{4} w_i - \sum_{i=1}^{4} z_i.$$

Similar but different decompositions are given in [1,3]. Note that v is defined in (4) and the boundary layer function w_1 associated with the edge $y = 0$ is defined as the solution of the problem

$$L_\varepsilon w_1 = -\varepsilon(1-y)s_1''(x), (x,y) \in \Omega, \tag{6a}$$
$$w_1(0,y) = q_4(y), \quad w_1(1,y) = q_2(y), \qquad 0 \le y \le 1, \tag{6b}$$
$$w_1(x,0) = s_1(x) := (u-v)(x,0), \quad w_1(x,1) = 0, 0 \le x \le 1, \tag{6c}$$
$$-\varepsilon q_4'' + b(0,y)q_4 = 0,\ y \in (0,1), \quad q_4(0) = s_1(0),\ q_4(1) = 0 \tag{6d}$$
$$-\varepsilon q_2'' + b(1,y)q_2 = 0,\ y \in (0,1), \quad q_2(0) = s_1(1),\ q_2(1) = 0. \tag{6e}$$

Lemma 1. *The solution of (6) satisfies the bounds*

$$|w_1(x,y)| \le Ce^{-y\sqrt{\beta/\varepsilon}} + C\varepsilon(1-y), \tag{7a}$$
$$\left\|\frac{\partial^{i+j} w_1}{\partial x^i \partial y^j}\right\| \le C\varepsilon^{-(i+j)/2}, 1 \le i+j \le 3,$$
$$\left\|\frac{\partial^j w_1}{\partial x^j}\right\| \le C,\ j = 1,2. \tag{7b}$$

Proof. Note that

$$|q_4(y)| \le C|s_1(0)|e^{-y\sqrt{\beta/\varepsilon}}, \quad |q_2(y)| \le C|s_1(1)|e^{-y\sqrt{\beta/\varepsilon}}.$$

Consider the following interpolant of the boundary data

$$\begin{aligned} h(x,y) =&(s_1(x) - s_1(0)(1-x))(1-y) + (q_4(y) - q_4(1)y)(1-x) \\ &+ (q_2(y) - q_2(0)(1-y))x. \end{aligned}$$

Then

$$L_\varepsilon h = -\varepsilon(1-y)s_1''(x) + T(x,y),$$

where $T(x,y) := bh - (1-x)b(0,y)q_4(y) - xb(1,y)q_2(y)$. Note that $T(x,y) = 0$ at each of the four corners. Then since $L_\varepsilon(w_1 - h) = T(x,y)$, we have sufficiently compatibility (3b) for $w_1 \in C^{3,\alpha}(\overline{\Omega})$ and

$$|(w_1 - h)(x,y)| \le Cx(1-x).$$

Using the maximum principle and classical bounds on the derivatives [3] we have that

$$|w_1(x,y)| \le Ce^{-y\sqrt{\beta/\varepsilon}} + C\varepsilon(1-y), \ \left\|\frac{\partial^{i+j} w_1}{\partial x^i \partial y^j}\right\| \le C\varepsilon^{-(i+j)/2},\ i+j \le 3.$$

Also $|T(x,y)| \le Cx(1-x)$, which implies that

$$\left|\frac{\partial w_1}{\partial x}(0,y)\right| \le C, \ \left|\frac{\partial w_1}{\partial x}(1,y)\right| \le C,$$

and using the differential equation (6a), we conclude that

$$\left|\frac{\partial^2 w_1}{\partial x^2}(0,y)\right| \le C(1-y)|s_1''(0)|, \ \left|\frac{\partial^2 w_1}{\partial x^2}(1,y)\right| \le C(1-y)|s_1''(1)|.$$

$$\text{Since,} \qquad L_\varepsilon \frac{\partial w_1}{\partial x} = -\varepsilon(1-y)s_1^{(3)}(x) - b_x w_1,$$

$$\text{and} \qquad L_\varepsilon \frac{\partial^2 w_1}{\partial x^2} = -\varepsilon(1-y)s_1^{(4)}(x) - b_{xx} w_1 - 2b_x \frac{\partial w_1}{\partial x},$$

we can use the maximum principle to establish the bounds $\|\frac{\partial^i w_1}{\partial x^i}\| \le C, \ i = 1,2$ on the derivatives orthogonal to the layer. □

Define the corner layer function z_1 associated with the corner $(0,0)$ as follows:

$$L_\varepsilon z_1 = 0, (x,y) \in \Omega, \tag{8a}$$

$$z_1(0,y) = w_1(0,y) = q_4(y), \ z_1(1,y) = 0, \ 0 \le y \le 1, \tag{8b}$$

$$z_1(x,0) = w_4(x,0) = q_1(x), \ z_1(x,1) = 0, 0 \le x \le 1, \tag{8c}$$

$$-\varepsilon q_1'' + b(x,0)q_1 = 0, \ x \in (0,1), \ q_1(0) = s_1(0), \ q_1(1) = 0. \tag{8d}$$

Then $z_1 \in C^{1,\alpha}(\overline{\Omega})$ and we have that

$$|z_1(x,y)| \le Ce^{-x\sqrt{\beta/\varepsilon}} e^{-y\sqrt{\beta/\varepsilon}}. \tag{9a}$$

Analogous bounds hold for the other boundary (corner) layer functions associated with the other three edges (corners).

4 Discrete Algorithm

We employ the standard central finite difference operator

$$L^N U^N := -\varepsilon(\delta_x^2 + \delta_y^2)U^N + bU^N = f,$$

which can also be generated from a standard finite element formulation on a structured tensor product grid with lumping as a quadrature rule. Here δ_x^2 denotes the classical three-point finite difference approximation to u_{xx} on a non-uniform mesh. We initially solve for an approximation $\bar{V}$ to the regular component v on a uniform coarse grid $\bar{\Omega}_u^N = \{(x_i, y_j) | x_i = i/N, y_j = j/N, 0 \le i, j \le N\}$. That is, the mesh function V^N is the solution of

$$L^N V^N = f, \ (x_i, y_j) \in \Omega_u^N; \ V^N = v, \ (x_i, y_j) \in \partial\Omega \cap \bar{\Omega}_u^N.$$

A global approximation to v is a simple interpolant of the form

$$\bar{V} = \sum_{i,j} V^N(x_i, y_j)\phi_i(x)\psi_j(y),$$

where $\phi_i(x)$ and $\psi_j(y)$ are the standard hat functions associated with x_i and y_j respectively. Define the following subdomains: $\Omega_1 = (0,1) \times (0,\sigma)$, $\Omega_2 = (1-\sigma, 1) \times (0,1)$, $\Omega_3 = (0,1) \times (1-\sigma, 1), \Omega_4 = (0,\sigma) \times (0,1)$. On each of these subdomains we define a tensor product of two uniform meshes. That is, $\bar{\Omega}_1^N := \{(x_i, y_j)|x_i = i/N, y_j = j\sigma/N, 0 \le i, j \le N\}$, where the Shishkin transition parameter σ is taken to be

$$\sigma := \min\left\{1, 2\sqrt{\frac{\varepsilon}{\beta}} \ln N\right\}. \tag{10}$$

The nodal values of an approximation $\bar{W}_1$ (defined solely on the layer region $\bar{\Omega}_1$) to the boundary layer function w_1 are computed by solving

$$L^N W_1^N = 0, (x_i, y_j) \in \Omega_1^N,$$
$$W_1^N(0, y_j) = s_1(0)e^{-y_j\sqrt{b(0,0)/\varepsilon}},\ W_1^N(1, y_j) = s_1(1)\sigma^{-1}(\sigma - y_j), 0 \le y_j \le \sigma,$$
$$W_1^N(x_i, 0) = s_1(x_i),\ W_1^N(x_i, \sigma) = W_1^N(0,\sigma)(1 - x_i) + W_1^N(1,\sigma)x_i, 0 < x_i < 1.$$

The nodal values of an approximation $\bar{Z}_1$ (defined solely on the corner layer region $\bar{\Omega}_5 \equiv \bar{\Omega}_1 \cap \bar{\Omega}_4$) to the corner layer function z_1 are computed by solving

$$L^N Z_1^N = 0, (x_i, y_j) \in \Omega_1^N \cap \Omega_4^N,$$
$$Z_1^N(0, y_j) = W_1^N(0, y_j),\ Z_1^N(\sigma, y_j) = \sigma^{-1} W_4^N(0,\sigma)(\sigma - y_j),\ 0 \le y_j \le \sigma,$$
$$Z_1^N(x_i, 0) = W_4^N(x_i, 0),\ Z_1^N(x_i, \sigma) = \sigma^{-1} W_1^N(0,\sigma)(\sigma - x_i),\ 0 < x_i < \sigma.$$

The approximations to the other six layer functions are defined analogously. The approximation $\bar{U}$ to the solution is patched together using the sum

$$\bar{U} = \bar{V} + \sum_{i=1}^{4} \bar{W}_i - \sum_{i=1}^{4} \bar{Z}_i.$$

5 Error Analysis

Theorem 1. *For the solution of (2a) and the approximation defined in Sect. 4*

$$\|u - \bar{U}\| \le CN^{-1} \ln N + C\sqrt{\varepsilon}.$$

Proof. Note that on the coarse uniform mesh Ω_u^N

$$\left|L^N(u_0 - V^N)(x_i, y_j)\right| = \left|(L^N - L_\varepsilon)u_0(x_i, y_j)\right| + C\varepsilon, \ (x_i, y_j) \in \Omega_u^N$$
$$\leq CN^{-2}\varepsilon + C\varepsilon \leq CN^{-2}\varepsilon + C\varepsilon.$$

Then

$$\|v - \bar{V}\| \leq \|v - u_0\| + \|u_0 - \bar{V}\| \leq CN^{-2}\varepsilon + C\varepsilon + CN^{-2}. \qquad (11)$$

Within the boundary layer region Ω_1^N, by (6) and the bounds in Lemma 1, we have that

$$\left|L^N(w_1 - W_1^N)(x_i, y_j)\right| = \left|(L^N - L_\varepsilon)w_1(x_i, y_j)\right| + C\varepsilon \leq CN^{-1}\ln N + C\varepsilon.$$

Note that, if $\Psi(y) := s_1(0)e^{-y\sqrt{b(0,0)/\varepsilon}}$ then $(\Psi - q_4)(0) = 0$,

$$-\varepsilon(\Psi(y) - q_4(y))'' + b(0, y)(\Psi(y) - q_4(y)) = (b(0, y) - b(0, 0))\Psi(y),$$

and $|(b(0, y) - b(0, 0))|\Psi(y) \leq C\sqrt{\varepsilon}$. From this, on the boundary $\partial\Omega_1^N$ we have

$$|(W_1^N - w_1)(0, y_j)| \leq C\sqrt{\varepsilon}, \ |(W_1^N - w_1)(1, y_j)| \leq C\sqrt{\varepsilon}, 0 \leq y_j \leq \sigma,$$
$$(W_1^N - w_1)(x_i, 0) = 0, \ (W_1^N - w_1)(x_i, \sigma)| \leq CN^{-2} + C\varepsilon, 0 < x_i < 1.$$

Then we can conclude that over the entire domain Ω

$$\|w_1 - \bar{W}_1\| \leq C(N^{-1}\ln N + \sqrt{\varepsilon}). \qquad (12)$$

Within the corner region, we follow closely the approach of Andreev [1]. We first further decompose the corner layer function z_1. Let $z_1 = q_1(x)q_4(y) + z_{00} + \sqrt{\varepsilon}R_2$, where

$$|L_\varepsilon R_2| = |(b(0, 0) - L_\varepsilon)q_1(x)q_4(y)| \leq C\sqrt{q_1(x)q_4(y)}, \ R_2 = 0, (x, y) \in \partial\Omega$$
$$L_\varepsilon z_{00} = b(0, 0)q_1(x)q_4(y), (x, y) \in \Omega, \qquad z_{00} = 0, (x, y) \in \partial\Omega.$$

Note that $|z_{00}(x, y)| \leq Cq_1(x)q_4(y)$. The discrete version of this secondary decomposition is

$$Z_1^N = q_1(x_i)q_4(y_j) + Z_{00}^N + \sqrt{\varepsilon}R_2^N$$
$$L^N Z_{00}^N = b(0, 0)q_1(x_i)q_4(y_j) + (L_\varepsilon - L^N)q_1(x_i)q_4(y_j), (x_i, y_j) \in \Omega_1^N \cap \Omega_4^N,$$
$$Z_{00}^N = 0, (x_i, y_j) \in \partial(\Omega_1^N \cap \Omega_4^N).$$

Hence $|R_2^N| \leq C$ and on the boundary of the corner patch we have that

$$|R_2^N(x_i, y_j)| \leq C(N^{-1}\ln N + \sqrt{\varepsilon}), \ (x_i, y_j) \in \partial(\Omega_1^N \cap \Omega_4^N).$$

It remains to estimate the error in $|z_{0,0} - Z_{0,0}^N|$. Set $\tau := \sum_{(x_i,y_j)\in\Omega_1^N\cap\Omega_4^N} |(L^N - L_\varepsilon)z_{0,0}(x_i, j_j)|$. We decompose z_{00} as in [1, Theorem 2.1], ($\chi_{1,1} = -b(0, 0)$, $\chi_{1,2} = 0$) and from [1, p. 962], we have that $\tau \leq C\ln N$. In the corner layer region, we then bound the nodal error using the discrete stability bound given in [1,

Theorem 3.1], as follows

$$
\begin{aligned}
|z_{0,0} - Z^N_{0,0}| &\le CN^{-2}(\ln N)^4 + CN^{-2}\ln N \\
&\quad \sum_{(x_i,y_j)\in\Omega_1^N\cap\Omega_4^N} |L^N(z_{0,0} - Z^N_{0,0})(x_i,y_j)| \\
&\le CN^{-2}(\ln N)^4 + CN^{-2}\ln N \\
&\quad \sum_{(x_i,y_j)\in\Omega_1^N\cap\Omega_4^N} |(L_\varepsilon - L^N)q_1(x_i)q_4(y_j)| \\
&\le CN^{-2}(\ln N)^4 + CN^{-2}\ln N \\
&\quad \sum_{(x_i,y_j)\in\Omega_1^N\cap\Omega_4^N} \frac{h^2}{\varepsilon} e^{-x_{i-1}\sqrt{\beta/\varepsilon}} e^{-y_{j-1}\sqrt{\beta/\varepsilon}} \\
&\le CN^{-2}(\ln N)^4 + CN^{-2}\ln N \Big(\frac{\rho}{1-e^{-\rho}}\Big)^2,\ \rho = h\sqrt{\beta/\varepsilon}, h = \sigma/N \\
&\le CN^{-2}(\ln N)^4.
\end{aligned}
$$

By explicitly differentiating the leading term in the representation given in [1, Theorem 2.1], we can deduce the following bound on the first derivatives:

$$
\Big\|\frac{\partial^{i+j} z_{0,0}}{\partial x^i \partial y^j}\Big\| \le C\varepsilon^{-1/2},\ i + j = 1.
$$

Use of the interpolation bound in [11, Lemma 4.1] completes the proof. □

Remark 2. It is worth noting that if the additional compatibility conditions (3b) are assumed to hold at all four corners, then $|s_1(0)| \le C\varepsilon$ and $|s_1(1)| \le C\varepsilon$. It follows that is not necessary to patch in the corners (i.e. it is not required to compute Z) in order to derive the following error bound

$$
\|u - \bar{U}\| \le CN^{-1}\ln N + C\varepsilon.
$$

6 Numerical Results

6.1 Test Example 1

We consider a particular example of problem (2a) with the following coefficients:

$$
b(x,y) = 1 + x^2y^2,\ f(x,y) = 1 + 2xy \tag{13}
$$

and boundary data

$$
g_1(x) \equiv g_4(y) \equiv 1,\ g_3 = 1 - x^2,\ g_2 = 1 - y^2. \tag{14}
$$

Table 1 Parameter-uniform global two-mesh differences D^N and rates ρ^N on a patched mesh for test example 1 over the range $R_\varepsilon = [2^{-40}, 2^{-7}]$

N	2^5	2^6	2^7	2^8	2^9
D^N	1.16×10^{-2}	2.92×10^{-3}	7.32×10^{-4}	1.83×10^{-4}	4.57×10^{-5}
ρ^N	1.99	2.00	2.00	2.00	

Note that in this particular example the zero level compatibility conditions (3a) are satisfied at all four corners, but the compatibility condition at the first level (corresponding to (3b)) is not satisfied at the corner $(1, 1)$. Tensor product meshes with N steps in each direction are used both for the boundary and corner patches, while a triangular mesh with N^2 degrees of freedom is used in computing an approximation to the regular component over the entire domain. The convergence behaviour of the numerical method is reported in Table 1 where the global two mesh differences D^N and the approximate uniform rates of convergence ρ^N were computed over a certain range R_ε of values for ε, using

$$D^N := \max_{\varepsilon \in R_\varepsilon} \|\bar{U}^N - \bar{U}^{2N}\|_{\Omega_S^{10N}}, \quad \rho^N := \log_2 \frac{D^N}{D^{2N}}.$$

Here Ω_S^{10N} is a tensor product piecewise-uniform Shishkin mesh [3] with $10N$ elements in each coordinate direction. We choose to measure the difference between the two interpolants on this finer mesh Ω_S^{10N}, as the maximum difference between the two interpolants may not occur over the set of mesh points $\Omega^N \cup \Omega^{2N}$. The computed uniform rate of convergence for this example is greater than what is established theoretically in Theorem 1.

6.2 Test Example 2

To assess the applicability of the patched mesh method to a problem posed on a non-rectangular domain, we consider a problem of the form (1) set in a domain $\Omega \equiv \Omega_1 \cup \Omega_2$ with $\Omega_1 \equiv (-1, 1) \times (-1, 0)$ and $\Omega_2 \equiv \{(x, y)|x^2 + y^2 < 1\}$. For this test example, the coefficients b, f and g are given by

$$\begin{cases} f(x, y) = b(x, y) = 1, & (x, y) \in \Omega \\ g(x, y) = \frac{2-\tanh(12y)-\tanh(12)}{2}, & (x, y) \in \partial\Omega. \end{cases}$$

For this choice of data no boundary layer occurs near the side $y = -1$. Let $\partial\Omega_L := \partial\Omega \setminus \{(x, -1), 0 < x < 1\}$ be the remainder of the boundary. The patch for this problem is taken to be $\bar{\Omega}_p := \{\mathbf{x} \in \bar{\Omega} | \text{dist}(\mathbf{x}, \partial\Omega_L) \leq \sigma\}$, where σ is as given in (10).

The solution to this second test example is shown in Fig. 1b, while Tables 2 and 3 show the performance of the patched mesh method and of a standard finite

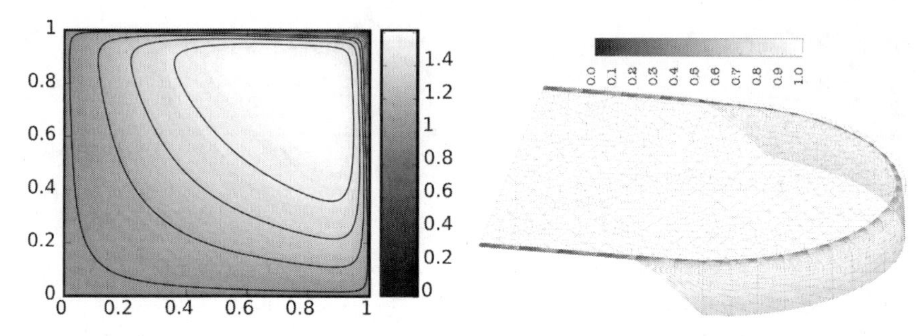

Fig. 1 Computed solutions to the two test examples using a patched mesh method with $N = 64$

Table 2 Parameter-uniform global two-mesh differences D^N and rates ρ^N on a patched mesh for test example 2 over the range $R_\varepsilon = 10^{-4}[2^{-20}, 1]$

N	2^4	2^5	2^6	2^7	2^8
D^N	1.47×10^{-2}	5.61×10^{-3}	1.99×10^{-3}	6.77×10^{-4}	2.22×10^{-4}
ρ^N	0.91	1.26	1.67	2.12	

Table 3 Parameter-uniform global two-mesh differences D^N and rates ρ^N of a standard finite element method on a quasi-uniform mesh for test example 2 over the range $R_\varepsilon = 10^{-4}[2^{-20}, 1]$

N	2^4	2^5	2^6	2^7	2^8
D^N	0.257	0.684	0.58	0.437	0.471
ρ^N	1.58	0.00	−0.04	0.44	

element method on a quasi uniform mesh respectively. The rates in Table 2 suggest that the patched method is parameter uniform for this problem, which contrasts with the apparent lack of uniform convergence displayed in Table 3 for a standard finite element method on a quasi-uniform mesh.

Acknowledgement This research was supported by the Mathematics Applications Consortium for Science and Industry in Ireland (MACSI) under the Science Foundation Ireland (SFI) mathematics initiative.

References

1. V.B. Andreev. On the accuracy of grid approximations to nonsmooth solutions of a singularly perturbed reaction–diffusion equation in a square. *Differential Equations*, 42(7):954–966, 2006.
2. F. Brezzi, J.-L. Lions, and O. Pironneau. Analysis of a chimera method. *Comptes Rendus de l'Academie des Sciences, Series I Mathematics*, 332(7):655–660, 2001.
3. C. Clavero, J.L. Gracia and E. O'Riordan. A parameter robust numerical method for a two-dimensional reaction–diffusion problem. *Mathematics of Computation*, 74:1743–1758, 2005.

4. M. Culpo, C. de Falco, and E. O' Riordan. Patches of finite elements for singularly-perturbed diffusion reaction equations with discontinuous coefficients. In *Proceedings of the ECMI 2008 Conference*, 2008.
5. P.A. Farrell, A.F. Hegarty, J.J.H. Miller, E. O'Riordan, and G.I. Shishkin. *Robust Computational Techniques for Boundary Layers.* Chapman and Hall/CRC, New York/Boca Raton, 2000.
6. R. Glowinski, J. He, A. Lozinski, J. Rappaz, and J. Wagner. Finite element approximation of multi-scale elliptic problems using patches of elements. *Numerische Mathematik*, 101(4):663–687, 2005.
7. H. Han and R.B. Kellogg. Differentiability properties of solutions of the equation $-\varepsilon^2 \Delta u + ru = f(x, y)$ in a square. *SIAM Journal of Mathematical Analysis*, 21:394–408, 1990.
8. N. Kopteva. Maximum norm error analysis of a 2D singularly perturbed semilinear reaction–diffusion problem. *Mathematics of Computation*, 76(258):631–646, 2007.
9. A. Quarteroni and A. Valli. *Domain Decomposition Methods for Partial Differential Equations*. Numerical Mathematics and Scientific Computation. Clarendon Press, Oxford, 1999.
10. A.H. Schatz and L.B. Wahlbin. On the finite element method for singularly perturbed reaction–diffusion problems in two and one dimensions. *Mathematics of Computation*, 40(161):47–89, 1983.
11. M. Stynes and E. O'Riordan. A uniformly convergent galerkin method on a shishkin mesh for a convection-diffusion problem. *Journal of Mathematical Analysis and Applications*, 214:36–54, 1997.

Singularly Perturbed Reaction–Diffusion Problem with a Boundary Turning Point

C. de Falco and E. O'Riordan

Abstract Parameter uniform numerical methods for singularly perturbed reaction diffusion problems have been examined extensively in the literature. By using layer adapted meshes of Bakhvalov or Shishkin type, it is now well established that one can achieve second order (or almost second order in the case of the simpler Shishkin meshes) parameter uniform convergence globally in the pointwise maximum norm. Note that, in proving such results, it is often assumed that the coefficient of the reactive term is strictly positive throughout the domain. In this paper, we examine a problem where the reaction coefficient is zero on parts of the boundary.

1 Introduction

Parameter-uniform [6] numerical methods for singularly perturbed reaction diffusion problems of the form

$$-\varepsilon \triangle u + bu = f, \mathbf{x} \in \Omega, \; u = g, \; \mathbf{x} \in \partial\Omega, \tag{1}$$

have been examined extensively [1–3, 10]. By using layer adapted meshes of Bakhvalov [2] or Shishkin [10] type, it is well established that one obtains second order (or almost second order in the case of the simpler Shishkin meshes) uniform convergence globally in the pointwise maximum norm. Note that, normally one assumes that

$$b(\mathbf{x}) \geq \beta > 0, \; \forall \mathbf{x} \in \overline{\Omega}. \tag{2}$$

In this paper, we examine a problem where b is zero on parts of the boundary $\partial\Omega$ and also depends on ε. We are interested in the necessary modifications required when using a piecewise-uniform Shishkin mesh and in the subsequent error analysis.

C. de Falco (✉)
School of Mathematical Sciences, Dublin City University, Dublin 9, Ireland,
E-mail: carlo.defalco@dcu.ie

A.F. Hegarty et al. (eds.), *BAIL 2008 - Boundary and Interior Layers.*
Lecture Notes in Computational Science and Engineering,
DOI: 10.1007/978-3-642-00605-0,

As for a possible application of the results presented here, in [5] a method was presented for computing the differential capacitance of Metal Oxide Semiconductor (MOS) structure and the advantage of using a layer-adapted mesh was shown to be non negligible for values of the coefficients within a physically reasonable range. The simple example presented in the numerical experiments of Sect. 4 demonstrates that similar benefits are to be expected if the model of [5] is extended to take into account quantization effects [4, 7].

Notation: Throughout this paper C (sometimes subscripted) denotes a generic constant that is independent of ε and N. We also use the following notation

$$|f|_k := \max_{x\in(0,1)} \left|\frac{d^k f}{dx^k}(x)\right| \quad \text{and} \quad \|f\| := \max_{x\in[0,1]} |f(x)|.$$

2 Continuous Problem

Consider the following two point boundary value problem

$$Lu := -\varepsilon u'' + b(x;\varepsilon)u = f(x), \; x \in \Omega = (0,1), \tag{3a}$$

$$b(x;\varepsilon) \geq 0, \; x \in \bar{\Omega}, \; u(0) = u_0, \; u(1) = u_1, \tag{3b}$$

where f, b are sufficiently smooth and the coefficient b satisfies the following

$$b(0;\varepsilon) = 0; \; |b|_k \leq C\varepsilon^{-k/2}, k \leq 2, \tag{3c}$$

$$(1-\gamma)b(x;\varepsilon) + \sqrt{\varepsilon}\sqrt{\gamma}(\sqrt{b})'(x;\varepsilon) \geq m > 0, \; 0 < \gamma < 1, \tag{3d}$$

$$|b(x;\varepsilon) - b(x;0)| \leq Me^{-x\sqrt{\frac{\theta}{\varepsilon}}}, \; \theta \geq \gamma\|b\|, \tag{3e}$$

$$b(x;0) := \lim_{\varepsilon\to 0} b(x;\varepsilon), \; \forall x \in \bar{\Omega}. \tag{3f}$$

For any specific b, we will need to identify the parameters m, M, θ, γ. We note in passing that we do not assume that $f(0) = 0$. We adopt the following notation for the following particular ordering of the two limits

$$b(0;0) := \lim_{x\to 0}\big(\lim_{\varepsilon\to 0} b(x;\varepsilon)\big).$$

From the above assumptions on b, the function $y := \sqrt{b}$ satisfies the following singularly perturbed nonlinear Riccati equation

$$\sqrt{\gamma}\sqrt{\varepsilon}y' + (1-\gamma)y^2 = g^2, x > 0, \; y(0) = 0.$$

We construct a lower solution $\underline{y}$ for y of the form $\underline{y} = C_1(1 - e^{-x\sqrt{\frac{\gamma}{\varepsilon}}})$, $C_1 \leq \|g\|$, where $C_1 > 0$ is such that

$$\sqrt{\gamma}\sqrt{\varepsilon}\underline{y}' + (1-\gamma)\underline{y}^2 = C_1\gamma(e^{-t} + K_1(1-e^{-t})^2) \leq m,$$

$$\text{and} \qquad t := x\sqrt{\frac{\gamma}{\varepsilon}}, \; \gamma K_1 := C_1(1-\gamma).$$

For the function $h(t) = e^{-t} + K_1(1-e^{-t})^2, \; t \geq 0$, we note that

$$\min\left\{K_1, 1 - \frac{1}{4K_1}\right\} \leq h(t) \leq \max\{1, K_1\}.$$

Hence,

$$\sqrt{\gamma}\sqrt{\varepsilon}\underline{y}' + (1-\gamma)\underline{y}^2 \leq C_1 \max\{\gamma, C_1(1-\gamma)\}.$$

Thus, the choice of

$$C_1 := \min\left\{\frac{m}{\gamma}, \sqrt{\frac{m}{1-\gamma}}\right\} \tag{4}$$

suffices for $\underline{y}$ to be a lower solution. It follows that, from (3c) and (3d) that

$$\sqrt{b(x;\varepsilon)} \geq C_1(1 - e^{-x\sqrt{\frac{\gamma}{\varepsilon}}}), \; x \in [0,1], \tag{5a}$$

$$b(x;\varepsilon) \geq \beta := C_1^2(1-e^{-1})^2 > 0, \; x \geq \sqrt{\frac{\varepsilon}{\gamma}}. \tag{5b}$$

Note that (5a) implies that $b(x;0) > 0, \; x \in [0,1]$.

The standard maximum principle [9] is still valid for the linear differential operator L. That is, if $w \in C^0(\bar{\Omega}) \cap C^2(\Omega), w(0) \geq 0, w(1) \geq 0$, and for all $x \in \Omega, \; Lw \geq 0$, then $w \geq 0$ for all $x \in \bar{\Omega}$.

Lemma 1. *For all* $k, \; 0 \leq k \leq 4, \; |u|_k \leq C\left(1 + \varepsilon^{-k/2}\right)$*, where* u *is the solution of problem (3).*

Proof. Consider the following barrier function

$$\phi(x) = A^* + B^*(1 - e^{-\sqrt{\frac{\eta}{\varepsilon}}x}) \geq A^*,$$

where A^*, B^* and η are positive constants specified below. Note that, outside the layer region, if $\beta A^* \geq \|f\|$, then

$$L\phi(x) \geq \beta\phi(x) \geq \|f\|, \; x \geq \sqrt{\frac{\varepsilon}{\gamma}}.$$

In the layer region, where $x \leq \sqrt{\frac{\varepsilon}{\gamma}}$, we have that

$$L\phi(x) = b\phi + B^*\eta e^{-x\sqrt{\frac{\eta}{\varepsilon}}} \geq B^*(\eta - b)e^{-x\sqrt{\frac{\eta}{\varepsilon}}} \geq B^*(\eta - \|b\|)e^{-\sqrt{\frac{\eta}{\gamma}}} \geq \|f\|,$$

if $\eta \geq \|b\|$ and $B^*(\eta - \|b\|) \geq \|f\|e^{\sqrt{\frac{\eta}{\gamma}}}$. We can choose $\eta = \gamma + \|b\|$, then by an appropriate choice of A^* and B^* we deduce the stability bound

$$\|u\| \leq \max\{\frac{\|f\|}{\beta}, |u_0|, |u_1|\} + \frac{\|f\|}{\gamma} e^{\sqrt{1+\frac{\|b\|}{\gamma}}} (1 - e^{-x\sqrt{\frac{\gamma+\|b\|}{\varepsilon}}}).$$

Recall that (3c) allows the derivatives of the coefficient b to depend adversely on ε. Hence, as we derive parameter explicit bounds on the derivatives of u below, we need to identify how the error constants in these bounds depend on b and its derivatives explicitly. To bound the first derivative of u we use an argument from [2]. Let $x \in \Omega$ and construct a neighbourhood $N_x = (a, a + \sqrt{\varepsilon})$ such that $x \in N_x$ and $N_x \subset \Omega$. Then, by the mean value theorem, for some $y \in \bar{N}_x$,

$$\frac{u(a + \sqrt{\varepsilon}) - u(a)}{\sqrt{\varepsilon}} = u'(y).$$

It follows that $|u'(y)| \leq 2\varepsilon^{-1/2}\|u\|_{N_x} \leq 2\|u\|\varepsilon^{-1/2}$. Now

$$u'(x) = u'(y) + \int_y^x u''(\xi)d\xi = u'(y) + \varepsilon^{-1} \int_y^x (f - bu)(\xi)\, d\xi$$

and so

$$\|u'\| \leq \big(\|f\| + (2 + \|b\|)\|u\|\big)\varepsilon^{-1/2}.$$

The bounds on the higher derivatives are obtained using the differential equation (3a) and (3c). Note that

$$\begin{aligned}
\varepsilon^{1/2}|u|_1 &\leq \|f\| + (2 + \|b\|)\|u\|; \qquad \varepsilon|u|_2 \leq \|f\| + \|b\|\|u\|;\\
\varepsilon^{3/2}|u|_3 &\leq \|b\|\|f\| + \sqrt{\varepsilon}|f|_1 + (\|b\|(2 + \|b\|) + \sqrt{\varepsilon}|b|_1)\|u\|;\\
\varepsilon^2|u|_4 &\leq \varepsilon\big(|f|_2 + \|b\||u|_2 + 2|b|_1|u|_1 + \|u\||b|_2\big)\\
&\leq \varepsilon|f|_2 + \|b\|\varepsilon|u|_2 + 2\sqrt{\varepsilon}|b|_1\sqrt{\varepsilon}|u|_1 + \|u\|\varepsilon|b|_2.
\end{aligned}$$

□

Define the associated operator

$$L_1\omega(x) := (-\varepsilon\omega'' + b(x;0)\omega),\ x \in \Omega,\ b(x;0) \geq \beta > 0.$$

Decompose the solution into three components of the form

$$u(x) = v(x) + w_L(x) + w_R(x),$$

where

$$\begin{aligned}
L_1 v &= f, x \in \Omega, & v(0) &= \frac{f(0)}{b(0;0)},\quad v(1) = \frac{f(1)}{b(1;0)},\\
Lw_L &= (L_1 - L)v, x \in \Omega, & w_L(0) &= u(0) - v(0),\ w_L(1) = 0,\\
Lw_R &= 0, x \in \Omega, & w_R(0) &= 0,\ w_R(1) = u(1) - v(1).
\end{aligned}$$

The boundary conditions for the regular component v have been selected [8] so that

$$|v|_k \leq C\frac{\|f\|}{\beta}(1+\varepsilon^{1-k/2}), \quad k \leq 4. \tag{6}$$

Lemma 2. *The layer components w_L, w_R in the solution of problem (3) satisfy the following bounds*

$$|w_L(x)| \leq Ce^{-\int_0^x \sqrt{\frac{\gamma b(t;\varepsilon)}{\varepsilon}}\,dt}, \ |w_R(x)| \leq Ce^{-\sqrt{\frac{\beta}{\varepsilon}}(1-x)}, \tag{7a}$$

$$|w_L|_k \leq C\varepsilon^{-k/2}, \ |w_R|_k \leq C\varepsilon^{-k/2}, 1 \leq k \leq 4. \tag{7b}$$

Proof. To obtain the pointwise bound on $w_L(x)$, use the barrier function

$$\Psi_L(x) = K_2 e^{-\int_0^x \sqrt{\frac{\gamma b(t;\varepsilon)}{\varepsilon}}\,dt}, \quad K_2 = \max\{\frac{M}{m}\|v\|, |u(0)-v(0)|\},$$

which satisfies

$$L\Psi_L = \big(b(x;\varepsilon)(1-\gamma) + \sqrt{\gamma\varepsilon}(\sqrt{b(x;\varepsilon)})'\big)\Psi_L \geq m\Psi_L$$

$$\geq M\|v\|e^{-x\sqrt{\frac{\theta}{\varepsilon}}} \geq |(L-L_1)v|.$$

To obtain the pointwise bound on $w_R(x)$ use the barrier function

$$\Psi_R(x) = K_3 e^{-\sqrt{\frac{\beta}{\varepsilon}}(1-x)} + K_4 e^{-\sqrt{\frac{\beta}{\varepsilon}}}(1-e^{-\sqrt{\frac{1+\|b\|}{\varepsilon}}x}).$$

Outside the left layer, we have that

$$L\Psi_R(x) \geq K_3(b-\beta)e^{-\sqrt{\frac{\beta}{\varepsilon}}(1-x)} \geq 0, \ x \geq \sqrt{\frac{\varepsilon}{\gamma}},$$

and within the left layer, for $x \leq \sqrt{\frac{\varepsilon}{\gamma}}$,

$$L\Psi_R(x) \geq K_4 e^{-\sqrt{\frac{\beta}{\varepsilon}}}(1+\|b\|-b)e^{-\sqrt{\frac{1+\|b\|}{\varepsilon}}x} - K_3\beta e^{-\sqrt{\frac{\beta}{\varepsilon}}(1-x)}$$

$$\geq K_4 e^{-\sqrt{\frac{\beta}{\varepsilon}}}e^{-\sqrt{\frac{1+\|b\|}{\gamma}}} - K_3\beta e^{-\sqrt{\frac{\beta}{\varepsilon}}}e^{\sqrt{\frac{\beta}{\gamma}}} \geq 0,$$

when we choose K_4 such that $K_4 \geq K_3\beta e^{\sqrt{\frac{1+\|b\|}{\gamma}}}e^{\sqrt{\frac{\beta}{\gamma}}}$ and $K_3 \geq |u(1)-v(1)|$. The bounds on the derivatives are derived as in the proof of Lemma 1. □

3 Discrete Problem

On the domain Ω a piecewise-uniform Shishkin mesh [1] of N mesh intervals is constructed in the usual way. The domain $\overline{\Omega}$ is subdivided into the three subintervals $[0,\sigma_1]$, $[\sigma_1, 1-\sigma_2]$ and $[1-\sigma_2, 1]$. On $[0,\sigma_1]$ and $[1-\sigma_2, 1]$ a uniform mesh with

$\frac{N}{4}$ mesh-intervals is placed, while $[\sigma_1, 1-\sigma_2]$ has a uniform mesh with $\frac{N}{2}$ mesh-intervals. The interior mesh points are denoted by Ω_ε^N and $h_i := x_i - x_{i-1}$ is the mesh step. Let

$$\sigma_1 := \min\left\{\frac{1}{4}, \tau\right\}, \quad \sigma_2 := \min\left\{\frac{1}{4}, 2\sqrt{\frac{\varepsilon}{\beta}}\ln N\right\}. \tag{8a}$$

Our transition point τ in our Shishkin mesh will be chosen such that

$$\int_0^\tau \sqrt{\gamma b(t;\varepsilon)}\, dt \geq 2\sqrt{\varepsilon}\ln N.$$

For example, based on the lower bound on $\sqrt{b}$ in (5) we take,

$$\tau \geq \frac{\sqrt{\varepsilon}}{\sqrt{\gamma}}(1 + \frac{2}{C_1}\ln N),\ C_1 = \min\left\{\sqrt{\frac{m}{1-\gamma}}, \frac{m}{\gamma}\right\}. \tag{8b}$$

The discrete problem is: Find U such that

$$L^N U(x_i) := -\varepsilon\delta^2 U(x_i) + b(x_i;\varepsilon)U(x_i) = f(x_i),\ x_i \in \Omega_\varepsilon^N, \tag{9a}$$

$$U(0) = u(0),\ U_\varepsilon(1) = u_\varepsilon(1), \tag{9b}$$

$$\text{where}\quad \delta^2 Z(x_i) := (\frac{Z(x_{i+1}) - Z(x_i)}{h_{i+1}} - \frac{Z(x_i) - Z(x_{i-1})}{h_i})\frac{2}{h_i + h_{i+1}}.$$

The finite difference operator L^N satisfies a discrete comparison principle. That is for any mesh function Z, if $L^N Z(x_i) \geq 0$ for all $x_i \in \Omega_\varepsilon^N, Z(0) \geq 0, Z(1) \geq 0$ then $Z(x_i) \geq 0$ for all $x_i \in \overline{\Omega}_\varepsilon^N$.

Lemma 3. *For any mesh function Z then*

$$\|Z\| \leq C(\|L^N Z\| + |Z(0)| + |Z(1)|).$$

Proof. Consider the following barrier function

$$\Phi(x_i) = B_1^*(1 - W(x_i)) + \max\{\frac{\|L^N Z\|}{\beta}, |u_0|, |u_1|\},\quad W(x_i) := \Pi_{j=1}^i (1 + h_j\sqrt{\frac{\zeta}{\varepsilon}})^{-1},$$

where B_1^*, ζ are specified below. Note that $W(x_i) \geq e^{-x_i\sqrt{\frac{\zeta}{\varepsilon}}}$ and

$$L^N\Phi(x_i) = b\Phi(x_i) + B_1^*\frac{2\zeta h_{i+1}}{h_i + h_{i+1}} W(x_{i+1}).$$

We note that outside the layer, $L^N\Phi(x_i) \geq \|L^N Z\|,\ x_i \geq \sqrt{\varepsilon\gamma^{-1}}$. In the layer region, where $x_i < \sqrt{\varepsilon\gamma^{-1}}$, we have that $h_{i+1} = h_i =: h$ and for sufficiently large N

$$h\sqrt{\frac{\zeta}{\varepsilon}} \leq \frac{8\sqrt{\zeta}}{C_1\sqrt{\gamma}}N^{-1}\ln N + \frac{4\sqrt{\zeta}}{\sqrt{\gamma}}N^{-1} \leq \ln 2.$$

Hence, we have that

$$\begin{aligned} L^N\Phi(x_i) &\geq b\Phi + B_1^*\frac{\zeta}{2}W(x_{i+1}) \geq B_1^*(\zeta W(x_{i+1}) - \|b\|W(x_i)) \\ &\geq B_1^*(\zeta(1+\ln 2)^{-1} - \|b\|)e^{-x_i\sqrt{\frac{\zeta}{\varepsilon}}} \\ &\geq B_1^*(\zeta(1+\ln 2)^{-1} - \|b\|)e^{-\sqrt{\frac{\zeta}{\gamma}}} \geq \|L^N Z\|, \end{aligned}$$

if we choose $\zeta = (1+\ln 2)\|b\| + \gamma$ and $\gamma B_1^* = (1+\ln 2)\|L^N Z\|e^{\sqrt{\frac{\zeta}{\gamma}}}$. □

The discrete solution is decomposed analogously to the continuous solution. That is

$$U(x_i) = V(x_i) + W_L(x_i) + W_R(x_i),$$

where $V(0) = v(0), V(1) = v(1),\ W_L(0) = w_L(0), W_L(1) = 0,\ W_R(0) = 0,$ $W_R(1) = w_R(1)$ and

$$L_1^N V = f,\ L^N W_L = (L_1^N - L^N)V,\ L^N W_R = 0, x_i \in \Omega_\varepsilon^N.$$

From the bounds on the derivatives of the components and Lemma 3, we can follow the argument in [8] to deduce that

$$\|\bar{U} - u\| \leq C(N^{-1}\ln N)^2, \tag{10}$$

where $\bar{U}$ is the piecewise linear interpolant of the discrete solution U of the discrete problem (9) and u is the solution of the continuous problem (3).

4 Numerical Experiments

As mentioned in the introduction, a physical problem whose numerical approximation requires relaxing the hypothesis (2) is that of computing the capacitance of an MOS structure where energy quantization in the inversion layer is to be taken into account. By choosing to model such quantization effects following the approach of [7] and performing the scaling and linearization procedure presented in [5], this problem leads to an equation of the form

$$-\varepsilon u'' + e^{A(x)}(1 - e^{-x^2/\lambda^2})u = f(x),$$

where $A(x)$ is the scaled electric potential, λ is the scaled electron wavelength and the semiconductor insulator interface is placed at $x = 0$. Rescaling we get

$$-\varepsilon e^{-A}u'' + (1 - e^{-x^2/\lambda^2})u = e^{-A}f(x),\ b(x;\lambda) = 1 - e^{-t}, t := x^2/\lambda^2. \tag{11a}$$

We set A to be constant and below we will see that it is necessary that

$$\lambda^2 = C\varepsilon. \tag{11b}$$

Let us check that finite values $m^*, M^*, \theta^*, \gamma^*$ exist for the parameters m, M, θ, γ so that the constraints (3c)–(3e) on the coefficient of the zero order term are satisfied, which are required by the theory in the preceding sections. Introduce the additional parameter α_0 defined by $e^A\lambda^2 =: \alpha_0\varepsilon$. Observe that

$$b(0;\varepsilon) = 0; \ |b|_k \le C(\lambda^2)^{-k/2}, k \le 2, \ \|b\| \le 1,$$

$$\begin{aligned}(1-\gamma)b + \sqrt{\gamma\varepsilon e^{-A}}(\sqrt{b})' &= (1-\gamma)(1-e^{-t}) + \sqrt{\frac{\gamma\varepsilon e^{-A}}{\lambda^2}}\sqrt{t}e^{-t}(1-e^{-t})^{-1/2} \\ &= \sqrt{\frac{\gamma}{\alpha_0}}\left(K(1-e^{-t}) + \sqrt{t}e^{-t}(1-e^{-t})^{-1/2}\right), \ K = (1-\gamma)\sqrt{\frac{\alpha_0}{\gamma}}.\end{aligned}$$

Note that $K(1-e^{-t}) + \sqrt{t}e^{-t}(1-e^{-t})^{-1/2} \ge \min\{1, K\}$. We then have that

$$(1-\gamma)b + \sqrt{\gamma\varepsilon e^{-A}}(\sqrt{b})' \ge \min\{1-\gamma, \frac{\sqrt{\gamma}}{\sqrt{\alpha_0}}\} =: m^*.$$

For all $\alpha_0 > 0$, we can choose $0 < \gamma^* < 1$ so that

$$1-\gamma^* = \frac{\sqrt{\gamma^*}}{\sqrt{\alpha_0}}.$$

Hence,

$$\sqrt{\gamma^*} := \frac{2\sqrt{\alpha_0}}{\sqrt{1+4\alpha_0}+1} \quad \text{and} \quad m^* := 1-\gamma^*. \tag{12}$$

Let us examine condition (3e)

$$b(x;\varepsilon) - b(x;0) = e^{-t} \le Me^{-\sqrt{\theta\alpha_0 t}}, \ \theta \ge \gamma.$$

Then, we choose M^* and θ^* so that, $\theta^* := \gamma^*$ and $M^* := e^{0.25\gamma^*\alpha_0}$. Under these constraints, we take the transition point in (8) to be

$$\tau = \sqrt{\frac{e^A\,\lambda^2}{\alpha_0\gamma^*}}\left[1 + \frac{2\,\ln(N)}{\min\left\{1, \dfrac{1-\gamma^*}{\gamma^*}\right\}}\right], \tag{13}$$

where

$$e^A\lambda^2 =: \alpha_0\varepsilon \quad \text{and} \quad \sqrt{\gamma^*} = \frac{2\sqrt{\alpha_0}}{\sqrt{1+4\alpha_0}+1}.$$

Table 1 The ε-uniform nodal differences D^N, nodal rates p^n, global differences $\bar{D}^N$ and global rates $\bar{p}^N$ for the numerical approximations to the solution of problem (14)

	$N = 32$	$N = 64$	$N = 128$	$N = 256$	$N = 512$	$N = 1,024$	$N = 2,048$	$N = 4,096$	$N = 8,192$	$N = 16,384$
D^N	0.01714	0.01196	0.004853	0.001692	0.0008845	0.00036	0.0001223	3.636e-05	1.049e-05	3.024e-06
p^N	0.6178	0.5424	1.287	1.529	0.9384	1.3	1.558	1.75	1.794	1.794
$\bar{D}^N$	0.1871	0.08985	0.04164	0.01445	0.004413	0.00135	0.0004053	0.0001205	3.49e-05	1.006e-05
$\bar{p}^N$	0.8217	1.108	1.097	1.536	1.716	1.714	1.736	1.75	1.788	1.795

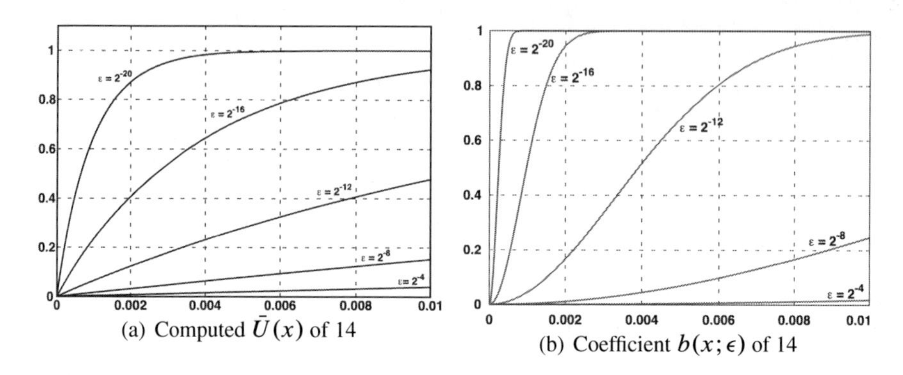

(a) Computed $\bar{U}(x)$ of 14

(b) Coefficient $b(x;\epsilon)$ of 14

Fig. 1 The computed approximations $\bar{U}(x)$ of (14), using $N = 4{,}096$, and the coefficient $b(x;\varepsilon)$ for several different values of ε over the interval $[0,0.01]$

Note that if λ^2 is not bounded above by $C\varepsilon$ then $M^* \to \infty$ as $\varepsilon \to 0$ and so the above error bounds are not uniformly bounded. Hence we require that $\lambda^2 = \mathcal{O}(\varepsilon)$. Below we present numerical results for the particular problem

$$-\varepsilon u'' + (1 - e^{-x^2/\lambda^2})u = 1, \ \lambda^2 = 0.09\varepsilon, \tag{14a}$$

$$u(0) = 0, \ u(1) = (1 - e^{-1/\lambda^2})^{-1}. \tag{14b}$$

The boundary condition at $x = 1$ means that there will be no layer in the vicinity of $x = 1$ and so it suffices to have $\sigma_2 = 0.25$, $\sigma_1 = \min\{0.5, \tau\}$ and to place $N/2$ mesh points in the intervals $[0, \sigma_1]$, $[\sigma_1, 1]$, in this case. In Fig. 1, plots of the layer and of the coefficient $b(x;\varepsilon)$ are displayed for several values of ε.

In Table 1, both the global and the nodal orders of convergence are estimated over the parameter range $\varepsilon = 2^0, 2^{-1}, \ldots, 2^{-20}$, using the double mesh principle [6]. The computed ε-uniform orders of convergence displayed are in line with the theoretical ε-uniform global convergence rate stated in the error bound (10).

Acknowledgement This research was supported by the Mathematics Applications Consortium for Science and Industry in Ireland (MACSI) under the Science Foundation Ireland (SFI) mathematics initiative.

References

1. V.B. Andreev, On the accuracy of grid approximations to nonsmooth solutions of a singularly perturbed reaction–diffusion equation in a square, *Diff. Eqs.*, **42**(7), 2006, 954–966.
2. N.S. Bakhvalov, On the optimization of methods for boundary-value problems with boundary layers. *J. Numer. Meth. Math. Phys.*, **9**, 1969, 841–859 (in Russian).
3. C. Clavero, J.L. Gracia, and E. O'Riordan, A parameter robust numerical method for a two dimensional reaction–diffusion problem, *Math. Comp.*, **74**, 2005, 1743–1758.
4. C. de Falco, E. Gatti, A.L. Lacaita, and R. Sacco, Quantum-corrected drift-diffusion models for transport in semiconductor devices, *J. Comp. Phys.*, **204**(2), 2005, 533–561.

5. C. de Falco, E. O'Riordan, Singularly perturbed reaction–diffusion equation with discontinuous diffusion coefficient, Preprint MS-08-01, DCU School of Math. Sc., 2008 (submitted).
6. P.A. Farrell, A.F. Hegarty, J.J.H. Miller, E. O'Riordan, and G.I. Shishkin, *Robust computational techniques for boundary layers*, Chapman and Hall/CRC, New York/Boca Raton, FL, (2000).
7. C. Jungermann, C. Nguyen, B. Neinhüs, S. Decker, and B. Meinerzhagen, Improved modified local density approximation for modeling of size quantization in NMOSFETs. In *Tech. Proc. of 2001 Intern. Conf. on Modeling and Simulation of Microsystems*, 2001, 458–461.
8. J.J.H. Miller, E. O'Riordan, G.I. Shishkin, and L.P. Shishkina, Fitted mesh methods for problems with parabolic boundary layers, *Math. Proc. R. Ir. Acad.*, **98A**(2), 1998, 173–190.
9. M.H. Protter and H.F. Weinberger, *Maximum Principles in Differential Equations*. Springer, Berlin (1984).
10. G.I. Shishkin, *Discrete approximation of singularly perturbed elliptic and parabolic equations*, Russian Academy of Sciences, Ural section, Ekaterinburg (1992).

Examination of the Performance of Robust Numerical Methods for Singularly Perturbed Quasilinear Problems with Interior Layers

P.A. Farrell and E. O'Riordan

Abstract Parameter-robust numerical methods for a particular class of singularly perturbed quasilinear boundary value problems were constructed and analysed in Farrell et al. (Math Comp 78:103–127, 2009). Certain constraints were imposed in Farrell et al. (Math Comp 78:103–127, 2009) on the data to establish the final theoretical error bound. In this companion paper to Farrell et al. (Math Comp 78:103–127, 2009), the parameter-uniform performance of the numerical method is examined (via numerical experiments) when one or more of these constraints are violated. The numerical results in this paper suggest that the numerical approximations converge for a wider class of problems to that covered by the theoretical convergence analysis in Farrell et al. (Math Comp 78:103–127, 2009).

1 Continuous Problem Class

Convection–diffusion equations of the form $(-\varepsilon u_x)_x + (g(u))_x = f(x)$, with a nonlinearity of the type $g(u) = u^2$, arise in numerous applications involving fluid dynamics. In this paper we examine the numerical performance of parameter-robust numerical methods [1] for the following class of quasilinear singularly perturbed boundary value problems: Let $\Omega^- := (0, d),\ \Omega^+ := (d, 1)$ and find $u_\varepsilon \in C^1(\bar{\Omega}) \cap C^2(\Omega^- \cup \Omega^+)$ such that

$$\varepsilon u_\varepsilon'' + b(x,u)u_\varepsilon' = f, \quad \text{for all} \quad x \in \Omega^- \cup \Omega^+, \tag{1a}$$

$$u_\varepsilon(0) = A,\ u_\varepsilon(1) = B, \tag{1b}$$

E. O'Riordan (✉)
School of Mathematical Sciences, Dublin City University, Ireland,
E-mail: eugene.oriordan@dcu.ie

A.F. Hegarty et al. (eds.), *BAIL 2008 - Boundary and Interior Layers.* Lecture Notes in Computational Science and Engineering,
DOI: 10.1007/978-3-642-00605-0,

$$b(x,u) = \begin{cases} b_1(u) = -1 + cu, \ x < d \\ b_2(u) = 1 + cu, \ x > d \end{cases}, \quad f(x) = \begin{cases} -\delta_1, \ x < d \\ \delta_2, \ x > d \end{cases} \tag{1c}$$

$$-1 < u_\varepsilon(0) < 0, \ 0 < u_\varepsilon(1) < 1, \ < c \leq 1, \tag{1d}$$

where c is a positive constant and δ_1, δ_2 are non-negative constants. Note the strict inequalities in (1d), which are imposed in order to ensure that the solution exhibits a standard convex–concave (or S-type) shock layer, as opposed to a concave–convex (or Z-type) layer (cf. [3, pp. 15–16]).

This paper is a companion paper to [2], where asymptotic error bounds for the numerical method examined in this paper were established. In order to guarantee existence and uniqueness of the solution of the continuous problem, additional conditions on the magnitudes of $\|f\|$ and the boundary values $|u_\varepsilon(0)|, \ |u_\varepsilon(1)|$ were imposed in [2]. Further restrictions are required in the theoretical analysis in [2] to prove uniform in ε convergence of the numerical method described below. These conditions are stated in (4) and (10).

The reduced solution $v_0 : [0,1] \to (-1,1)$ is defined to be the solution of the following nonlinear *first order* problem

$$b(v_0, x)v_0' = f, \ x \in \Omega^- \cup \Omega^+, \ v_0(0) = u_\varepsilon(0), \ v_0(1) = u_\varepsilon(1). \tag{2}$$

A unique reduced solution v_0 with the additional sign-pattern property of $v_0(x) < 0, \ x \in \Omega^-; \quad v_0(x) > 0, \ x \in \Omega^+$ exists if the conditions [2]

$$\delta_1 d < -u_\varepsilon(0) + 0.5cu_\varepsilon^2(0), \ \delta_2(1-d) < u_\varepsilon(1) + 0.5cu_\varepsilon^2(1), \tag{3}$$

are satisfied by the data. For a unique solution of the full continuous problem to exist it suffices [2] that

$$\delta_1 d < -u_\varepsilon(0), \ \delta_2(1-d) < u_\varepsilon(1), \tag{4a}$$

$$u_\varepsilon(1) - u_\varepsilon(0) < 1/c + \min\{\frac{\delta_1 d}{1 - cu_\varepsilon(0)}, \frac{\delta_2(1-d)}{1 + cu_\varepsilon(1)}\}. \tag{4b}$$

Let $\mathbf{C}_1$ be the class of problems defined by (1), (3); $\mathbf{C}_2$ be the class of problems defined by (1), (4) and $\mathbf{C}_3$ be the class of problems defined by (1), (4) and (10). Note that (4a) implies (3) and hence $\mathbf{C}_3 \subset \mathbf{C}_2 \subset \mathbf{C}_1$. The proof of parameter uniform convergence of the numerical approximations given in [2, Theorem 6.2] restricts the problem to the smallest of these three classes $\mathbf{C}_3$. Figure 1 displays some typical solutions for two problems in $\mathbf{C}_3$, with $\varepsilon = 0.000001, d = 0.25, \delta_2 = 0.13, u_\varepsilon(0) = -0.09$ and $u_\varepsilon(1) = 0.098$. The left one is for a problem with $\delta_1 = 0.1$ and the right one for a problem with $\delta_1 = 0.35$. In this paper, we examine (via numerical experiments) the parameter-uniform performance of the numerical method when one or more of the conditions (3), (4) or (10) are violated.

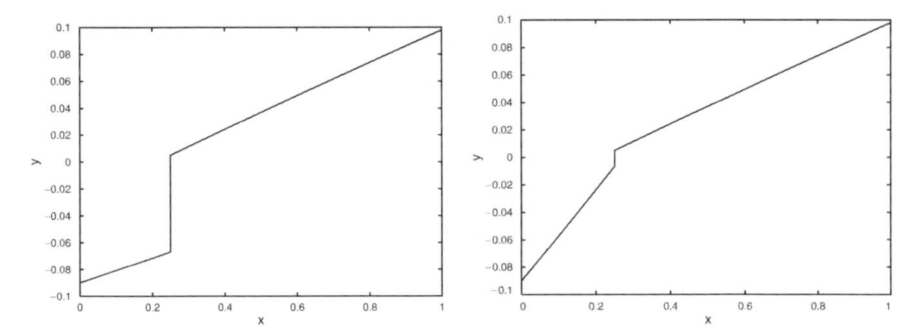

Fig. 1 Solution of (1) for sample problems in $\mathbf{C}_3$

Furthermore, we deduce in [2] that for the solution to a problem in $\mathbf{C}_2$ we have that

$$|b_1(u_\varepsilon)| > \theta_1 := \max\{-cu_\varepsilon(0), 1 - cu_\varepsilon(1)\},\ x \le d; \tag{5a}$$

$$b_2(u_\varepsilon) > \theta_2 := \max\{cu_\varepsilon(1), 1 + cu_\varepsilon(0)\},\ x \ge d. \tag{5b}$$

Lemma 1 ([2]). *Assume the problem is in* $\mathbf{C}_2$. *The solution can be written as a linear sum of the form* $u_\varepsilon = v_\varepsilon + w_\varepsilon$, *where for each integer k, satisfying* $1 \le k \le 3$, *these components satisfy the following bounds,*

$$\|v_\varepsilon\| \le C,\ \|v_\varepsilon^{(k)}\|_{\Omega^- \cup \Omega^+} \le C(1 + \varepsilon^{2-k}),$$

$$|[v_\varepsilon](d)| \le C, |[v_\varepsilon'](d)| \le C, |[v_\varepsilon''](d)| \le C,$$

$$|w_\varepsilon^{(k)}(x)| \le \begin{cases} C\varepsilon^{-k} e^{-(d-x)\theta_1/\varepsilon}, & x \in \Omega^-, \\ C\varepsilon^{-k} e^{-(x-d)\theta_2/\varepsilon}, & x \in \Omega^+, \end{cases}$$

where C *is a constant independent of* ε.

2 Numerical Method

The domain $\overline{\Omega}$ is subdivided into the four subintervals

$$[0, d - \sigma_1] \cup [d - \sigma_1, d] \cup [d, d + \sigma_2] \cup [d + \sigma_2, 1], \tag{6a}$$

for some σ_1, σ_2 that satisfy $0 < \sigma_1 \le \frac{d}{2}$, $0 < \sigma_2 \le \frac{1-d}{2}$. On each of the four subintervals a uniform mesh with $\frac{N}{4}$ mesh-intervals is placed. The interior mesh points are denoted by

$$\Omega_\varepsilon^N := \{x_i : 1 \le i \le N - 1,\ i \ne N/2\}. \tag{6b}$$

Clearly $x_{\frac{N}{2}} = d, \overline{\Omega}_\varepsilon^N = \{x_i\}_0^N$ and σ_1, σ_2 are taken to be the following

$$\sigma_1 := \min\left\{\frac{d}{2}, 2\frac{\varepsilon}{\theta_1}\ln N\right\}, \quad \sigma_2 := \min\left\{\frac{1-d}{2}, 2\frac{\varepsilon}{\theta_2}\ln N\right\}, \tag{6c}$$

whose choice can be motivated from (5) and the earlier bounds on $w_\varepsilon^{(k)}$. Then the fitted mesh method for problem (1) is: Find a mesh function U_ε such that

$$\varepsilon\delta^2 U_\varepsilon(x_i) + b(x_i, U_\varepsilon(x_i))DU_\varepsilon(x_i) = f(x_i) \quad \text{for all} \quad x_i \in \Omega_\varepsilon^N, \tag{7a}$$

$$U_\varepsilon(0) = u_\varepsilon(0), \quad U_\varepsilon(1) = u_\varepsilon(1), \tag{7b}$$

$$D^- U_\varepsilon(x_{\frac{N}{2}}) = D^+ U_\varepsilon(x_{\frac{N}{2}}), \tag{7c}$$

where

$$\delta^2 Z_i = \frac{D^+ Z_i - D^- Z_i}{(x_{i+1} - x_{i-1})/2}, \quad DZ_i = \begin{cases} D^- Z_i, & i < N/2, \\ D^+ Z_i, & i > N/2, \end{cases}$$

D^+ and D^- are the standard forward and backward finite difference operators, respectively. In order to solve this nonlinear finite difference scheme we use a variant of the continuation method from [1, Sect. 10.3].

$$(\varepsilon\delta_x^2 + b(x_i, U_\varepsilon(x_i, t_{j-1}))D - D_t^-)U_\varepsilon(x_i, t_j) = f(x_i), x_i \neq d, j = 1, \dots K, \tag{8a}$$

$$D_x^- U_\varepsilon(d, t_j) = D_x^+ U_\varepsilon(d, t_j), j = 1, \dots K, \tag{8b}$$

$$U_\varepsilon(0, t_j) = u_\varepsilon(0),\ U_\varepsilon(1, t_j) = u_\varepsilon(1) \text{ for all } j, \tag{8c}$$

$$U_\varepsilon(x, 0) = u(0) + (u(1) - u(0))x, \tag{8d}$$

and D_t^- is the standard backward finite difference operator in time. The choices of the uniform time-like step $k = t_j - t_{j-1}$ and the number of iterations K are determined as follows. Defining

$$e(j) := \max_{1 \le i \le N} |U_\varepsilon(x_i, t_j) - U_\varepsilon(x_i, t_{j-1})|/k, \quad \text{for } j = 1, 2, \cdots, K \tag{9a}$$

the time-like step k is chosen sufficiently small so that

$$e(j) \le e(j-1), \quad \text{for all } j \text{ satisfying } 1 < j \le K. \tag{9b}$$

Then the number of iterations K is chosen such that

$$e(K) \le \text{ TOL } := 10^{-7}. \tag{9c}$$

The numerical solution is computed using the following algorithm. Start from t_0 with the initial timestep $k = 1.0$. If, at some value of j, (9b) is not satisfied, then discard the timestep from t_{j-1} to t_j and restart from t_{j-1} with half the time step, that is $k^{new} = k/2$, and continue halving the timestep until one finds a k for which (9b) is satisfied. Assuming that (9b) is satisfied at each timestep, continue until either (9c) is satisfied or $t_j = 1{,}000$. If (9c) is not satisfied, we repeat the entire process again from t_0, halving the initial timestep k to $k = 0.5$. If the process still stalls, we restart from t_0 again halving the initial timestep. If (9c) is satisfied the resulting values of $U_\varepsilon(x, K)$ are taken as the approximations to the solution of the continuous problem.

The same conditions required for existence of the solution of the full continuous problem are also sufficient for the existence (but not uniqueness) of the solution of the discrete nonlinear problem.

In [2], it is established that, providing N is sufficiently large and ε is sufficiently small, independently of each other, under the further implicit restriction that

$$b^2(x_i, U_\varepsilon) - 4\varepsilon c u'_\varepsilon > 0, \quad x_i \neq d, \tag{10}$$

we can prove a uniform in ε error bound at all the mesh points of the form

$$\|U_\varepsilon - u_\varepsilon\|_\Omega \leq CN^{-1}(\ln N)^2, \tag{11}$$

where u_ε is the continuous solution, U_ε is a discrete solution of (7), and C is a constant independent of N and ε. The condition (10) is implicit as the exact solution u_ε is, in general, unknown.

3 Robustness of the Solution Method

Example 1. For the uniform convergence result (11) to be valid, [2] requires that (4) and (10) must be satisfied. For example, if

$$c = 1, \quad \delta_1 d < -u_\varepsilon(0) < 0.1 \quad \text{and} \quad \delta_2(1-d) < u_\varepsilon(1) < 0.1$$

then the data constraints (4) and (10) in $\mathbf{C}_3$ are both satisfied. Thus a problem with

$$d = 0.25,\ \delta_2 = 0.13,\ \delta_1 < 0.4,\ 0.0975 < u_\varepsilon(1) < 0.1,\ -0.1 < u_\varepsilon(0) < -\delta_1/4$$

satisfies these constraints. We consider a problem with $\mathrm{u}(0) = -0.09$, $\mathrm{u}(1) = 0.098$, $\delta_2 = 0.13$ and δ_1 varying from 0.1 to 0.35. This choice for the data satisfies all three assumptions including the implicit one (10). We verify this assertion numerically by computing

$$T_\varepsilon^N(x_i) = \begin{cases} b^2(x_i, U_\varepsilon^N) - 4\varepsilon D^- U_\varepsilon^N, & x_i < d \\ b^2(x_i, U_\varepsilon^N) - 4\varepsilon D^+ U_\varepsilon^N, & x_i > d \end{cases} \tag{12}$$

and observing that $T_\varepsilon^N = \min_i T_\varepsilon^N(x_i) > 0$ for all values of ε and N used. The computed uniform rates of convergence p_N, using the double mesh principle and the uniform fine mesh errors E_N (see [1, pp. 104, 190] for details on how these quantities are calculated) are computed over the range $\varepsilon = 2^{-j}, j = 1, 2, \ldots 25$ and are presented in Table 1. These results confirm uniform convergence in this range of the data.

Table 1 Maximum errors E_N and computed rates of convergence p_N for the numerical method (6), (7) for problems within $\mathbf{C}_3$ in the case of Example 1

N	32	64	128	256	512	1,024
			$\delta_1 = 0.1$			
E_N	0.004962	0.003227	0.002017	0.001175	0.000637	0.000313
p_N	0.46	0.75	0.63	0.72	0.68	0.84
			$\delta_1 = 0.2$			
E_N	0.003583	0.002245	0.001346	0.000771	0.000413	0.000201
p_N	0.57	0.76	0.72	0.72	0.72	0.85
			$\delta_1 = 0.3$			
E_N	0.002549	0.001403	0.000809	0.000457	0.000243	0.000117
p_N	0.70	0.90	0.79	0.76	0.73	0.86
			$\delta_1 = 0.35$			
E_N	0.002205	0.001151	0.000584	0.000295	0.000155	0.000075
p_N	0.90	0.94	0.96	0.93	0.72	0.88

Table 2 Maximum errors E_N and computed rates of convergence p_N for a problem outside $\mathbf{C}_1$, but satisfying (10), in the case of Example 1

			$\delta_1 = 0.39$			
N	32	64	128	256	512	1,024
E_N	0.002282	0.001154	0.000578	0.000283	0.000133	0.000057
p_N	0.98	0.96	0.98	0.99	0.99	1.00

Now consider the same problem with $u(0) = -0.09$, $u(1) = 0.098$, $\delta_2 = 0.13$ and $\delta_1 = 0.39$. This does not satisfy (3) and hence is not in $\mathbf{C}_1$. However, this scheme does numerically satisfy the implicit condition (10).

The results presented in Table 2 imply that the scheme is still convergent uniformly in ε.

Example 2. For the existence of a continuous solution we have the sufficient conditions (4). As an example, take

$$c = 1,\ u_\varepsilon(1) = 0.7,\ u_\varepsilon(0) = -0.5\ d = 0.25.$$

Then (3) is satisfied when $\delta_1 < 2.5$ and $\delta_2 < 1.26$. Also (4a) is satisfied when

$$\delta_1 < 2 \text{ and } \delta_2 < \frac{2.8}{3} \approx 0.933333$$

and (4b) is satisfied when

$$\delta_1 > 1.2 \text{ and } \delta_2 > \frac{1.36}{3} \approx 0.453333.$$

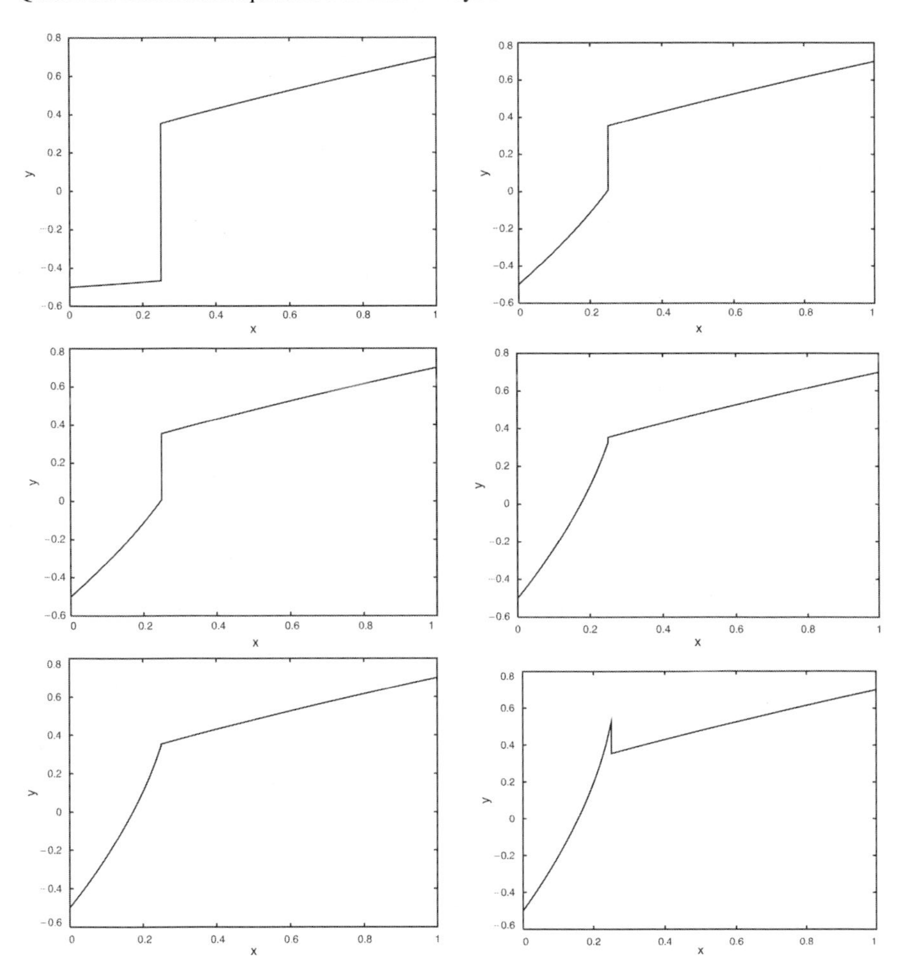

Fig. 2 Solution of (1) for problems which do not satisfy $\mathbf{C}_3$. In all these figures, $\delta_2 = 0.7,\ u(0) = -0.5,\ u(1) = 0.7,\ N = 64$ and $\varepsilon = 0.000001$. From *top left* to *bottom right:* δ_1 = 0.2, 2.4999, 2.5, 3.5, 3.55, 3.9

We fix $\delta_2 = 0.7$ and consider various values of δ_1, in particular ones which violate one or more of the conditions (3), (4a) or (4b). For the problems examined in this example, it has been observed numerically, using condition (12), that the implicit condition (10) is not satisfied for any of the values of δ_1 considered. That is, these problems lie outside the class $\mathbf{C}_3$. Problems are in the class $\mathbf{C}_2 \backslash \mathbf{C}_3$ if $1.2 < \delta_1 < 2$, in the class $\mathbf{C}_1 \backslash \mathbf{C}_2$ if $2 \le \delta_1 < 2.5$ or if $\delta_1 \le 1.2$ and finally the problem lies outside $\mathbf{C}_1$ if $\delta_1 \ge 2.5$.

Illustrations of the corresponding solutions are given in Fig. 2, and the convergence results are given in Tables 3–5. They show that provided the reduced solution

Table 3 Maximum errors E_N and computed rates of convergence p_N for the numerical method (6), (7) applied to problems in $\mathbf{C}_2$, where (10) is violated, that is within $\mathbf{C}_2\backslash\mathbf{C}_3$ in the case of Example 2 with $\delta_2 = 0.7$

N	32	64	128	256	512	1,024
			$\delta_1 = 1.3$			
E_N	0.067928	0.053165	0.033076	0.020709	0.011692	0.005732
p_N	0.09	0.65	0.71	0.57	0.74	0.71
			$\delta_1 = 1.8$			
E_N	0.058642	0.047114	0.029970	0.018685	0.010404	0.005133
p_N	0.13	0.66	0.73	0.56	0.70	0.71

Table 4 Maximum errors E_N and computed rates of convergence p_N for the numerical method (6), (7) applied to problems in $\mathbf{C}_1$, where (4) and (10) are violated, that is within $\mathbf{C}_1\backslash\mathbf{C}_2$ in the case of Example 2 with $\delta_2 = 0.7$

N	32	64	128	256	512	1,024
			$\delta_1 = 0.2$			
E_N	0.085977	0.070653	0.045129	0.028786	0.016281	0.008038
p_N	0.01	0.62	0.70	0.55	0.70	0.70
			$\delta_1 = 0.5$			
E_N	0.081286	0.063318	0.039899	0.025084	0.014299	0.007035
p_N	0.00	0.62	0.70	0.56	0.74	0.70
			$\delta_1 = 1.1$			
E_N	0.071339	0.055289	0.034691	0.021476	0.012067	0.005918
p_N	0.08	0.65	0.71	0.57	0.76	0.71
			$\delta_1 = 2.1$			
E_N	0.052495	0.042713	0.027518	0.016995	0.009474	0.004675
p_N	0.16	0.68	0.73	0.57	0.69	0.71
			$\delta_1 = 2.4$			
E_N	0.045858	0.037679	0.024406	0.014925	0.008380	0.004132
p_N	0.21	0.68	0.74	0.59	0.67	0.72
			$\delta_1 = 2.4999$			
E_N	0.043529	0.035851	0.023213	0.014147	0.007960	0.003927
p_N	0.23	0.67	0.74	0.60	0.68	0.72

of the problem remains monotonic increasing, the method is robust in the sense that the numerical method remains uniformly in ε convergent. When the problem ceases to be monotonic the layer type changes from a standard shock layer to a Z-layer. As the Z-layer grows in amplitude the nonlinear solver does not converge and thus the method ceases to be robust.

Table 5 Maximum errors E_N and computed rates of convergence p_N for the numerical method (6), (7) applied to problems outside $\mathbf{C}_1$, that is where (3), (4) and (10) are violated, in the case of Example 2 with $\delta_2 = 0.7$

N	32	64	128	256	512	1,024
			$\delta_1 = 2.8$			
E_N	0.041487	0.029870	0.019123	0.011529	0.006529	0.003246
p_N	0.39	0.64	0.77	0.65	0.68	0.71
			$\delta_1 = 3.0$			
E_N	0.043328	0.025441	0.015947	0.009703	0.005490	0.002714
p_N	0.83	0.63	0.79	0.69	0.68	0.71
			$\delta_1 = 3.5$			
E_N	0.075558	0.032340	0.015213	0.007286	0.003408	0.001470
p_N	1.32	1.12	1.04	1.00	0.99	0.98
			$\delta_1 = 3.8$			
E_N	0.168256	0.056174	0.024782	0.011446	0.005227	0.002217
p_N	1.84	1.24	1.10	1.05	1.02	1.01

4 Sensitivity to the Position of the Transition Points

We examine the effect of varying the fine mesh width by incorporating a constant C_* in a revised formula for σ_1 and σ_2 given by

$$\sigma_1 = \min\left\{\frac{d}{2},\ C_*\frac{\varepsilon}{\theta_1}\ln N\right\},\quad \sigma_2 = \min\left\{\frac{1-d}{2},\ C_*\frac{\varepsilon}{\theta_2}\ln N\right\},\tag{13}$$

where C_* is a parameter and θ_1, θ_2 are specified in (5).

Table 6 give the results for Example 2 with $\delta_1 = 1.20010$. For the range of C_* tested, it was observed that the number of iterations varied by at most a factor of two.

Thus the method is not particularly sensitive to the fine mesh width and, in fact, a choice of a value of C_* less than that of $C_* = 2$ used in [2] seems to give better performance. In the example considered here, the errors are smallest and the rate of convergence best for $C_* = 0.5$.

Remark 1. The theoretical rate of convergence given in (11) can be compared to the observed rates of convergence given in Tables 1–6, by using Table 7. For example, Table 1 exhibits rates close to $N^{-1}\ln N$ and Tables 3–6 mainly exhibit rates close to $N^{-1}(\ln N)^2$.

Table 6 Maximum errors E_N and computed rates of convergence p_N for various choices of the transition point in the case of Example 2 with $\delta_1 = 1.20010,\ \delta_2 = 0.7$

N	32	64	128	256	512	1,024
			$C_* = 0.125$			
E_N	0.077109	0.063909	0.052342	0.040499	0.028576	0.017859
p_N	0.37	0.34	0.27	0.24	0.26	0.27
			$C_* = 0.25$			
E_N	0.055713	0.034658	0.020660	0.011906	0.006556	0.003274
p_N	0.70	0.68	0.71	0.71	0.71	0.70
			$C_* = 0.5$			
E_N	0.039241	0.021406	0.012181	0.006681	0.003483	0.001645
p_N	0.81	0.89	0.79	0.80	0.82	0.78
			$C_* = 1.0$			
E_N	0.052324	0.033291	0.020706	0.011990	0.006454	0.003099
p_N	0.23	0.79	0.68	0.73	0.77	0.76
			$C_* = 2.0$			
N	32	64	128	256	512	1,024
E_N	0.069652	0.054194	0.033899	0.021033	0.011889	0.005824
p_N	0.08	0.65	0.71	0.57	0.75	0.71

Table 7 Orders of local convergence p^N corresponding to different theoretical error bounds for various values of N

N	32	64	128	256	512	1,024
$N^{-1} \ln N$	0.68	0.74	0.78	0.81	0.83	0.85
$N^{-1}(\ln N)^2$	0.28	0.44	0.53	0.60	0.65	0.69

5 Conclusions

The numerical results in this paper indicate a possible gap between the theory in [2] and what is observed in practice. As was proven in [2] the scheme (6), (7) is a parameter-uniform scheme under the conditions (4) and (10). However these sufficient conditions appear to be overly restrictive, since, in practice, the numerical approximations appear to converge for a wider range of data. In any attempt to extend the theory in [2] to a wider class of problems, a reasonable constraint on the data to aim for (in place of (4)) would be that the reduced solution is monotonic increasing, which is a necessary condition to exclude Z-layers from appearing in the solution of (1).

The implicit condition (10) is not satisfied for some of the examples presented here, while the numerical approximations still converge uniformly in ε. When the constraint (10) is violated it appears that $T_\varepsilon^N(x_i) < 0$ in a particular neighborhood of the point d and not at the transition points between the fine and coarse mesh. Proving convergence without (10) being satisfied would require a method of proof other than the maximum principle arguments used in [2]. These numerical results also suggest that a different finite difference equation (other than continuity of the discrete first derivative) at the point of the discontinuity d may ensure that $T_\varepsilon^N > 0$, which in turn might improve the performance of the scheme and also assist in extending the scope of the current theory.

References

1. P.A. Farrell, A.F. Hegarty, J.J.H. Miller, E. O'Riordan, and G.I. Shishkin, *Robust Computational Techniques for Boundary Layers*, Chapman and Hall/CRC, New York/Boca Raton, (2000)
2. P.A. Farrell, E. O'Riordan, and G.I. Shishkin, A class of singularly perturbed quasilinear differential equations with interior layers, *Mathematics of Computation* 78(265):103–127 (2009)
3. F.A. Howes, Boundary-interior layer interactions in nonlinear singular perturbation theory, *Memoirs of the AMS* 15:203 (1978)

Glycolysis as a Source of "External Osmoles": The Vasa Recta Transient Model

M. Gonzalez, A.F. Hegarty, and S.R. Thomas

Abstract The kidney is one of the most important organs in our body, responsible for regulating the volume and composition of the extracellular fluid; excreting metabolic waste (as urine) and foreign substances; and also producing some hormones.

The mechanisms that contribute to the urine concentrating mechanism are not completely understood. Some ideas have been proposed over the last years and this paper is based on the hypothesis of Thomas (Am J Physiol Renal Physiol 279:468–481, 2000), that glycolysis as a source of external osmoles could contribute to the urine concentrating mechanism. Based on the steady state model developed by Thomas and also on the model developed by Zhang and Edwards (Am J Physiol Renal Physiol 290:87–102, 2005) (a model focused on microcirculation), we have developed a time-dependent model where, besides verifying some of the steady state results of Thomas (Am J Physiol Renal Physiol 279:468–481, 2000), we can also study some time dependent issues, such as the time that it will take to wash out the gradient created by glycolysis if an increase in blood inflow occurs.

1 The Kidney

In a normal human adult (Fig. 1), each kidney is about 11 cm long and about 5 cm thick, weighing 150 g. If the kidney is bisected from top to bottom, the two major regions that can be visualized are the outer *cortex* and the inner region referred as the *medulla*, where we can also distinguish two regions, the *outer medulla* (OM) and the *inner medulla*(IM).

The nephron (Fig. 2) is the functional unit of the kidney. There are more than a million in each normal adult human kidney. Each nephron contains a tuft of

A.F. Hegarty (✉)
University of Limerick, Limerick, Ireland, E-mail: alan.hegarty@ul.ie

A.F. Hegarty et al. (eds.), *BAIL 2008 - Boundary and Interior Layers.* Lecture Notes in Computational Science and Engineering,
DOI: 10.1007/978-3-642-00605-0,

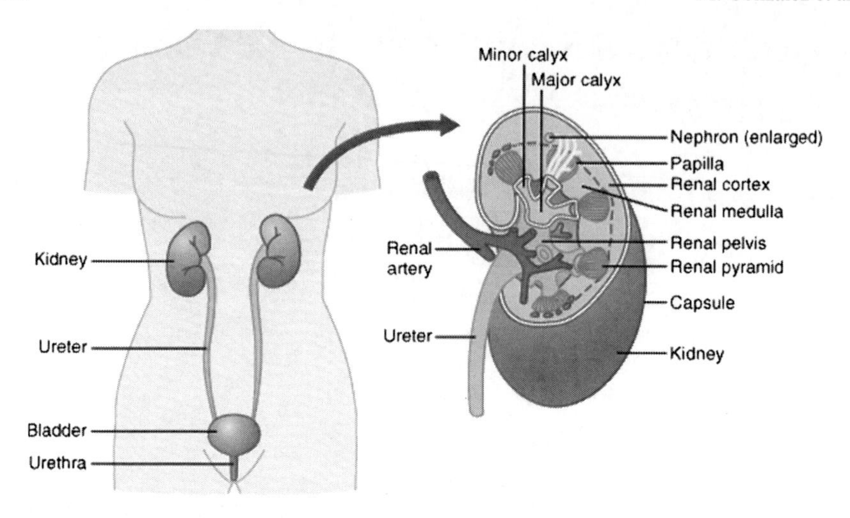

Fig. 1 Urinary system and the kidney [1]

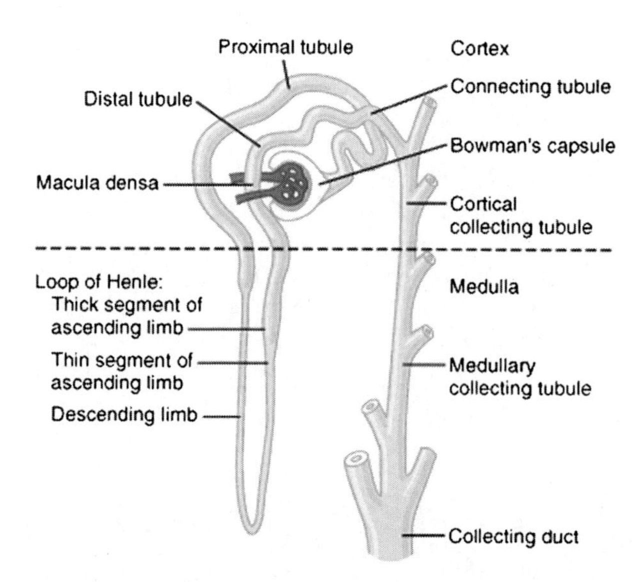

Fig. 2 Parts of the nephron [1]

capillaries called the *glomerulus*, through which large amounts of fluid are filtered from the blood, and a long *tubule* in which the filtered fluid is converted into urine on its way to the pelvis of the kidney.

Depending on how deep they lie into the medulla we can distinguish two types of nephrons: *Cortical nephrons* and *Juxtamedullary nephrons* (Fig. 4).

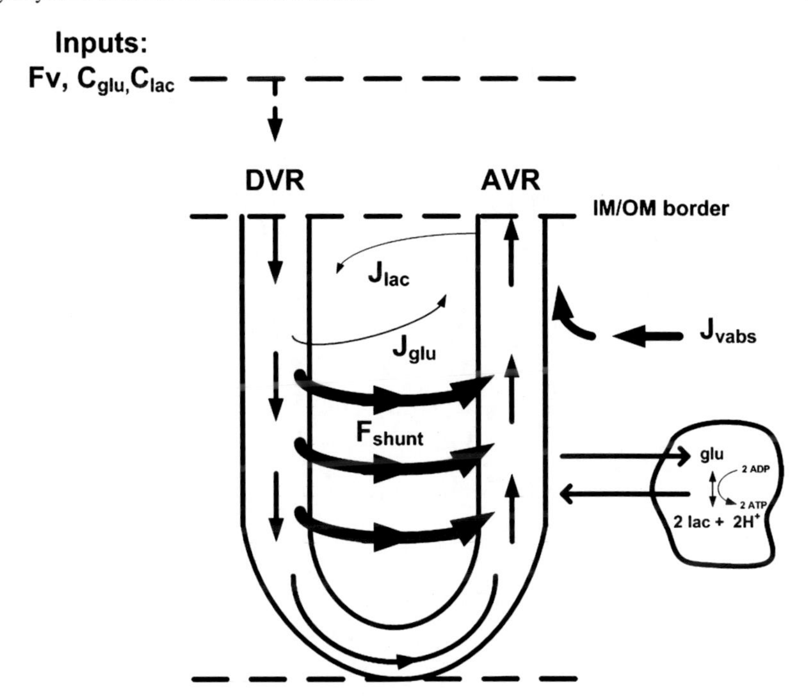

Fig. 3 Process of glycolysis in the IM cells

2 The Urine Concentrating Mechanism

Looking at urine osmolalities, mammals can produce urine that has a much higher osmolality than that of blood plasma (270–300 mOsm). This capability to concentrate their urine allows them to excrete metabolic and other waste products without compromising their water balance.

Since the 1950s, in renal physiology, the explanation for the capability of producing hypertonic urine has been a major open question. The following model was developed to study the possibility that the process of glycolysis, taking place in the IM cells (see Fig. 3) to obtain a large fraction of the energy for cell metabolism, could contribute significantly to the build-up of this gradient. During glycolysis one molecule of glucose is converted into two lactates.

3 Model Description

In the present work, a transient version of the previous paper published by Thomas [6] is developed. We consider this time the vasa recta (see Vasa Recta location in Fig. 4) and also the interstitium, where only solute movement by diffusion is considered.

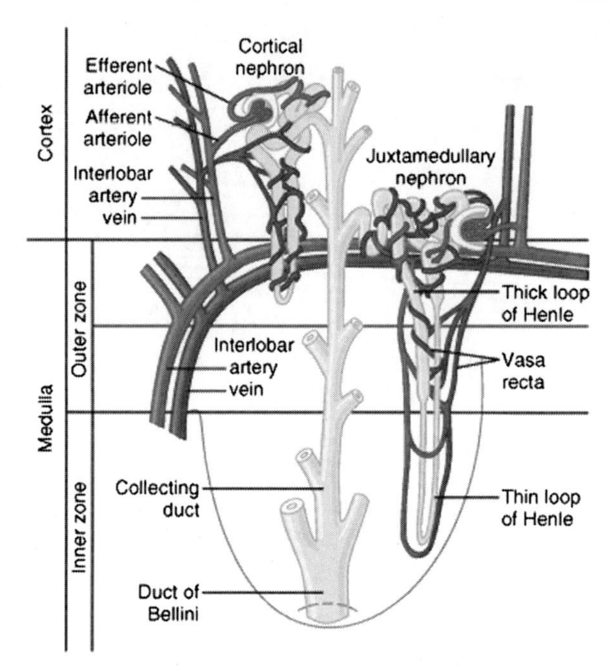

Fig. 4 Nephron blood supply showing cortical and juxtamedullary nephrons

As in previous models of the kidney we will consider a population of vasa recta represented by a single composite structure, where a fraction of the descending limb flow is shunted directly into the ascending limb at every node in the discretization where a single vasa recta turns in the inner medulla (see Fig. 3). The variable x denotes distance along the medulla, $x = 0$ at the OM/IM border and $x = L$ ($L =$ 4 mm) at the papillary tip. The numbers of DVR (descending vasa recta) and AVR (ascending vasa recta) are assumed to diminish exponentially in number along the IM toward the tip of the papilla according to the same relation as in earlier models and in conformity with reported rat anatomy:

$$N(x) = N(0)e^{-k_{sh}x} \tag{1}$$

where $N(0)$ is the number of VR at the OM/IM border. We let $N(0) = 128$ and the species dependent factor $k_{sh} = 1.213\,\text{mm}^{-1}$ which gives us a system with a single vasa recta at the papillary tip.

4 Equations

4.1 Volume Flow Equations

As in [5] we will consider the renal parenchyma indistensible, so all fluid reabsorbed flows immediately into the AVR. With such an assumptionthe equations describing water movement are the following:

$$\frac{dF_v^d}{dx} = -J_v^d - F_{shunt} \tag{2}$$

$$\frac{dF_v^a}{dx} = -J_v^a + F_{shunt} + J_v^{ABS} \tag{3}$$

where F_v represents volume flow in descending (d) and ascending (a) vasa recta and J_v represents transmural flux of volume. As J_v depends on forces not represented in this model, we will take this as an explicit fraction ($\bar{v}$) of the entering flow.

$$J_v^d(x) = \bar{v}\frac{F_v^d(0)}{N(0)L}N(x) \tag{4}$$

$$J_v^a(x) = -J_v^d(x) \tag{5}$$

F_{shunt} is shunt transfer of volume (or solute for the equations in the next section) from DVR to AVR and is calculated as follows [10]:

$$F_{shunt}(x) = \frac{F_d(x)}{N(x)}\frac{dN(x)}{dx} \tag{6}$$

Also included in the above equations is net volume reabsorption into the AVR from LDL (long descending limb) and CD (collecting duct), designated as J_v^{ABS}.

4.2 Solute Equations

For solute flow equations we will consider the following assumptions:

1. Axial movement of solutes is by convection in the VR.
2. In the interstitium solute movement occurs by diffusion only.
3. Glucose consumed by cellular glycolysis is supplied from AVR and the resulting lactate is recovered into interstitium.
4. Interstitial cross-sectional area is taken as 40% of the total tubular luminal cross-sectional area.

Considering this we will write a PDE coupled system where we have three equations for each solute considered (glucose = g, lactate = l). As before we will use the superindexes d for DVR, a for AVR and i for interstitium equations.

$$\frac{\partial C_g^d}{\partial t} = \frac{1}{A}\left(-\frac{\partial(F_v^d C_g^d)}{\partial x} - J_g^d - F_{shunt}\right) \tag{7}$$

$$\frac{\partial C_g^a}{\partial t} = \frac{1}{A}\left(-\frac{\partial(F_v^a C_g^a)}{\partial x} - J_g^a + F_{shunt} - J_{gly}\right) \tag{8}$$

$$\frac{\partial C_g^i}{\partial t} = D_g^i \frac{\partial^2 C_g^i}{\partial x^2} + \frac{1}{A_{int}} J_g^i \tag{9}$$

$$\frac{\partial C_l^d}{\partial t} = \frac{1}{A}\left(-\frac{\partial (F_v^d C_l^d)}{\partial x} - J_l^d - F_{shunt}\right) \tag{10}$$

$$\frac{\partial C_l^a}{\partial t} = \frac{1}{A}\left(-\frac{\partial (F_v^a C_l^a)}{\partial x} - J_l^a + F_{shunt}\right) \tag{11}$$

$$\frac{\partial C_l^i}{\partial t} = D_l^i \frac{\partial^2 C_l^i}{\partial x^2} + \frac{1}{A_{int}} (J_l^i + 2J_{gly}) \tag{12}$$

where C is concentration of solute in each tube and A represents the cross-sectional area of each tube.

The relation between axial solute and axial volume flow is given by (see [9])

$$F_{ik} = F_{iv} C_{ik} - D_k \frac{\partial C_{ik}}{\partial x} \tag{13}$$

4.2.1 Membrane Flux Equations

$$J_k^j(x,t) = 2\pi r P_k (C_k^j(x,t) - C_k^i(x,t)) N(x) + (1-\sigma_k) J_v(x) \frac{C_k^j(x,t) + C_k^i(x,t)}{2} \tag{14}$$

$$J_k^i(x,t) = \sum_{j=DVR,AVR} J_k^j \tag{15}$$

In (14) the first term refers to membrane diffusion (where P_k are permeabilities to glucose and lactate). The second term refers to solvent drag (where a piecewise lineralization of the Kedem–Katchalscky equation is taken [7]) and σ_k are reflection coefficients.

4.2.2 Glycolysis

Glycolytic rate is described simply with a first degree Michaelis–Menten equation saturable as a function of AVR glucose consumption

$$J_{gly}(x,t) = N(x) \frac{V_{max} C_g^a(x,t)}{K_m + C_g^a(x,t)} \tag{16}$$

4.2.3 Initial and Boundary Conditions

Flows and concentrations in the descending structures and interstitium are known at the OM/IM boundary ($x = 0$), continuity conditions are applied at the papillary tip from the ascending structures and also $(\partial C^i/\partial x)(L) = 0$. The initial concentrations throughout the tubes are set to their known values at $x = 0$.

4.3 Numerical Method

Since transmural flux of volume depends on forces not included in the model, the volume flow equations were solved analytically. The solute flow equations were solved numerically using the Method of Lines (MOL) (previously applied by Moore and Marsh [4,5]). Finite difference approximations for each of the partial derivatives with respect to the distance along the corticopapillary axis were used for the first order partial derivatives in space:

$$C_x \approx \frac{C_i - C_{i-1}}{h}, \tag{17}$$

while the second order partial derivatives were approximated with a three-point centred difference expression

$$C_{xx} \approx \frac{C_{i-1} - 2C_i + C_{i+1}}{h^2} \tag{18}$$

It might be more efficient to solve the PDE by a method specially constructed to suit the problem [2, 3], but the MOL usually enables us to solve quite general and complicated PDEs relatively easily and with acceptable efficiency. It is also attractive since powerful ODE solvers are readily available, as in our case, the ODE Matlab solver *ode15s*.

4.4 Simulations

During all simulations parameters not indicated were set as their baseline values. Twenty per cent consumption was adopted as the baseline value for glycolysis in all simulations when this is not tested. J_v^{ABS} and J_v baseline values were set at 30%.

It has been shown in previous studies [8] that 20–100 mOsm/kgH_2O of an unspecified external interstitial osmolytes could improve the concentration ability. Figure 5 shows different glycolysis consumptions; note that for the highest values (as was shown by Thomas [6]) the lower bound of the interval above is reached, which suggests that glycolysis should be considered in models of the urine concentrating mechanism.

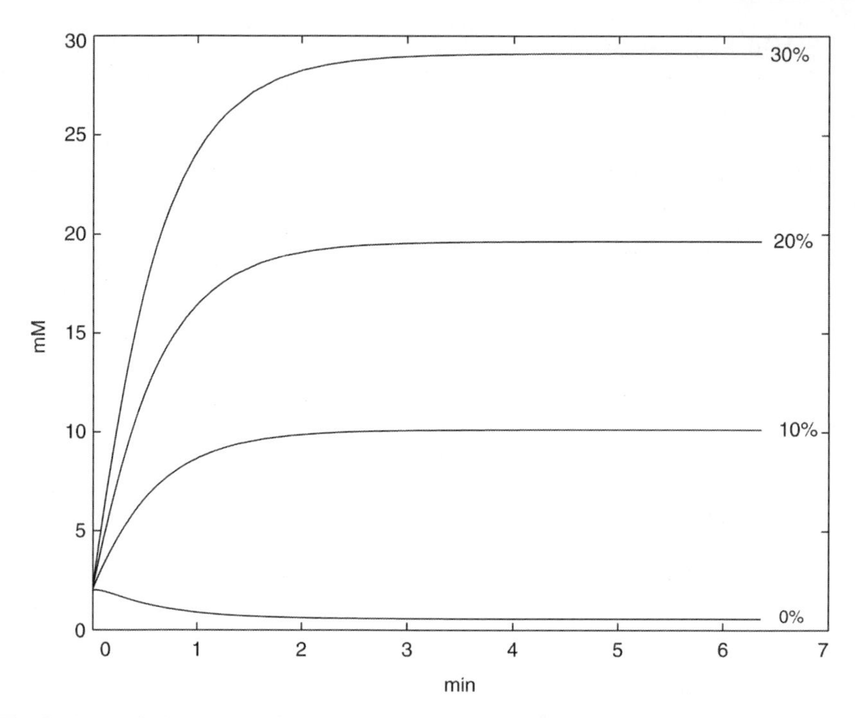

Fig. 5 The graph shows the time that it takes the lactate gradient to build up at the papillary tip for different glycolysis consumptions

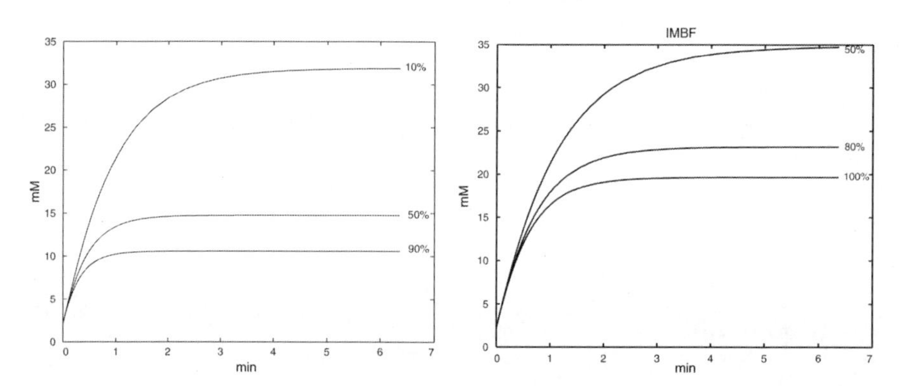

Fig. 6 *Left:* Accumulation of lactate for different volume reabsorption from nephrons at the papillary tip. *Right:* Lactate concentration when IMBF is reduced from its baseline value (absolute glucose consumption was held constant)

The effect of varying J_v^{ABS} is shown in Fig. 6. Absorption rates of 10% , 50% and 90% of DVR inflow are shown here. Increasing volume reabsorption affect significantly lactate accumulation. Also this figure shows that lactate accumulation increases dramatically as IMBF falls to one-half its baseline value, as may occur in antidiuresis. The predicted lactate profiles clearly suggest that IMBF may play an important role in the extent of lactate accumulation.

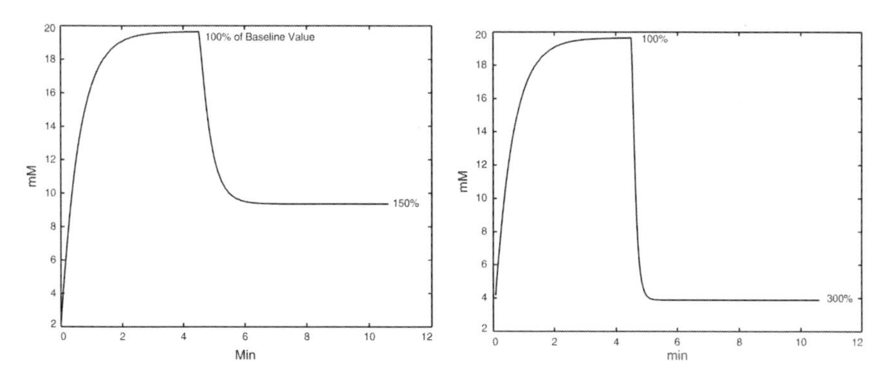

Fig. 7 Different situations of the lactate gradient being washed out after increasing IMBF at the papillary tip

Finally Fig. 7 shows different situations of the gradient being washed out after increasing the inner medullary blood flow. It can be seen that the time it takes the gradient to disappear is considerably less than the time it takes to be built up.

References

1. A.C. Guyton and J.E. Hall, *Textbook of Medical Physiology, Eleventh edition*. Elsevier Saunders, Philadelphia, 2006.
2. A.T. Layton, *A methodology for tracking solute distribution in a mathematical model of the kidney*, J. Biol. Syst. **13** (2005), 1–21.
3. A.T. Layton and H.E. Layton, *An efficient numerical method for distributed-loop models of the urine concentrating mechanism*, Math. Biosci. **181** (2003), 111–132.
4. L.C. Moore and D.J. Marsh, *How descending limb of Henle's loop permeability affects hypertonic urine formation*, Am. J. Physiol. Renal Physiol. **239** (1980), F57–F71.
5. L.C. Moore, D.J. Marsh, and C.M. Martin, *Loop of Henle during the water-to-antidiuresis transition in Brattleboro rats*, Am. J. Physiol. Renal Physiol. **239** (1980), F72–F83.
6. S.R. Thomas, *Inner medullary lactate production and accumulation: a vasa recta model*, Am. J. Physiol. Renal Physiol. **279** (2000), 468–481.
7. S.R. Thomas and D.C. Mikulecky, *Transcapillay solute exhange: a comparison of the Kedem–Katchalsky convection–diffusion equations with the rigorous nonlinear equations for this special case*, Microvasc. Res. **15** (1978), 207–220.
8. S.R. Thomas and A.S. Wexler, *Inner medullary external osmotic driving force in a 3-D model of the renal concentrating mechanism*, Am. J. Physiol. Renal Physiol. **269** (1995), F159–F171.
9. J.L. Stephenson, *Urinary concentration and dilution: models*, in Handbook of Physiology – Renal Physiology, sect. 8, vol. 2. Oxford University Press, Oxford (1992), 1349–1408.
10. A.S. Wexler, R.E. Kalaba, and D.J. Marsh, *Three-dimensional anatomy and renal concentrating mechanism. I. Modeling results*, Am. J. Physiol. Renal Physiol. **260**, (1991) F368–F383.
11. W. Zhang and A. Edwards, *A model of glucose transport and conversion to lactate in the renal medullary microcirculation*, Am. J. Physiol. Renal Physiol. **290** (2005), 87–102.

A System of Singularly Perturbed Semilinear Equations

J.L. Gracia, F.J. Lisbona, M. Madaune-Tort, and E. O'Riordan

Abstract In this paper systems of singularly perturbed semilinear reaction-diffusion equations are examined. A numerical method is constructed for these systems which involves an appropriate layer–adapted piecewise-uniform mesh. The numerical approximations generated from this method are shown to be uniformly convergent with respect to the singular perturbation parameters.

1 Introduction

In this paper we consider semilinear systems of the form

$$\mathbf{Tu} := -E\mathbf{u}'' + \mathbf{b}(x,\mathbf{u}) = \mathbf{0},\ x \in \Omega = (0,1),\ \mathbf{u}(0) = \mathbf{a},\ \mathbf{u}(1) = \mathbf{b}, \quad (1a)$$

$$\mathbf{b}(x,\mathbf{u}) = (b_1(x,\mathbf{u}),\dots,b_m(x,\mathbf{u}))^T \in C^4(\bar{\Omega}\times\mathbb{R}^m), \quad (1b)$$

and $\forall(x,\mathbf{y}) \in \bar{\Omega}\times\mathbb{R}^m$ we assume that the nonlinear terms satisfy

$$\frac{\partial b_i}{\partial u_j}(x,\mathbf{y}) \le 0, \forall i \ne j, \text{ and } \sum_{j=1}^{m} \frac{\partial b_i}{\partial u_j}(x,\mathbf{y}) > \beta^2 > 0,\ \beta > 0,\ \forall i = 1,\dots,m, \quad (1c)$$

where $E = \text{diag}\{\varepsilon_1^2,\dots,\varepsilon_m^2\}$ is a diagonal matrix, $0 < \varepsilon_1 \le \dots \le \varepsilon_m \le 1$ and $\mathbf{u} = (u_1,\dots,u_m)^T$.

In [1, 3], information about the layer structure for linear singularly perturbed reaction–diffusion systems was obtained via linear decompositions of the solution into regular and singular components. Here we show that these techniques are applicable to a semilinear system. The preprint [2] is available to the reader to supplement this paper with some additional details.

J.L. Gracia (✉)
Departamento de Matemática Aplicada, Universidad de Zaragoza, Spain,
E-mail: jlgracia@unizar.es

A.F. Hegarty et al. (eds.), *BAIL 2008 - Boundary and Interior Layers.*
Lecture Notes in Computational Science and Engineering,
DOI: 10.1007/978-3-642-00605-0,

For any $\mathbf{v}, \mathbf{w} \in \mathbb{R}^m$, we write $\mathbf{v} \le \mathbf{w}$ if $v_i \le w_i$, $\forall i$ and $|\mathbf{v}| := (|v_1|, |v_2|, \ldots, |v_m|)^T$; $\|f\|_\infty := \max_x |f(x)|$ and $\|\mathbf{f}\|_\infty := \max_i \|f_i\|_\infty$; $\mathbf{C} := C(1, 1, \ldots, 1)^T$ is a constant vector and C denotes a generic positive constant independent of $(\varepsilon_1, \varepsilon_2, \ldots, \varepsilon_m)$ and the discretization parameter.

2 Singularly Perturbed Semilinear Systems

Conditions (1b), (1c) and the implicit function theorem ensure that there exists a unique solution $\mathbf{u} \in (C^4(\bar{\Omega}))^m$ to (1a), and that the corresponding reduced problem $\mathbf{b}(x, \mathbf{r}) = \mathbf{0}$, $x \in \bar{\Omega}$, also has a unique solution in $\mathbf{r} \in (C^4(\bar{\Omega}))^m$. Note that the conditions (1c) on the Jacobian matrix J where

$$J(x, \mathbf{y}) := \left(\frac{\partial b_i}{\partial u_j}\right)(x, \mathbf{y}),$$

are the natural extension of the linear case [6] for the coupling matrix. These conditions guarantee that J is an M–matrix for all $(x, \mathbf{y}) \in \bar{\Omega} \times \mathbb{R}^m$.

To deduce the asymptotic behaviour of the solution, we consider the following decomposition $\mathbf{u} = \mathbf{v} + \mathbf{w} + \mathbf{w}_R$, where the regular component $\mathbf{v}$ is the solution of the problem

$$-\mathbf{E}\mathbf{v}'' + \mathbf{b}(x, \mathbf{v}) = \mathbf{0}, \; x \in \Omega, \; \mathbf{v}(0) = \mathbf{r}(0), \; \mathbf{v}(1) = \mathbf{r}(1), \tag{2}$$

and the singular components $\mathbf{w}, \mathbf{w}_R$ are the solutions of

$$\begin{aligned} &-\mathbf{E}\mathbf{w}'' + (\mathbf{b}(x, \mathbf{v} + \mathbf{w}) - \mathbf{b}(x, \mathbf{v})) = \mathbf{0}, \; x \in \Omega, \\ &\quad \mathbf{w}(0) = (\mathbf{u} - \mathbf{v})(0), \; \mathbf{w}(1) = \mathbf{0}, \end{aligned} \tag{3}$$

$$\begin{aligned} &-\mathbf{E}\mathbf{w}_R'' + (\mathbf{b}(x, \mathbf{v} + \mathbf{w} + \mathbf{w}_R) - \mathbf{b}(x, \mathbf{v} + \mathbf{w})) = \mathbf{0}, \; x \in \Omega, \\ &\quad \mathbf{w}_R(0) = \mathbf{0}, \; \mathbf{w}_R(1) = (\mathbf{u} - \mathbf{v})(1). \end{aligned} \tag{4}$$

Note (1c) guarantees existence and uniqueness of $\mathbf{v}, \mathbf{w}, \mathbf{w}_R$ and it will also be used below to establish existence and uniqueness for several further decompositions of these components. Below we state bounds on the derivatives of the left layer component $\mathbf{w}$. The corresponding bounds on the right layer component $\mathbf{w}_R$ are obtained by simply replacing x with $1 - x$.

Lemma 1. *The regular component* $\mathbf{v}$ *satisfies*

$$\left\|\frac{d^k \mathbf{v}}{dx^k}\right\|_\infty \le C, \; k = 0, 1, 2, \; \left\|\frac{d^k v_i}{dx^k}\right\|_\infty \le C\varepsilon_i^{2-k}, \; k = 3, 4, \; i = 1, \ldots, m. \tag{5}$$

Proof. Consider the secondary decomposition of $\mathbf{v} = \sum_{i=1}^{m} \mathbf{q}^{[i]}$, where

$$-E_m \frac{d^2\mathbf{q}^{[m]}}{dx^2} + \mathbf{b}(x, \mathbf{q}^{[m]}) = \mathbf{0},\ (\mathbf{q}^{[m]})_m(0) = r_m(0),\ (\mathbf{q}^{[m]})_m(1) = r_m(1), \quad (6)$$

$$-E_j \frac{d^2\mathbf{q}^{[j]}}{dx^2} + \mathbf{b}(x, \sum_{i=j}^{m} \mathbf{q}^{[i]}) - \mathbf{b}(x, \sum_{i=j+1}^{m} \mathbf{q}^{[i]}) = \varepsilon_j^2 \sum_{i=j+1}^{m} (\mathbf{q}^{[i]})_j \mathbf{e_j},\ x \in \Omega,$$

$$(\mathbf{q}^{[j]})_i(0) = (\mathbf{q}^{[j]})_i(1) = 0,\ j \leq i \leq m,\ 1 \leq j < m, \quad (7)$$

with the matrix E_i is the zero matrix except that on the main diagonal $(E_i)_{jj} = \varepsilon_j^2, j \geq i$, (note that in this notation $E_1 = E$) and $\mathbf{e_i}$ is the ith vector of the canonical basis. Conditions (1c) imply that $\mathbf{q}^{[m]}(0) = \mathbf{r}(0)$, $\mathbf{q}^{[m]}(1) = \mathbf{r}(1)$, and $\mathbf{q}^{[j]}(0) = \mathbf{q}^{[j]}(1) = \mathbf{0}$, for $1 \leq j < m$.

To obtain estimates for the component $\mathbf{q}^{[m]}$, we introduce the function $\mathbf{z} = \mathbf{q}^{[m]} - \mathbf{r}$, which is the solution of the problem

$$-E_m\mathbf{z}'' + \int_{s=0}^{1} J(x, \mathbf{r} + s\mathbf{z})ds\, \mathbf{z} = E_m\mathbf{r}'', \quad \mathbf{z}(0) = \mathbf{z}(1) = \mathbf{0}.$$

The conditions (1c) ensure that a maximum principle holds for this system. Thus $\|\mathbf{z}\|_\infty \leq C\varepsilon_m^2$ and $\|z_m''\|_\infty \leq C$ and follows that $\|z_m'\|_\infty \leq C$. We conclude that

$$\left\| \frac{d^k(\mathbf{q}^{[m]})_m}{dx^k} \right\|_\infty \leq C,\ k = 0, 1, 2, \quad \text{and} \quad \left\| \mathbf{q}^{[m]} \right\|_\infty \leq C.$$

In addition, from the nonlinear system $b_1\left(x, \mathbf{q}^{[m]}\right) = \cdots = b_{m-1}\left(x, \mathbf{q}^{[m]}\right) = 0$, we have that

$$\left\| \frac{d^k(\mathbf{q}^{[m]})_i}{dx^k} \right\|_\infty \leq C,\ k = 1, 2,\ 1 \leq i < m.$$

Differentiating the mth equation of (6) twice and using the above bound we conclude that $\|d^4(\mathbf{q}^{[m]})_m/dx^4\|_\infty \leq C\varepsilon_m^{-2}$. Hence $\|d^3(\mathbf{q}^{[m]})_m/dx^3\|_\infty \leq C\varepsilon_m^{-1}$ and, using the first $m-1$ equations of (6), we have that

$$\left\| \frac{d^k(\mathbf{q}^{[m]})_i}{dx^k} \right\|_\infty \leq C\varepsilon_m^{2-k},\ k = 3, 4,\ 1 \leq i < m.$$

Now consider the component $\mathbf{q}^{[j]}$ with $1 \leq j < m$. It is the solution of

$$-E_j \frac{d^2\mathbf{q}^{[j]}}{dx^2} + \int_0^1 J(x, \sum_{i=j+1}^{m} \mathbf{q}^{[i]} + s\mathbf{q}^{[j]})ds\, \mathbf{q}^{[j]} = \varepsilon_j^2 \sum_{i=j+1}^{m} (\mathbf{q}^{[i]})_j\, \mathbf{e_j},\ x \in \Omega,$$

$$(\mathbf{q}^{[j]})_i(0) = (\mathbf{q}^{[j]})_i(1) = 0,\ j \leq i \leq m.$$

The maximum principle yields $\|\mathbf{q}^{[j]}\|_\infty \le C\varepsilon_j^2$, and then $\|d^2(\mathbf{q}^{[j]})_i/dx^2\|_\infty \le C(\varepsilon_j/\varepsilon_i)^2 \le C,\ j \le i \le m$. Then, $\|d(\mathbf{q}^{[j]})_i/dx\|_\infty \le C,\ j \le i \le m$, and hence (if $j > 1$) we have $\|d(\mathbf{q}^{[j]})_i/dx\|_\infty \le C,\ \|d^2(\mathbf{q}^{[j]})_i/dx^2\|_\infty \le C,\ 1 \le i \le j-1$.

Differentiating the differential equation (7) twice, using the bounds for $\mathbf{q}^{[j]}$ and its derivatives, we deduce that $\|d^4(\mathbf{q}^{[j]})_i/dx^4\|_\infty \le C\varepsilon_i^{-2},\ i = 1,\dots,m$. Hence $\|d^3(\mathbf{q}^{[j]})_i/dx^3\|_\infty \le C\varepsilon_i^{-1},\ i = 1,\dots,m$. □

To establish first order error bounds in the case of an arbitrary number of equations, we consider a further decomposition of the singular component **w**, which is similar to that used in [1] for linear systems. For simplicity, we present the main ideas for the particular case of two equations and these decompositions can be extended to the general case of m semilinear equations using the arguments in [1,3].

In the case of $m = 2$, consider the following decomposition of the left singular component **w**

$$\mathbf{w} = \mathbf{w}^{[1]} + \mathbf{w}^{[2]}, \tag{8a}$$

where $\mathbf{w}^{[2]}(1) = \mathbf{w}^{[1]}(1) = \mathbf{0}$, and

$$-\mathbf{E}\frac{d^2\mathbf{w}^{[2]}}{dx^2} + (\mathbf{b}(x,\mathbf{v}+\mathbf{w}^{[2]}) - \mathbf{b}(x,\mathbf{v})) = \mathbf{0}, \quad x \in \Omega, \tag{8b}$$

$$b_1(0,\mathbf{v}(0)+\mathbf{w}^{[2]}(0)) - b_1(0,\mathbf{v}(0)) = 0,\ w_2^{[2]}(0) = w_2(0), \tag{8c}$$

$$-\mathbf{E}\frac{d^2\mathbf{w}^{[1]}}{dx^2} + (\mathbf{b}(x,\mathbf{v}+\mathbf{w}) - \mathbf{b}(x,\mathbf{v}+\mathbf{w}^{[2]})) = \mathbf{0},\ x \in \Omega, \tag{8d}$$

$$w_1^{[1]}(0) = w_1(0) - w_1^{[2]}(0),\ w_2^{[1]}(0) = 0. \tag{8e}$$

Below we see that the components $\mathbf{w}^{[2]}$ depend weakly on ε_1 and the appearance of $\mathbf{w}^{[1]}$ requires that $w_1(0) - w_1^{[2]}(0) \neq 0$. Moreover, if $\varepsilon_1 = \varepsilon_2$, it is not necessary to decompose **w** into these subcomponents to perform the numerical analysis. We introduce the following notation

$$B_\varepsilon(x) := e^{-x\beta/\varepsilon}, \text{ where } \beta \text{ is defined by (1c).}$$

Lemma 2. *For any $x \in \Omega$, the component $\mathbf{w}^{[2]}$, satisfies the bounds*

$$\left|\frac{d^k\mathbf{w}^{[2]}}{dx^k}(x)\right| \le \mathbf{C}\varepsilon_2^{-k} B_{\varepsilon_2}(x),\ k = 0,1,2,$$

$$\left|\frac{d^3\mathbf{w}^{[2]}}{dx^3}(x)\right| \le C\left(\varepsilon_1^{-2},\varepsilon_2^{-2}\right)^T \varepsilon_2^{-1} B_{\varepsilon_2}(x).$$

Proof. Note that

$$-\mathbf{E}\frac{d^2\mathbf{w}^{[2]}}{dx^2} + \int_{s=0}^{1} J\left(x,\mathbf{v}+s\mathbf{w}^{[2]}\right) ds\ \mathbf{w}^{[2]} = \mathbf{0},$$

from which it follows that $|\mathbf{w}^{[2]}(x)| \leq \mathbf{C}B_{\varepsilon_2}(x)$. Then, from the second equation in (8b) we deduce that

$$\left|\frac{d^k w_2^{[2]}}{dx^k}(x)\right| \leq C\varepsilon_2^{-k} B_{\varepsilon_2}(x),\ k = 0, 1, 2. \tag{9}$$

To obtain bounds for the first component, consider the decomposition $\mathbf{w}^{[2]} = \mathbf{p}^{[2]} + \mathbf{r}^{[2]}$, $r_2^{[2]} \equiv 0$, where

$$b_1\left(x, \mathbf{v} + \mathbf{p}^{[2]}\right) - b_1(x, \mathbf{v}) = 0, \tag{10a}$$

$$-\varepsilon_1^2 \frac{d^2 r_1^{[2]}}{dx^2} + b_1\left(x, \mathbf{v} + \mathbf{p}^{[2]} + \mathbf{r}^{[2]}\right) - b_1\left(x, \mathbf{v} + \mathbf{p}^{[2]}\right) = \varepsilon_1^2 \frac{d^2 p_1^{[2]}}{dx^2}. \tag{10b}$$

As $p_2^{[2]} \equiv w_2^{[2]}$ this is simply a decomposition of the first component $w_1^{[2]}$. Note that the condition on the coefficients (1c) means that $p_1^{[2]}(0) = w_1^{[2]}(0)$, and $p_1^{[2]}(1) = w_1^{[2]}(1)$. Therefore $r_1^{[2]}(0) = r_1^{[2]}(0) = 0$. Writing (10a) in the form

$$\sum_{i=1}^{2} \int_0^1 \frac{\partial b_1}{\partial u_i}\left(x, \mathbf{v} + s\mathbf{p}^{[2]}\right) p_i^{[2]} ds = 0,$$

and using (1b) and (9), we deduce that $|p_1^{[2]}(x)| \leq CB_{\varepsilon_2}(x)$ for any $x \in \Omega$.

Differentiating (10a) and grouping terms, we have

$$\frac{\partial}{\partial x}\left(b_1(x, \mathbf{v} + \mathbf{p}^{[2]}) - b_1(x, \mathbf{v})\right) + \left(\nabla_u b_1(x, \mathbf{v} + \mathbf{p}^{[2]}) - \nabla_u b_1(x, \mathbf{v})\right) \frac{d\mathbf{v}}{dx}$$

$$+ \nabla_u b_1\left(x, \mathbf{v} + \mathbf{p}^{[2]}\right) \frac{d\mathbf{p}^{[2]}}{dx} = 0, \quad \text{where} \quad \nabla_u b_1 := \left(\frac{\partial b_1}{\partial u_1}, \frac{\partial b_1}{\partial u_2}\right)^T.$$

Note if $b_1(x, \mathbf{u} + \mathbf{v}) - b_1(x, \mathbf{u}) = Q(x)$, then

$$\frac{\partial}{\partial x}\left[b_1(x, \mathbf{u} + \mathbf{v}) - b_1(x, \mathbf{u})\right] = \frac{\partial}{\partial x}\left[\sum_{i=1}^{2} \int_0^1 \frac{\partial b_1}{\partial u_i}(x, \mathbf{u} + t\mathbf{v}) v_i\, dt\right],$$

which implies that $\left|\frac{\partial v_1}{\partial x}\right| \leq C\left|\frac{\partial Q}{\partial x}\right| + C\left|\frac{\partial v_2}{\partial x}\right| + C|v_1| + C|v_2|$.

From these expressions, (1b) and (9), we have that $\left|\frac{dp_1^{[2]}}{dx}(x)\right| \leq C\varepsilon_2^{-1} B_{\varepsilon_2}(x)$. Use the same argument to prove $\left|\frac{d^2 p_1^{[2]}}{dx^2}(x)\right| \leq C\varepsilon_2^{-2} B_{\varepsilon_2}(x)$.

The remainder is the solution of the following problem

$$-\varepsilon_1^2 \frac{d^2 r_1^{[2]}}{dx^2} + \left(\int_0^1 \frac{\partial b_1}{\partial u_1}(x, \mathbf{v} + \mathbf{p}^{[2]} + s\mathbf{r}_1^{[2]}) ds \right) r_1^{[2]}$$
$$= \varepsilon_1^2 \frac{d^2 p_1^{[2]}}{dx^2}, \; r_1^{[2]}(0) = r_1^{[2]}(1) = 0.$$

The maximum principle proves that $|r_1^{[2]}(x)| \le C\varepsilon_1^2 \varepsilon_2^{-2} B_{\varepsilon_2}(x)$. Hence,

$$\left| \frac{d^k r_1^{[2]}}{dx^k}(x) \right| \le C\varepsilon_2^{-k} B_{\varepsilon_2}(x), \; k = 1, 2.$$

To obtain the bound on the third derivatives, differentiate (8b) and use the bounds on the lower derivatives. □

Lemma 3. *For any $x \in \Omega$ and for $i = 1, 2$, the component $\mathbf{w}^{[1]}$ satisfies the bounds*

$$\left| w_1^{[1]}(x) \right| \le C(B_{\varepsilon_1}(x) + \frac{\varepsilon_1^2}{\varepsilon_2^2} B_{\varepsilon_2}(x)), \;\; \left| w_2^{[1]}(x) \right| \le C \frac{\varepsilon_1^2}{\varepsilon_2^2} B_{\varepsilon_2}(x),$$

$$\left| \frac{dw_i^{[1]}}{dx}(x) \right| \le C(\varepsilon_1^{-1} B_{\varepsilon_1}(x) + \varepsilon_2^{-1} B_{\varepsilon_2}(x)),$$

$$\varepsilon_i^2 \left| \frac{d^2 w_i^{[1]}}{dx^2}(x) \right| \le C(B_{\varepsilon_1}(x) + \frac{\varepsilon_1^2}{\varepsilon_2^2} B_{\varepsilon_2}(x)),$$

$$\varepsilon_i^2 \left| \frac{d^3 w_i^{[1]}}{dx^3}(x) \right| \le C(\varepsilon_1^{-1} B_{\varepsilon_1}(x) + \varepsilon_2^{-1} B_{\varepsilon_2}(x)).$$

Proof. Decompose $\mathbf{w}^{[1]}$ further into the following sum $\mathbf{w}^{[1]} = \mathbf{z}^{[1]} + \mathbf{s}^{[1]}$, where $\mathbf{z}^{[1]}(0) = \mathbf{w}^{[1]}(0)$, $\mathbf{z}^{[1]}(1) = \mathbf{w}^{[1]}(1) = \mathbf{s}^{[1]}(0) = \mathbf{s}^{[1]}(1) = \mathbf{0}$, and for $x \in \Omega$

$$-\varepsilon_1^2 \frac{d^2 z_1^{[1]}}{dx^2} + \left(\int_0^1 \frac{\partial b_1}{\partial u_1}(x, \mathbf{v} + \mathbf{w}^{[2]} + s(z_1^{[1]}, 0)^T) ds \right) z_1^{[1]} = 0,$$

$$-\varepsilon_2^2 \frac{d^2 z_2^{[1]}}{dx^2} + \left(\int_0^1 \frac{\partial b_2}{\partial u_2}(x, \mathbf{v} + \mathbf{w}^{[2]} + t\mathbf{z}^{[1]}) dt \right) z_2^{[1]}$$
$$= -\left(\int_0^1 \frac{\partial b_2}{\partial u_1}(x, \mathbf{v} + \mathbf{w}^{[2]} + t\mathbf{z}^{[1]}) dt \right) z_1^{[1]},$$

$$-\mathbf{E} \frac{d^2 \mathbf{s}^{[1]}}{dx^2} + \int_0^1 J(x, \mathbf{v} + \mathbf{w}^{[2]} + \mathbf{z}^{[1]} + t\mathbf{s}^{[1]}) dt \; \mathbf{s}^{[1]}$$
$$= \left(b_1(x, \mathbf{v} + \mathbf{w}^{[2]} + (z_1^{[1]}, 0)^T) - b_1(x, \mathbf{v} + \mathbf{w}^{[2]} + \mathbf{z}^{[1]}), 0 \right)^T.$$

From the maximum principle, we have $|\frac{d^k z_1^{[1]}}{dx^k}| \leq C\varepsilon_1^{-k} B_{\varepsilon_1}(x),\ k = 0,1,2$. If $\varepsilon_2^2 \leq 2\varepsilon_1^2$, then the maximum principle proves $|z_2^{[1]}(x)| \leq CB_{\varepsilon_2}(x)$. For the case $\varepsilon_2^2 \geq 2\varepsilon_1^2$, to obtain appropriate bounds of $z_2^{[1]}$, we observe that

$$\left| \int_0^1 \frac{\partial b_2}{\partial u_1} \left(x, \mathbf{v} + \mathbf{w}^{[2]} + t\mathbf{z}^{[1]} \right) dt z_1^{[1]} \right| \leq C_1 B_{\varepsilon_1}(x).$$

Consider the barrier function Z [1], which is the solution of the problem

$$-\varepsilon_2^2 Z'' + \beta^2 Z = C_1 B_{\varepsilon_1}(x),\ Z(0) = Z(1) = 0.$$

This allows one to prove that $|z_2^{[1]}(x)| \leq Z(x) \leq C\varepsilon_1^2\varepsilon_2^{-2}B_{\varepsilon_2}(x)$, if $2\varepsilon_1^2 \leq \varepsilon_2^2$. Thus, for all $\varepsilon_1 \leq \varepsilon_2$, we have $|z_2^{[1]}(x)| \leq C\varepsilon_1^2\varepsilon_2^{-2}B_{\varepsilon_2}(x)$. Hence,

$$\varepsilon_2^2 |\frac{d^2 z_2^{[1]}}{dx^2}(x)| \leq C(B_{\varepsilon_1}(x) + \varepsilon_1^2\varepsilon_2^{-2}B_{\varepsilon_2}(x))$$
$$\leq CB_{\varepsilon_2}(x),\ |\frac{dz_2^{[1]}}{dx}(x)| \leq C\varepsilon_2^{-1}B_{\varepsilon_2}(x).$$

To obtain bounds for the remainder $\mathbf{s}^{[1]}$, note that the first component of the right–hand–side can be written as

$$b_1(x, \mathbf{v} + \mathbf{w}^{[2]} + (z_1^{[1]}, 0)^T) - b_1(x, \mathbf{v} + \mathbf{w}^{[2]} + \mathbf{z}^{[1]})$$
$$= -\int_0^1 \frac{\partial b_1}{\partial u_2}(x, \mathbf{v} + \mathbf{w}^{[2]} + (z_1^{[1]}, tz_2^{[1]})^T) dt z_2^{[1]}.$$

Then, the maximum principle proves that $|\mathbf{s}^{[1]}(x)| \leq \mathbf{C}\varepsilon_1^2\varepsilon_2^{-2}B_{\varepsilon_2}(x)$. Hence,

$$|\frac{d^k \mathbf{s}^{[1]}}{dx^k}(x)| \leq C(\varepsilon_1^{-k}, \varepsilon_2^{-k})^T \frac{\varepsilon_1^2}{\varepsilon_2^2} B_{\varepsilon_2}(x),\ k = 0,1,2.$$

Differentiate (8d) and use above arguments to bound the third derivatives. □

3 Discrete Problem and Analysis of Uniform Convergence

The domain is divided into the subintervals $[0, \tau_{\varepsilon_1}], [\tau_{\varepsilon_1}, \tau_{\varepsilon_2}], \ldots, [\tau_{\varepsilon_m}, 1-\tau_{\varepsilon_m}], \ldots, [1-\tau_{\varepsilon_1}, 1]$. Distribute half the mesh points uniformly within $(\tau_{\varepsilon_m}, 1-\tau_{\varepsilon_m}]$ and the other half in the remaining intervals, distributing $N/(4m)+1$ mesh points uniformly in each $(\tau_{\varepsilon_i}, \tau_{\varepsilon_{i+1}}]$. The transition points are defined as

$$\tau_{\varepsilon_m} = \min\{0.25, 2\varepsilon_m/\beta \ln N\},\ \tau_{\varepsilon_i} = \min\left\{0.5\tau_{\varepsilon_{i+1}}, 2\varepsilon_i/\beta \ln N\right\},\ 1 \leq i < m. \tag{11}$$

On the mesh $\bar{\Omega}^N = \{x_i\}_{i=0}^N$, consider the following finite difference scheme

$$(\mathbf{T}_N\mathbf{U})(x_j) := -(\mathbf{E}\delta^2\mathbf{U})(x_j) + \mathbf{b}(x_j, \mathbf{U}(x_j)) = \mathbf{0},\ x_j \in \Omega^N = \bar{\Omega}^N \cap \Omega, \quad (12)$$

with $\mathbf{U}(0) = \mathbf{u}(0)$, $\mathbf{U}(1) = \mathbf{u}(1)$ and δ^2 is the classical three–point finite difference approximation of the second derivative on a non–uniform mesh.

From (1c), the Frechet–derivative $\mathbf{T}'_N$ is an M–matrix and then for any two mesh functions $\mathbf{Y}$ and $\mathbf{Z}$ with $\mathbf{Y}(0) = \mathbf{Z}(0)$ and $\mathbf{Y}(1) = \mathbf{Z}(1)$, we have that

$$\|\mathbf{Y} - \mathbf{Z}\|_\infty \leq \|(\mathbf{T}'_N)^{-1}\|_\infty \|\mathbf{T}_N\mathbf{Y} - \mathbf{T}_N\mathbf{Z}\|_\infty \leq \frac{1}{\min\{1, \beta^2\}} \|\mathbf{T}_N\mathbf{Y} - \mathbf{T}_N\mathbf{Z}\|_\infty.$$

This implies the uniqueness of the solution to problem (12). In bounding the truncation error, we must bound the same terms

$$|\mathbf{T}_N\mathbf{u}(x)| = |\mathbf{T}_N\mathbf{u}(x) - \mathbf{T}\mathbf{u}(x)| \leq \mathbf{E}|(\delta^2\mathbf{v} - \mathbf{v}'')(x)| + \mathbf{E}|(\delta^2\mathbf{w} - \mathbf{w}'')(x)|,$$

as in the linear problem [1]. The derivatives of both the regular and singular components have a similar behaviour to their linear counterparts, and thus we can deduce that $\|\mathbf{T}_N\mathbf{u}\|_\infty \leq CN^{-1}$.

Theorem 1. *Let* $\mathbf{u}$ *be the solution of the problem (1) and* $\mathbf{U}$ *the solution of problem (12) on the Shishkin mesh* $\bar{\Omega}^N$. *Then,*

$$\|\mathbf{U} - \mathbf{u}\|_\infty \leq CN^{-1}.$$

Remark 1. In the particular case of equal diffusion parameters $\varepsilon_i = \varepsilon, i = 1, \ldots m$, it is possible [2] to prove essentially second order uniform convergence. In the linear case of $m = 2$, Linss and Madden [4] have established second order (up to logarithmic factors). To achieve this higher order, Linss and Madden [4] employ a decomposition (based on the decomposition in Madden and Stynes [6]) of the solutions, which is different to the decomposition presented in this paper. In the linear case of $m \geq 2$ Linß and Madden [5] have established second order convergence for arbitrary ε_i, under the assumption that the elements in the coefficient matrix $B(x)$ of the zero order terms satisfy

$$b_{ii}(x) > 0,\ \sum_{k \neq i}^{m} \|b_{ik}(x)/b_{ii}(x)\| < 1,\ 1 \leq i \leq m.$$

For variable coefficients and $m > 2$, these conditions will only be satisfied by a subset of problems from the class (1). Hence, the question of proving second order convergence for the class of problems in (1) for $m > 2$ and arbitrary ε_i remains open.

4 Numerical Experiments

Example 1. Consider a nonlinear problem of type (1) where $m = 2$, $\mathbf{u}(0) = \mathbf{u}(1) = (0, 0)^T$, and

$$b_1(x,\mathbf{u}) = u_1 - 1 - (1-u_1)^3 + e^{u_1-u_2},\ b_2(x,\mathbf{u}) = u_2 - 0.5 - (0.5-u_2)^5 + e^{u_2-u_1}.$$

The corresponding nonlinear systems of equations associated with the discrete problem are solved using Newton's method with zero as an initial guess. We iteratively compute $\mathbf{U}^k(x_j)$, for $k = 1, 2, \ldots, K$, until

$$\|\mathbf{U}^K(x_j) - \mathbf{U}^{K-1}(x_j)\|_\infty \leq N^{-2}.$$

To estimate the pointwise errors $|\mathbf{U}^K(x_j) - \mathbf{u}(x_j)|$ we calculate a new approximation $\{\hat{\mathbf{U}}^K(x_j)\}$ on the mesh $\{\hat{x}_j\}$ that contains the mesh points of the original mesh and its midpoints. At the coarse mesh points we calculate the uniform two-mesh differences and the orders of convergence

$$d_i^{N,K} = \max_{S_\varepsilon} \max_{0\leq j\leq N} |U_i^K(x_j) - \hat{U}_i^K(x_{2j})|,\ p_{i,uni}^{N,K} = \log_2(d_i^{N,K}/d_i^{2N,K}),\ i = 1, 2,$$

where the singular perturbation parameters take values in the set

$$S_\varepsilon = \{(\varepsilon_1, \varepsilon_2) \mid \varepsilon_2^2 = 2^0, 2^{-1}, \ldots, 2^{-30},\ \varepsilon_1^2 = \varepsilon_2^2, 2^{-1}\varepsilon_2^2, \ldots, 2^{-59}, 2^{-60}\}.$$

In Table 1 we display the uniform two-mesh differences and the approximate orders of convergence for both components u_1 and u_2. Finally, we report that $K \leq 4$ for all $(\varepsilon_1, \varepsilon_2) \in S_\varepsilon$ and all $N = 2^{-j},\ j = 5, \ldots, 12$.

Example 2. Consider a linear problem of the type (1) where $m = 3,\ \mathbf{u}(0) = \mathbf{u}(1) = (1, 1, 1)^T$, and

$$\begin{aligned} b_1(x,\mathbf{u}) &= 2.1u_1 - (1-x)u_2 - (1+x)u_3 - x, \\ b_2(x,\mathbf{u}) &= -xu_1 + (1.1+x)u_2 - xu_3 + x, \\ b_3(x,\mathbf{u}) &= -(2+x)u_1 - (1-x)u_2 + (3.1+x)u_3 - 1. \end{aligned}$$

This linear problem is not covered by the theory in [5], but is covered by the theory in this paper. In Table 2 the uniform two-mesh differences and the approximate orders of uniform convergence are displayed, where the values of the singular perturbation parameters vary over the range

$$\varepsilon_3 = 2^0, 2^{-2}, \ldots, 2^{-30},\ \varepsilon_2 = \varepsilon_3, 2^{-2}\varepsilon_3, \ldots, 2^{-40},\ \varepsilon_1 = \varepsilon_2, 2^{-2}\varepsilon_2, \ldots, 2^{-60}.$$

Table 1 Uniform two-mesh differences $\mathbf{d}^{N,K}$ and orders of convergence $\mathbf{p}_{uni}^{N,K}$ for Example 1

$(\varepsilon_1, \varepsilon_2) \in S_\varepsilon$	N = 32	N = 64	N = 128	N = 256	N = 512	N = 1,024	N = 2,048	N = 4,096
$d_1^{N,K}$	6.861E−3	6.222E−3	3.568E−3	1.313E−3	4.486E−4	1.423E−4	4.327E−5	1.291E−5
$p_{1,uni}^{N,K}$	0.141	0.802	1.443	1.549	1.656	1.718	1.745	
$d_2^{N,K}$	8.130E−3	3.915E−3	1.523E−3	5.343E−4	1.736E−4	5.375E−5	1.644E-5	4.943E−6
$p_{2,uni}^{N,K}$	1.054	1.362	1.511	1.622	1.691	1.709	1.733	

Table 2 Uniform two-mesh differences $\mathbf{d}^N$ and approximate uniform orders of convergence $\mathbf{p}_{uni}^N$ for Example 2

	N = 16	N = 32	N = 54	N = 128	N = 256	N = 512	N = 1,024	N = 2,048
$[\mathbf{d}^N]_1$	0.151E+00	0.135E+00	0.113E+00	0.747E−01	0.378E−01	0.145E−01	0.484E−02	0.154E−02
$[\mathbf{p}_{uni}^N]_1$	0.159	0.256	0.599	0.982	1.381	1.586	1.655	
$[\mathbf{d}^N]_2$	0.159E+00	0.147E+00	0.119E+00	0.778E−01	0.381E−01	0.145E−01	0.472E−02	0.150E−02
$[\mathbf{p}_{uni}^N]_2$	0.115	0.303	0.613	1.030	1.391	1.620	1.656	
$[\mathbf{d}^N]_3$	0.158E+00	0.142E+00	0.119E+00	0.784E−01	0.397E−01	0.152E−01	0.508E−02	0.161E-02
$[\mathbf{p}_{uni}^N]_3$	0.157	0.256	0.598	0.982	1.381	1.586	1.655	

For both examples, we observe uniform convergence of the finite difference approximations, which is in agreement with Theorem 1. However, orders greater than one are observed in both Tables.

Acknowledgement This research was partially supported by the project MEC/FEDER MTM2007-63204 and the Diputacion General de Aragon.

References

1. J.L. Gracia, F.J. Lisbona, E. O'Riordan, A system of singularly perturbed reaction-diffusion equations, Dublin City University School of Mathematical Sciences preprint MS-07-10, (2007).
2. J.L. Gracia, F.J. Lisbona, M. Madaune-Tort, E. O'Riordan, A coupled system of singularly perturbed semilinear reaction-diffusion equations, Dublin City University School of Mathematical Sciences Preprint, MS-08-11, (2008).
3. J.L. Gracia, F.J. Lisbona, E. O'Riordan, A coupled system of singularly perturbed parabolic reaction-diffusion equations, *Adv. Comp. Math.*, DOI 10.1007/s10444-008-9086-3.
4. T. Linß, N. Madden, Accurate solution of a system of coupled singularly perturbed reaction-diffusion equations, *Computing*, **73** (2004), 121–133.
5. T. Linß, N. Madden, Layer-adapted meshes for a system of coupled singularly perturbed reaction-diffusion equations, *IMA J. Numer. Anal.*, **29** (2009), 109–125.
6. N. Madden, M. Stynes, A uniformly convergent numerical method for a coupled system of two singularly perturbed linear reaction-diffusion problems, *IMA J. Numer. Anal.*, **23** (2003), 627–644.

On Finite Element Methods for 3D Time-Dependent Convection–Diffusion–Reaction Equations with Small Diffusion

Volker John and Ellen Schmeyer

Abstract The paper studies finite element methods for the simulation of time-dependent convection-diffusion-reaction equations with small diffusion: the SUPG method, a SOLD method and two types of FEM–FCT methods. The methods are assessed, in particular with respect to the size of the spurious oscillations in the computed solutions, at a 3D example with nonhomogeneous Dirichlet boundary conditions and homogeneous Neumann boundary conditions.

1 Introduction

The simulation of various applications requires the numerical solution of time-dependent convection–diffusion–reaction equations. Processes which involve a chemical reaction in a flow field are a typical example [5]. Such a reaction can be modeled with a coupled system of time-dependent nonlinear convection–diffusion–reaction equations for the concentrations of the reactants and the products.

Typically, the solution of these equations possesses layers. A numerical method for the simulation of these equations, whose results can be considered to be useful, should meet the following requirements:

- The layers should be correctly localized,
- Sharp layers (with respect to the used mesh size) should be computed,
- Spurious oscillations in the solution must not occur.

The third requirement means in particular that the computed solution should not have negative values if, for instance, the behavior of concentrations is simulated. A number of finite element methods have been developed for the simulation of convection–diffusion–reaction equations with small diffusion. One of the most popular ones is the Streamline Upwind Petrov–Galerkin (SUPG) method from [1, 2].

V. John (✉)
FR 6.1 – Mathematik, Universität des Saarlandes, Postfach 15 11 50, 66041 Saarbrücken, Germany,
E-mails: john@math.uni-sb.de, schmeyer@math.uni-sb.de

A.F. Hegarty et al. (eds.), *BAIL 2008 - Boundary and Interior Layers.*
Lecture Notes in Computational Science and Engineering,
DOI: 10.1007/978-3-642-00605-0,

This method leads to solutions with correctly located and sharp layers, however also with sometimes considerable spurious oscillations. To reduce these oscillations, a number of so-called Spurious Oscillations at Layers Diminishing (SOLD) schemes have been proposed, see the reviews [3,4]. SOLD schemes add additional, in general nonlinear, stabilization terms to the SUPG method. A completely different finite element approach for treating equations with small diffusion is used in Finite Element Method Flux–Corrected–Transport (FEM–FCT) schemes [8, 10]. These methods do not modify the bilinear form but manipulate the matrix and the right-hand side of a Galerkin finite element method.

A first comparison of finite element methods for time-dependent convection-diffusion-reaction equations was presented in [6]. The numerical examples of [6] studied problems in 2D with homogeneous Dirichlet boundary conditions. The present paper extends the studies of [6] to 3D problems with inhomogeneous Dirichlet and homogeneous Neumann boundary conditions. This is a realistic situation in applications.

2 Finite Element Methods for Time-Dependent Convection–Diffusion–Reaction Equations

We consider a linear time-dependent convection–diffusion–reaction equation

$$u_t - \varepsilon \Delta u + \mathbf{b} \cdot \nabla u + cu = f \text{ in } (0, T] \times \Omega, \tag{1}$$

where $\varepsilon > 0$ is the diffusion coefficient, $\mathbf{b} \in L^\infty(0, T; (W^{1,\infty}(\Omega))^3)$ is the convection field, $c \in L^\infty(0, T; L^\infty(\Omega))$ is the non-negative reaction coefficient, $f \in L^2(0, T; L^2(\Omega))$ describes sources, $T > 0$ is the final time and $\Omega \subset \mathbb{R}^3$ is a bounded domain. This equation has to be equipped with an initial condition $u_0 = u(0, \mathbf{x})$ and with appropriate boundary conditions. Since the isothermal reaction considered in [5] leads to equations with non-negative reaction rates, we are particularly interested in the case $c(t, \mathbf{x}) \geq 0$ in $[0, T] \times \Omega$.

In the numerical studies, (1) is discretized in time with the Crank–Nicolson scheme using equidistant time steps Δt. This leads at the discrete time t_k to the equation

$$\begin{aligned} &u_k + 0.5\Delta t \left(-\varepsilon \Delta u_k + \mathbf{b}_k \cdot \nabla u_k + c_k u_k\right) \\ &\quad = u_{k-1} - 0.5\Delta t \left(-\varepsilon \Delta u_{k-1} + \mathbf{b}_{k-1} \cdot \nabla u_{k-1} + c_{k-1} u_{k-1}\right) \\ &\qquad + 0.5\Delta t f_{k-1} + 0.5\Delta t f_k. \end{aligned} \tag{2}$$

Equation (2) can be considered as a steady-state convection–diffusion–reaction equation, with the diffusion, convection and reaction, respectively, given by

$$D = 0.5\Delta t \varepsilon, \quad \mathbf{C}_k = 0.5\Delta t \mathbf{b}_k, \quad R_k = 1 + 0.5\Delta t c_k.$$

The Galerkin finite element method for (2) reads as follows: Find $u_k^h \in V_{\text{ans}}^h$ such that

$$
\begin{aligned}
&(u_k^h, v^h) + 0.5\Delta t \left((\varepsilon \nabla u_k^h, \nabla v^h) + (\mathbf{b}_k \cdot \nabla u_k^h + c_k u_k^h, v^h) \right) \\
&= (u_{k-1}^h, v^h) - 0.5\Delta t \left((\varepsilon \nabla u_{k-1}^h, \nabla v^h) + (\mathbf{b}_{k-1} \cdot \nabla u_{k-1}^h + c_{k-1} u_{k-1}^h, v^h) \right) \\
&\quad + 0.5\Delta t (f_{k-1}, v^h) + 0.5\Delta t (f_k, v^h)
\end{aligned} \tag{3}
$$

for all $v^h \in V_{\text{test}}^h$, where V_{ans}^h and V_{test}^h are appropriate finite element spaces. Here, $(\cdot, \cdot)$ denotes the inner product in $L^2(\Omega)$.

The SUPG method adds a consistent diffusion term in streamline direction

$$
\sum_{K \in \mathcal{T}^h} \tau_K \left(R^h(u_k^h), \mathbf{C}_k \cdot \nabla v^h \right)_K
$$

to the left-hand side of (3), where $\mathcal{T}^h$ is the given triangulation of Ω, $\{\tau_K\}$ is a set of parameters depending on the mesh cells $\{K\}$ and $(\cdot, \cdot)_K$ is the inner product in $L^2(K)$. The residual $R^h(u_k^h)$ is defined by the difference of the left-hand side and the right-hand side of (2). Different proposals for the choice of the parameters $\{\tau_K\}$ can be found in the literature. In the numerical studies of [6], the choice from [7]

$$
\tau_K = \min \left\{ \frac{h_K}{\Delta t \|\mathbf{b}_k\|_2}, \frac{1}{1 + 0.5\Delta t c_k}, \frac{2h_K^2}{\Delta t \varepsilon} \right\} \tag{4}
$$

has been proven to be the best one. In (4), $\|\cdot\|_2$ denotes the Euclidean norm of a vector and h_K is an appropriate measure of the size of the mesh cell K. For time-dependent problems which are discretized with small time steps, the second term in (4) dominates and the actual choice h_K is of minor importance. In the computations presented below, the diameter of the mesh cell K was chosen. It is well known that numerical solutions which are computed with the SUPG method often possess non-negligible spurious oscillations at the layers.

SOLD methods try to reduce the spurious oscillations of the SUPG method by adding another stabilization term to this method. This stabilization term is in general nonlinear. There are several classes of SOLD methods, see [3, 4]. It was found in the numerical studies of [6] that the best results among the SOLD methods were obtained with a method that adds an anisotropic diffusion term

$$
(\tilde{\varepsilon} \mathbb{C}_{\text{os},k} \nabla u_k^h, \nabla v^h) \quad \text{with} \quad \mathbb{C}_{\text{os},k} = \begin{cases} I - \dfrac{\mathbf{C}_k \otimes \mathbf{C}_k}{\|\mathbf{C}_k\|_2^2} & \text{if } \mathbf{C}_k \neq \mathbf{0}, \\ 0 & \text{else,} \end{cases}
$$

and the parameter

$$
\tilde{\varepsilon}|_K = \max \left\{ 0, C \frac{\operatorname{diam}(K) |R^h(u_k^h)|}{2\|\nabla u_k^h\|_2} - D \right\}, \tag{5}
$$

where $\mathrm{diam}(K)$ is the diameter of a mesh cell K. This type of parameter was proposed in [7] and modified to the form (5) in [3]. The SOLD parameter (5) contains a free parameter C which has to be chosen by the user. In analogy to [6], this SOLD method will be called KLR02.

The last approaches which will be studied in our numerical tests are FEM–FCT schemes. They start with the algebraic equation corresponding to the Galerkin finite element method (3)

$$(M_C + 0.5\Delta t A_k)\underline{u}_k = (M_C - 0.5\Delta t A_{k-1})\underline{u}_{k-1} + 0.5\Delta t \underline{f}_{k-1} + 0.5\Delta t \underline{f}_k, \quad (6)$$

where $\{\varphi_i\}$ is the basis of the finite element space and $(M_C)_{ij} = (m_{ij}) = (\varphi_j, \varphi_i)$ is the consistent mass matrix. The matrix representation of the second term of the left-hand side of (3) is denoted by $(A_k)_{ij} = (a_{ij})$. Vectors are indicated by an underline. The first idea of FEM–FCT schemes is to manipulate (6) so that a stable but low order scheme is represented. To this end, define $L_k = A_k + D_k$ with

$$D_k = (d_{ij}), \quad d_{ij} = -\max\{0, a_{ij}, a_{ji}\} \text{ for } i \neq j, \ d_{ii} = -\sum_{j=1, j\neq i}^{N} d_{ij},$$

and $M_L = \mathrm{diag}(m_i)$ with $m_i = \sum_{j=1}^{N} m_{ij}$, where N is the number of degrees of freedom. M_L is called lumped mass matrix. The low order scheme reads

$$(M_L + 0.5\Delta t L_k)\underline{u}_k = (M_L - 0.5\Delta t L_{k-1})\underline{u}_{k-1} + 0.5\Delta t \underline{f}_{k-1} + 0.5\Delta t \underline{f}_k. \quad (7)$$

The second idea of FEM–FCT schemes is to modify the right-hand side of (7) in such a way that diffusion is removed where it is not needed but spurious oscillations are still suppressed

$$\begin{aligned}(M_L + 0.5\Delta t L_k)\underline{u}_k = {} & (M_L - 0.5\Delta t L_{k-1})\underline{u}_{k-1} + 0.5\Delta t \underline{f}_{k-1} + 0.5\Delta t \underline{f}_k \\ & + \underline{f}^*(\underline{u}_k, \underline{u}_{k-1}). \end{aligned} \quad (8)$$

The computation of the anti-diffusive flux vector $\underline{f}^*(\underline{u}_k, \underline{u}_{k-1})$ is somewhat involved and we refer to [6, 8–10] for details. Its computation relies on a predictor step which uses an explicit and stable low order scheme. Thus, a stability issue arises in FEM–FCT schemes which leads to the CFL-like condition $\Delta t < 2 \min_i m_i / l_{ii}$. This condition was fulfilled in the numerical tests presented in Sect. 3. We will consider a nonlinear approach for computing $\underline{f}^*(\underline{u}_k, \underline{u}_{k-1})$ [9, 10] and a linear approach [8] (in the form which is presented in [6]).

3 Numerical Studies

We consider a situation which has some typical features of a chemical reaction in applications. First, the domain is three dimensional, $\Omega = (0,1)^3$. There is an inlet at $\{0\} \times (5/8, 6/8) \times (5/8, 6/8)$ and an outlet at $\{1\} \times (3/8, 4/8) \times (4/8, 5/8)$.

The convection is given by $\mathbf{b} = (1, -1/4, -1/8)^T$, which corresponds to the vector pointing from the center of the inlet to the center of the outlet. Thus, the convection will not be aligned to the mesh. The diffusion is given by $\varepsilon = 10^{-6}$ and the reaction by

$$c(\mathbf{x}) = \begin{cases} 1 \text{ if } \|\mathbf{x} - g\|_2 \leq 0.1, \\ 0 \text{ else,} \end{cases}$$

where g is the line through the center of the inlet and the center of the outlet. That means, a reaction takes place only where the solution (concentration) is expected to be transported. The inlet boundary condition is

$$u_{\text{in}}(t) = \begin{cases} \sin(\pi t/2) & \text{if } t \in [0, 1], \\ 1 & \text{if } t \in (1, 2], \\ \sin(\pi(t-1)/2) & \text{if } t \in (2, 3]. \end{cases}$$

At the outlet, homogeneous Neumann boundary conditions are prescribed. Apart from inlet and outlet, the solution should obey homogeneous Dirichlet conditions on the boundary. The right-hand side was set to be $f = 0$ in Ω for all times and the final time in our numerical studies was $T = 3$. The initial condition was set to be $u_0 = 0$. The orders of magnitude for diffusion, convection, reaction and concentration correspond to the situation of [5].

Results will be presented for the P_1 finite element on a tetrahedral mesh and the Q_1 finite element on a hexahedral mesh. The number of degrees of freedom on both meshes is 35 937, including Dirichlet nodes. The diameter of the mesh cells is about 0.054 for the hexahedral mesh and between 0.054 and 0.076 for the tetrahedral mesh. The Crank–Nicolson scheme was applied with $\Delta t = 0.001$.

From the construction of the problem, it is expected that the solution is transported from the inlet to the outlet with a little smearing due to the diffusion. It should take values in [0, 1]. The size of the spurious oscillations in the numerical schemes will be illustrated with the size of the undershoots $u^h_{\min}(t)$, see Fig. 1. The undershoots are particularly dangerous in applications since they represent non-physical situations, like negative concentrations. Figure 2 shows the distribution of

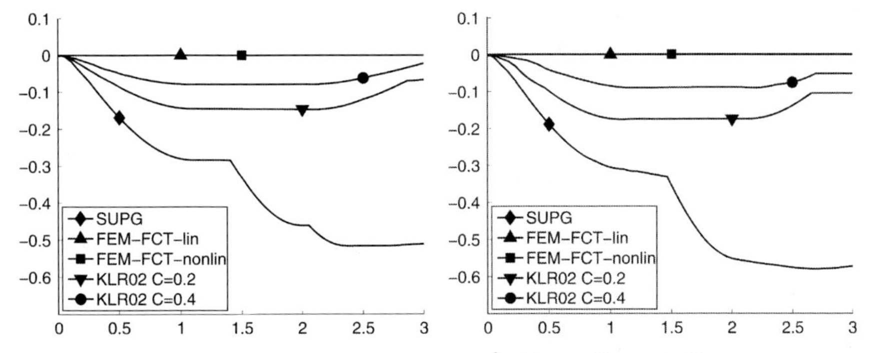

Fig. 1 Minimal value of the finite element solutions $u^h_{\min}(t)$, *left* Q_1, *right* P_1

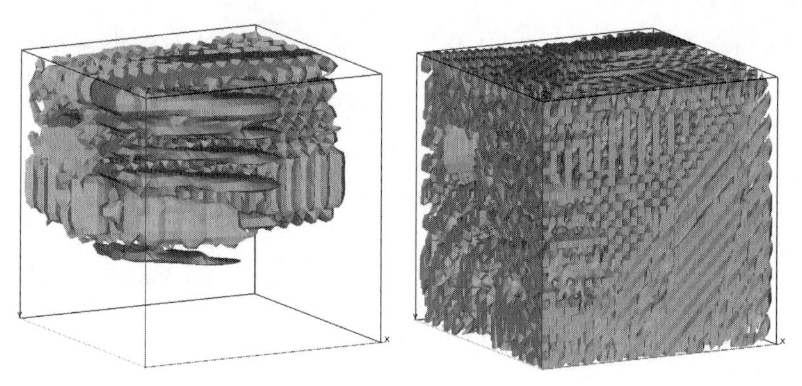

Fig. 2 Distribution of negative oscillations $u^h_{\min}(t) \leqslant 0.01$ for the SUPG method at $t = 2$, *left* Q_1, *right* P_1

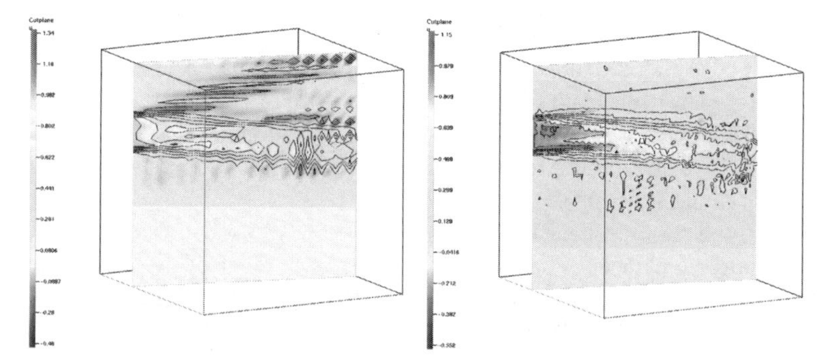

Fig. 3 Cut of the solution, SUPG method at $t = 2$, *left* Q_1, *right* P_1

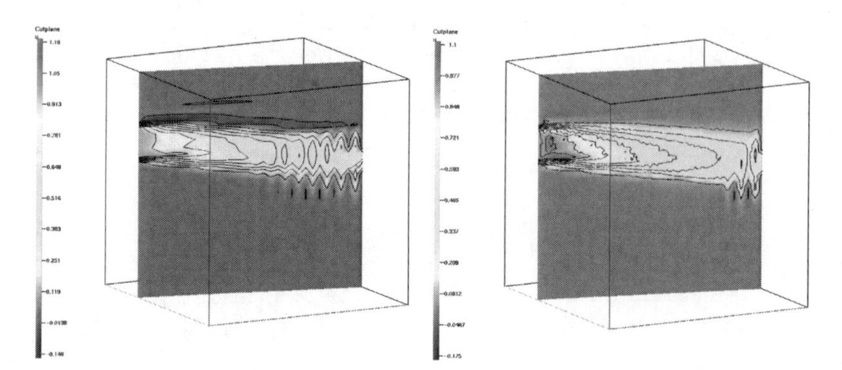

Fig. 4 Cut of the solution, SOLD method (5), $C = 0.2$ at $t = 2$, *left* Q_1, *right* P_1

the undershoots with $u^h_{\min}(t) \leqslant 0.01$ for the SUPG method at $t = 2$. Cut planes of the solutions at $t = 2$ are given in Figs. 3–7. These cut planes contain the centers of the inlet and the outlet and they are parallel to the z-axis. Note, some wiggles which can be seen in the contour lines might be due to the rather coarse meshes. For illustrating the spurious oscillations, a color bar is given for each cut plane.

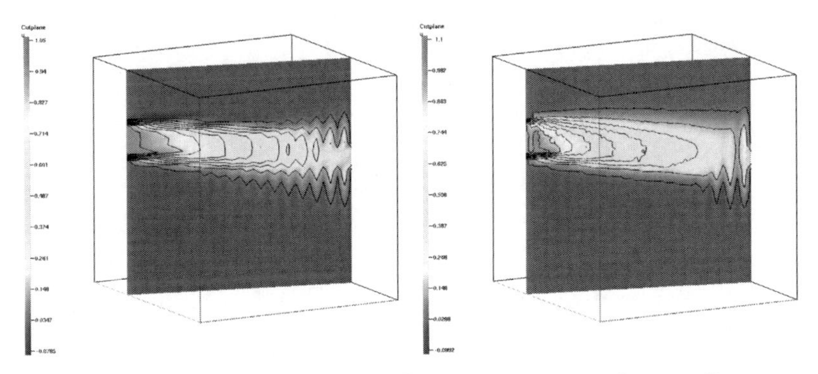

Fig. 5 Cut of the solution, SOLD method (5), $C = 0.4$ at $t = 2$, *left* Q_1, *right* P_1

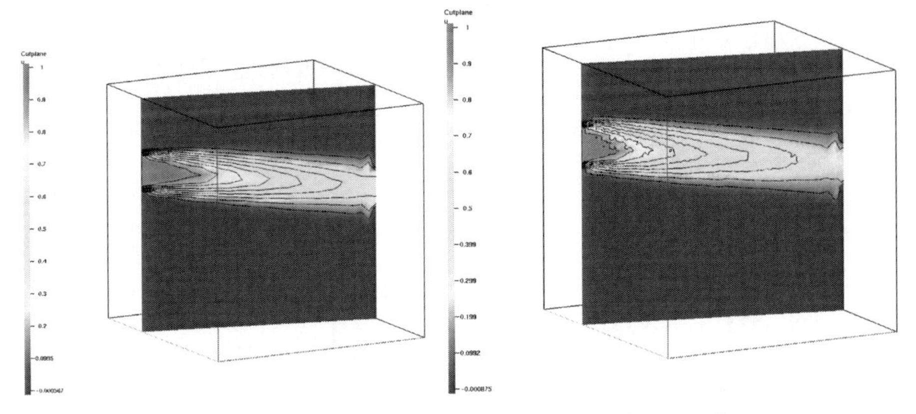

Fig. 6 Cut of the solution linear FEM–FCT method at $t = 2$, *left* Q_1, *right* P_1

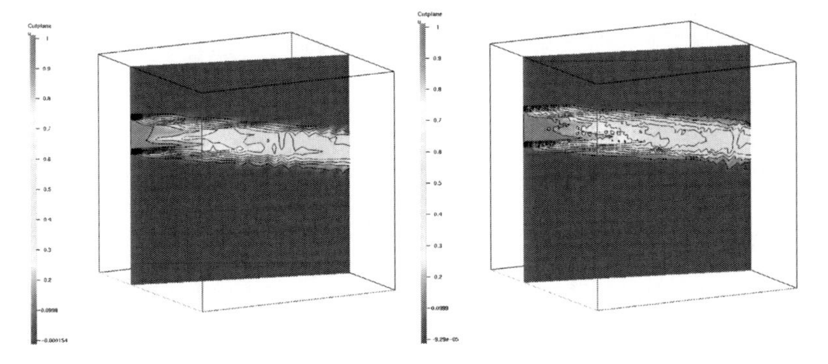

Fig. 7 Cut of the solution, nonlinear FEM–FCT method at $t = 2$, *left* Q_1, *right* P_1

The numerical results show the large amount of spurious oscillations in the solutions computed with the SUPG method. Figure 2 demonstrates that the solutions are globally polluted with spurious oscillations. The oscillations were considerably reduced and localized (not shown here) with the SOLD method KLR02. Increasing

Table 1 Computing times in seconds

Method	Q_1	P_1
SUPG	5,989	9,473
SOLD (5), $C = 0.2$	24,832	25,050
SOLD (5), $C = 0.4$	33,688	30,932
FEM–FCT linear	5,920	6,509
FEM–FCT nonlinear	9,768	10,398

the constant in (5) leads to a decrease of the spurious oscillations, Fig. 1. From the numerical studies of [3, 4] it is known that an increase of the constant in (5) results to somewhat more smearing of the solutions. However, this is rather tolerable in applications compared with spurious oscillations. The solutions obtained with the FEM–FCT methods are almost free of spurious oscillations. The smoother solutions of the linear FEM–FCT scheme, compared with the nonlinear FEM–FCT scheme, reflect that the linear scheme introduces more diffusion. This leads generally to a stronger smearing of the layers, see [6]. Altogether, the FEM–FCT schemes gave the best results in the numerical studies.

Computing times for the methods are given in Table 1. For solving the algebraic systems corresponding to the nonlinear schemes, the same fixed point iteration as described in [4, 6] was used. The iterations were stopped when the Euclidean norm of the residual was less than 10^{-8}. The computations were performed on a computer with Intel Xeon CPU with 2.66 GHz. It can be observed that the nonlinear schemes are considerably more expensive than the linear methods. For KLR02, the computing times increase with increasing size of the user-chosen parameter. All observations correspond to the results obtained in [6] for 2D problems.

4 Summary and Conclusions

The paper studied several finite element methods for solving time-dependent convection–diffusion–reaction equations in a 3D domain with inhomogeneous Dirichlet and homogeneous Neumann boundary conditions. The SUPG method led to solutions globally polluted with large spurious oscillations. These oscillations were reduced considerably with a SOLD method, but at the expense of much larger computing times. FEM–FCT methods led to almost oscillation-free solutions. From the aspects of solution quality and computing time, the linear FEM–FCT scheme seems to be, among the methods studied, the most appropriate method to be used in applications.

Acknowledgement The research of E. Schmeyer was supported by the Deutsche Forschungsgemeinschaft (DFG) by grant No. Jo 329/8–1 within the DFG priority program 1276 MetStröm: Multiple Scales in Fluid Mechanics and Meteorology.

References

1. A.N. Brooks and T.J.R. Hughes. Streamline upwind/Petrov-Galerkin formulations for convection dominated flows with particular emphasis on the incompressible Navier–Stokes equations. *Comput. Methods Appl. Mech. Eng.*, 32:199–259, 1982.
2. T.J.R. Hughes and A.N. Brooks. A multidimensional upwind scheme with no crosswind diffusion. In T.J.R. Hughes, editor, *Finite Element Methods for Convection Dominated Flows, AMD vol. 34*, pages 19–35. ASME, New York, 1979.
3. V. John and P. Knobloch. A comparison of spurious oscillations at layers diminishing (sold) methods for convection–diffusion equations: Part I – a review. *Comput. Methods Appl. Mech. Eng.*, 196:2197–2215, 2007.
4. V. John and P. Knobloch. A comparison of spurious oscillations at layers diminishing (sold) methods for convection–diffusion equations: Part II – analysis for P_1 and Q_1 finite elements. *Comput. Methods Appl. Mech. Eng.*, 197:1997–2014, 2008.
5. V. John, M. Roland, T. Mitkova, K. Sundmacher, L. Tobiska, and A. Voigt. Simulations of population balance systems with one internal coordinate using finite element methods. *Chem. Eng. Sci.*, 64:733–741, 2009.
6. V. John and E. Schmeyer. Stabilized finite element methods for time-dependent convection-diffusion–reaction equations. *Comput. Methods Appl. Mech. Eng.*, 198:475–494, 2008.
7. T. Knopp, G. Lube, and G. Rapin. Stabilized finite element methods with shock capturing for advection-diffusion problems. *Comput. Methods Appl. Mech. Eng.*, 191:2997–3013, 2002.
8. D. Kuzmin. Explicit and implicit FEM–FCT algorithms with flux linearization. Ergebnisberichte Angew. Math. 358, University of Dortmund, 2008. *J. Comput. Phys.*, 228:2517–2534, 2009.
9. D. Kuzmin and M. Möller. Algebraic flux correction I. Scalar conservation laws. In R. Löhner D. Kuzmin and S. Turek, editors, *Flux-corrected transport: Principles, algorithms and applications*, pages 155–206. Springer, Berlin, 2005.
10. D. Kuzmin, M. Möller, and S. Turek. High–resolution FEM–FCT schemes for multidimensional conservation laws. *Comput. Methods Appl. Mech. Eng.*, 193:4915–4946, 2004.

On the Application of Local Projection Methods to Convection–Diffusion–Reaction Problems

Petr Knobloch

Abstract We apply the local projection stabilization to finite element discretizations of scalar convection–diffusion–reaction equations with mixed boundary conditions. We derive general error estimates and discuss the choice of the stabilization parameter. Numerical results illustrate some drawbacks of the local projection stabilization in comparison to the SUPG method.

1 Introduction

Local projection stabilizations of finite element discretizations have become very popular during the last decade. First they were introduced by Becker and Braack [BB01] for the Stokes problem and later they have been applied to many other problems including transport problems, convection–diffusion–reaction equations, Oseen equations and Navier–Stokes equations, see, e.g., [BB06, BR06, MST07, MST08, RLL08]. In this paper we shall consider the convection–diffusion–reaction problem

$$-\varepsilon\,\Delta\,u+\mathbf{b}\cdot\nabla u+c\,u=f\text{ in }\Omega,\quad u=u_b\text{ on }\Gamma^D,\quad \varepsilon\,\frac{\partial u}{\partial\mathbf{n}}=g\text{ on }\Gamma^N,\qquad(1)$$

where $\Omega\subset\mathbb{R}^d$, $d=2,3$, is a bounded domain with a polyhedral Lipschitz-continuous boundary $\partial\Omega$ and $\Gamma^D,\Gamma^N\subset\partial\Omega$ are two relatively open disjoint sets satisfying $\overline{\Gamma^D\cup\Gamma^N}=\partial\Omega$ and $\text{meas}_{d-1}(\Gamma^D)>0$. We denote by $\mathbf{n}$ the outer unit normal vector to $\partial\Omega$. We assume that ε is a positive constant and $\mathbf{b}\in W^{1,\infty}(\Omega)^d$, $c\in L^\infty(\Omega)$, $f\in L^2(\Omega)$, $u_b\in H^{1/2}(\Gamma^D)$ and $g\in H^{-1/2}(\Gamma^N)$ are given functions satisfying

$$\sigma:=c-\tfrac{1}{2}\,\text{div}\,\mathbf{b}\geq\sigma_0\geq0\,,$$

P. Knobloch
Charles University, Faculty of Mathematics and Physics, Sokolovská 83, 186 75 Praha 8, Czech Republic, E-mail: knobloch@karlin.mff.cuni.cz

A.F. Hegarty et al. (eds.), *BAIL 2008 - Boundary and Interior Layers.* Lecture Notes in Computational Science and Engineering,
DOI: 10.1007/978-3-642-00605-0,

where σ_0 is a constant. Moreover, we assume that the inflow boundary is a part of the Dirichlet boundary, i.e.,

$$\{\mathbf{x} \in \partial\Omega\,;\ (\mathbf{b} \cdot \mathbf{n})(\mathbf{x}) < 0\} \subset \Gamma^D\,.$$

The plan of the paper is as follows. In Sect. 2, we introduce a local projection discretization of (1) and formulate assumptions which will be needed for the error analysis carried out in Sect. 3. Here, in contrast to, e.g., [MST07, MST08, RLL08], we do not construct any special interpolation operator but derive a general error estimate of the type of Strang's lemmas. Then, in Sect. 4, we discuss the choice of the local projection stabilization parameter with respect to the data of (1) based on the data dependence of the estimate from Sect. 3. This discussion reveals that a choice of a stabilization parameter possessing reasonable scaling properties does not allow to obtain optimal convergence results in some cases. Finally, in Sect. 5, we present numerical results illustrating this deterioration of the convergence order and demonstrating some drawbacks of the local projection stabilization in comparison to the SUPG method. Throughout the paper we use standard notation which can be found, e.g., in [Cia91]. Moreover, we use the notation $a \lesssim b$ if $a \leq C\,b$ with $C > 0$ independent of all relevant parameters like mesh size, finite element spaces and the parameter ε.

2 Discrete Problem

The discrete problem we will introduce in this section is based on the standard weak formulation of (1) which reads: Find $u \in H^1(\Omega)$ such that $u = u_b$ on Γ^D and

$$a(u,v) = (f,v) + \langle g,v\rangle_{\Gamma^N} \qquad \forall\ v \in V := \{v \in H^1(\Omega)\,;\ v = 0 \ \text{ on } \ \Gamma^D\}\,,$$

where

$$a(u,v) := \varepsilon\,(\nabla u, \nabla v) + (\mathbf{b} \cdot \nabla u, v) + (c\,u, v)\,,$$

$(\cdot,\cdot)$ denotes the inner product in $L^2(\Omega)$ or $L^2(\Omega)^d$ and $\langle\cdot,\cdot\rangle_{\Gamma^N}$ is the duality pairing between $H^{-1/2}(\Gamma^N)$ and $H^{1/2}(\Gamma^N)$. Since $a(v,v) \geq \varepsilon\,|v|^2_{1,\Omega}$ for any $v \in V$, the weak formulation has a unique solution.

Let $\mathcal{T}_h$ be a triangulation of Ω consisting of closed shape-regular cells K possessing the usual compatibility properties. We assume that all cells of $\mathcal{T}_h$ are of the same type (simplices, quadrilaterals or hexahedra) and are images of a reference cell under a (multi)linear regular mapping. We set $h_K = \mathrm{diam}(K)$ for any $K \in \mathcal{T}_h$ and assume that $h_K \leq h$ for all $K \in \mathcal{T}_h$. We introduce a coarse triangulation $\mathcal{M}_h$ constructed by coarsening the triangulation $\mathcal{T}_h$ such that each macro-element $M \in \mathcal{M}_h$ is the union of one or more neighboring cells $K \in \mathcal{T}_h$. The diameter of $M \in \mathcal{M}_h$ is denoted by h_M. We assume that the decomposition $\mathcal{M}_h$ of Ω is non-overlapping and shape-regular. Additionally, each cell is supposed to be of the same size as the macro-cell it belongs to:

$$\exists C > 0 : \quad h_M \leq C h_K \qquad \forall\, K \in \mathcal{T}_h,\ M \in \mathcal{M}_h \text{ with } K \subset M\,.$$

Using the triangulation $\mathcal{T}_h$, we define a finite element space $W_h \subset H^1(\Omega)$, see, e.g., [Cia91], and we set $V_h = W_h \cap V$. In addition, we introduce a discontinuous finite element space $D_h \subset L^2(\Omega)$ on the macro-partition $\mathcal{M}_h$. We denote by π_h a projection operator which maps $L^2(\Omega)$ onto D_h, resp. $L^2(\Omega)^d$ onto D_h^d, and we define the fluctuation operator $\kappa_h = id - \pi_h$ where id denotes the identity operator on $L^2(\Omega)$, resp. $L^2(\Omega)^d$. For any $M \in \mathcal{M}_h$, we define the local projection stabilization term

$$s_M(u,v) = (\kappa_h(\mathbf{b}\cdot\nabla u), \kappa_h(\mathbf{b}\cdot\nabla v))_M \tag{2}$$

or

$$s_M(u,v) = (\kappa_h \nabla u, \kappa_h \nabla v)_M \tag{3}$$

and we denote

$$s_h(u,v) = \sum_{M\in\mathcal{M}_h} \tau_M\, s_M(u,v),$$

where τ_M is a nonnegative stabilization parameter. Finally, we introduce a function $\widetilde{u}_{bh} \in W_h$ such that its trace approximates the boundary condition u_b.

Now the local projection discretization of (1) reads: Find $u_h \in W_h$ such that $u_h - \widetilde{u}_{bh} \in V_h$ and

$$a_h(u_h, v_h) = (f, v_h) + \langle g, v_h\rangle_{\Gamma^N} \qquad \forall v_h \in V_h\,,$$

where $a_h(u,v) = a(u,v) + s_h(u,v)$.

If we introduce the local projection norm

$$|||v|||_{LP} = \left(\varepsilon\, |v|_{1,\Omega}^2 + \|\sigma^{1/2}\, v\|_{0,\Omega}^2 + \frac{1}{2}\, \|(\mathbf{b}\cdot\mathbf{n})^{1/2}\, v\|_{0,\Gamma^N}^2 + s_h(v,v)\right)^{1/2},$$

then $a_h(v,v) = |||v|||_{LP}^2$ for any $v \in V$ and hence the discrete problem is uniquely solvable. To estimate the error of the discrete solution, we have to make several assumptions on the finite element spaces. First, we assume that, for some positive integer k and $2 \leq l \leq k+1$, we have

$$\inf_{v_h\in V_h} \left\{ \|v - v_h\|_{0,K} + \frac{h_K}{k}\, |v - v_h|_{1,K} \right\} \lesssim \frac{h_K^l}{k^l}\, |v|_{l,K} \quad \forall\ v \in V \cap H^l(\Omega),\ K \in \mathcal{T}_h.$$

Moreover, we assume that the space W_h satisfies the inverse inequality

$$|v_h|_{1,M} \leq \mu_k\, h_M^{-1}\, \|v_h\|_{0,M} \qquad \forall\ v_h \in W_h,\ \ M \in \mathcal{M}_h. \tag{4}$$

If W_h consists of piecewise polynomial functions of degree k, then this inequality holds with $\mu_k \sim k^2$ [Geo08]. For simplicity, we assume that $\mu_k \geq k$. Approximation properties of the space D_h are expressed by the assumption that, for $0 \leq l \leq k$,

$$\|\kappa_h q\|_{0,M} \lesssim \frac{h_M^l}{k^l} |q|_{l,M} \qquad \forall\, q \in L^2(\Omega),\ q|_M \in H^l(M),\ M \in \mathcal{M}_h. \quad (5)$$

Furthermore, we assume that the spaces W_h and D_h satisfy the inf–sup conditions

$$\exists\, \beta > 0 : \qquad \inf_{q_h \in D_h(M)} \sup_{v_h \in Y_h(M)} \frac{(v_h, q_h)_M}{\|v_h\|_{0,M}\, \|q_h\|_{0,M}} \geq \beta \qquad \forall M \in \mathcal{M}_h, \quad (6)$$

where $D_h(M) := \{q_h|_M\,;\ q_h \in D_h\}$ and $Y_h(M) := H_0^1(M) \cap \{v_h|_M\,;\ v_h \in W_h\}$. For suitable pairs of finite element spaces D_h, W_h satisfying the above assumptions we refer to [MST07].

Finally, we assume that

$$\begin{aligned} &\sigma_0 > 0 \quad \text{or} \quad \operatorname{div} \mathbf{b} = 0 \text{ in } \Omega \qquad && \text{if } s_M \text{ are given by (2),} \\ &\sigma_0 > 0 && \text{if } s_M \text{ are given by (3).} \end{aligned}$$

To enable a simultaneous analysis for both definitions of s_M, we set (with $l \in \mathbb{N}_0$)

$$\begin{aligned} &\nu_k = \mu_k, \quad \gamma_{M,l}(\mathbf{b}) = \|\mathbf{b}\|_{l,\infty,M}^2 \quad && \text{if } s_M \text{ are given by (2),} \\ &\nu_k = 1, \quad\ \ \gamma_{M,l}(\mathbf{b}) = 1 && \text{if } s_M \text{ are given by (3).} \end{aligned}$$

3 Error Analysis

Let $u \in H^1(\Omega)$ be the weak solution of (1). The local projection discretization is not consistent and we have $a_h(u - u_h, v_h) = s_h(u, v_h)$ for any $v_h \in V_h$. Denoting $W_h^b = \{w_h \in W_h\,;\ w_h - \widetilde{u}_{bh} \in V_h\}$, we obtain similarly as in the proof of the first Strang lemma (see, e.g., [Cia91])

$$\begin{aligned} |||u - u_h|||_{LP} \leq \inf_{w_h \in W_h^b} &\left\{ |||u - w_h|||_{LP} + \sup_{v_h \in V_h} \frac{a_h(u - w_h, v_h)}{|||v_h|||_{LP}} \right\} \\ &+ \sup_{v_h \in V_h} \frac{s_h(u, v_h)}{|||v_h|||_{LP}}\,. \end{aligned}$$

Lemma 1. *Let $D_h^\perp$ be the orthogonal complement of D_h in $L^2(\Omega)$. Then for any $w \in H^1(\Omega) \cap D_h^\perp$ and any $v_h \in V_h$, we have*

$$a_h(w, v_h) \lesssim \left(\|(\mathbf{b} \cdot \mathbf{n})^{1/2}\, w\|_{0,\Gamma^N}^2 + \sum_{M \in \mathcal{M}_h} C_M\, \|w\|_{1,M,*}^2 \right)^{1/2} |||v_h|||_{LP},$$

where $\|w\|_{1,M,*} = |w|_{1,M} + \mu_k\, h_M^{-1}\, \|w\|_{0,M}$ *and*

$$C_M = \lambda_M + h_M^2\, \nu_k^{-2} + \|\mathbf{b}\|_{0,\infty,M}^2\, h_M^2\, \mu_k^{-2}\, \lambda_M^{-1}\,, \qquad \lambda_M = \max\{\varepsilon, \tau_M\, \gamma_{M,0}(\mathbf{b})\}\,.$$

Proof. Consider any $w \in H^1(\Omega) \cap D_h^\perp$ and $v_h \in V_h$. Then

$$(\mathbf{b}\cdot\nabla w, v_h) + (c\, w, v_h) = -(w, \mathbf{b}\cdot\nabla v_h) + (\sigma\, w, v_h) - \tfrac{1}{2}((\operatorname{div}\mathbf{b})\, w, v_h) + \langle(\mathbf{b}\cdot\mathbf{n})\, w, v_h\rangle_{\Gamma^N}\,.$$

Furthermore, for any $M \in \mathcal{M}_h$, we derive

$$(w, \mathbf{b}\cdot\nabla v_h)_M \leq \|\mathbf{b}\|_{0,\infty,M}\, \|w\|_{0,M}\, |v_h|_{1,M}\,.$$

If s_M are defined by (2), we may also estimate

$$(w, \mathbf{b}\cdot\nabla v_h)_M = (w, \kappa_h(\mathbf{b}\cdot\nabla v_h))_M \leq \|w\|_{0,M}\, \|\kappa_h(\mathbf{b}\cdot\nabla v_h)\|_{0,M}\,.$$

If s_M are defined by (3), we define $\mathbf{b}_M = (1, \mathbf{b})_M/|M|$ and obtain

$$\begin{aligned}(w, \mathbf{b}\cdot\nabla v_h)_M &= (w, (\mathbf{b}-\mathbf{b}_M)\cdot\nabla v_h)_M + (w, \kappa_h(\mathbf{b}_M\cdot\nabla v_h))_M \\ &\lesssim \mu_k\, |\mathbf{b}|_{1,\infty,M}\, \|w\|_{0,M}\, \|v_h\|_{0,M} + \|\mathbf{b}\|_{0,\infty,M}\, \|w\|_{0,M}\, \|\kappa_h \nabla v_h\|_{0,M}\,.\end{aligned}$$

Therefore, for both definitions of s_M, we have

$$\begin{aligned}(w, \mathbf{b}\cdot\nabla v_h)_M &\lesssim \mu_k\, \|w\|_{0,M}\, \|\sigma^{1/2} v_h\|_{0,M} \\ &\quad + \|\mathbf{b}\|_{0,\infty,M}\, \lambda_M^{-1/2}\, \|w\|_{0,M}\, \big(\varepsilon\, |v_h|_{1,M}^2 + \tau_M\, s_M(v_h, v_h)\big)^{1/2}\end{aligned}$$

(the first term on the right-hand side can be dropped for s_M defined by (2)). Furthermore, we have

$$\begin{aligned}&(\sigma\, w, v_h)_M - \tfrac{1}{2}((\operatorname{div}\mathbf{b})\, w, v_h)_M \lesssim \|w\|_{0,M}\, \|\sigma^{1/2} v_h\|_{0,M}\,, \\ &\varepsilon\, (\nabla w, \nabla v_h)_M + \tau_M\, s_M(w, v_h) \lesssim \lambda_M^{1/2}\, |w|_{1,M}\, \big(\varepsilon\, |v_h|_{1,M}^2 + \tau_M\, s_M(v_h, v_h)\big)^{1/2}\,.\end{aligned}$$

Thus, in all the above inequalities, the right-hand sides can be estimated by

$$\big(\lambda_M\, |w|_{1,M}^2 + (\mu_k^2\, \nu_k^{-2} + \|\mathbf{b}\|_{0,\infty,M}^2\, \lambda_M^{-1})\, \|w\|_{0,M}^2\big)^{1/2} \times \Big(\varepsilon\, |v_h|_{1,M}^2 + \|\sigma^{1/2} v_h\|_{0,M}^2 + \tau_M\, s_M(v_h, v_h)\Big)^{1/2}\,,$$

which leads to the desired estimate. □

It is easy to show that, for any $w \in H^1(\Omega)$,

$$|||w|||_{LP} \lesssim \left(\|(\mathbf{b}\cdot\mathbf{n})^{1/2}\, w\|_{0,\Gamma^N}^2 + \sum_{M\in\mathcal{M}_h} (\lambda_M + h_M^2\, \nu_k^{-2}) \|w\|_{1,M,*}^2 \right)^{1/2}.$$

Therefore, denoting $W_h^b(u) = \{w_h \in W_h^b;\ u - w_h \in D_h^\perp\}$, it follows from Lemma 1 and the estimate before Lemma 1 that

$$|||u-u_h|||_{LP} \lesssim \inf_{w_h\in W_h^b(u)} \left(\|u-w_h\|^2_{0,\Gamma^N} + \sum_{M\in\mathcal{M}_h} C_M \|u-w_h\|^2_{1,M,*} \right)^{1/2} + \sup_{v_h\in V_h} \frac{s_h(u,v_h)}{|||v_h|||_{LP}}. \tag{7}$$

Lemma 2. *For any $w \in H^1(\Omega)$ there exists $z_h \in W_h \cap H_0^1(\Omega)$ such that $w - z_h \in D_h^\perp$ and $\|w-z_h\|_{1,M,*} \le (1+2/\beta)\|w\|_{1,M,*}$ for any $M \in \mathcal{M}_h$.*

Proof. Consider any $w \in H^1(\Omega)$ and $M \in \mathcal{M}_h$. The inf–sup conditions (6) imply that there exists $z_M \in Y_h(M)$ such that $(w - z_M, q_h)_M = 0\ \forall\, q_h \in D_h(M)$ and $\beta\|z_M\|_{0,M} \le \|w\|_{0,M}$. Since $|z_M|_{1,M} \le \mu_k\, h_M^{-1}\|z_M\|_{0,M}$, we obtain the lemma. □

In view of Lemma 2, the estimate (7) can be replaced by

$$|||u-u_h|||_{LP} \lesssim \inf_{w_h\in W_h^b} \left(\|u-w_h\|^2_{0,\Gamma^N} + \left(1+\frac{1}{\beta}\right)^2 \sum_{M\in\mathcal{M}_h} C_M \|u-w_h\|^2_{1,M,*} \right)^{1/2} + \sup_{v_h\in V_h} \frac{s_h(u,v_h)}{|||v_h|||_{LP}}.$$

It remains to estimate the consistency error.

Lemma 3. *Let $u \in H^{l+1}(M)$ for some $l \in \{0,\dots,k\}$ and for all $M \in \mathcal{M}_h$. If s_M are defined by* (2), *let $\mathbf{b}|_M \in W^{l,\infty}(M)^d$ for all $M \in \mathcal{M}_h$. Then*

$$s_h(u,v_h) \lesssim \left(\sum_{M\in\mathcal{M}_h} C_M^s\, \tau_M\, \gamma_{M,l}(\mathbf{b})\, \frac{h_M^{2l}}{k^{2l}} \|u\|^2_{l+1,M} \right)^{1/2} |||v_h|||_{LP} \qquad \forall v_h \in V_h$$

with

$$C_M^s = \min\left\{1, \frac{\tau_M\, \gamma_{M,0}(\mathbf{b})\, \mu_k^2}{\sigma_0\, h_M^2}\right\}.$$

Proof. Consider any $M \in \mathcal{M}_h$ and $v_h \in V_h$. Then the Cauchy–Schwarz inequality and (5) yield $s_M(u,v_h) \lesssim h_M^l\, k^{-l}\, \|u\|_{l+1,M}\, [\gamma_{M,l}(\mathbf{b})\, s_M(v_h,v_h)]^{1/2}$. Furthermore, we deduce using the L^2 stability of κ_h in (5) and the inverse inequality (4) that $s_M(v_h,v_h) \lesssim \gamma_{M,0}(\mathbf{b})\, \mu_k^2\, h_M^{-2}\, \|v_h\|^2_{0,M}$. Thus,

$$\tau_M\, s_M(u,v_h) \lesssim [C_M^s\, \tau_M\, \gamma_{M,l}(\mathbf{b})]^{1/2} \frac{h_M^l}{k^l} \|u\|_{l+1,M} \left(\|\sigma^{1/2} v_h\|^2_{0,M} + \tau_M\, s_M(v_h,v_h) \right)^{1/2},$$

which proves the lemma. □

The above estimates lead to the following result:

Theorem 1. *Let all assumptions made in Sects. 1 and 2 be satisfied and let the approximation* $\widetilde{u}_{bh}$ *of the Dirichlet boundary condition be sufficiently accurate,* $u \in H^{l+1}(\Omega)$ *for some* $l \in \{1, \dots, k\}$ *and, in case of* s_M *given by (2), let* $\mathbf{b}|_M \in W^{l,\infty}(M)^d$ *for all* $M \in \mathcal{M}_h$*. Then the solution of the local projection discretization satisfies the error estimate*

$$|||u - u_h|||_{LP} \lesssim \frac{h^{l+1/2}}{k^l} \|u\|_{l+1,\Omega} + \left(1 + \frac{1}{\beta}\right) \left(\sum_{M \in \mathcal{M}_h} C_M \frac{\mu_k^2}{k^2} \frac{h_M^{2l}}{k^{2l}} \|u\|_{l+1,M}^2 \right)^{1/2}$$

$$+ \left(\sum_{M \in \mathcal{M}_h} C_M^s \, \tau_M \, \gamma_{M,l}(\mathbf{b}) \frac{h_M^{2l}}{k^{2l}} \|u\|_{l+1,M}^2 \right)^{1/2} .$$

4 Choice of the Parameter τ_M

The estimate from Theorem 1 considerably depends on the choice of the parameter τ_M. Some authors simply set $\tau_M \sim h_M$, which leads to an optimal error estimate with respect to h. However, such definition of τ_M is not reasonable from the practical point of view since the parameter τ_M should possess certain scaling properties with respect to the solved problem. For example, if the data ε, $\mathbf{b}$, c, f and g are multiplied by a positive number α, the solution of (1) does not change. This property should be preserved by the discrete problem but this requires that τ_M changes to τ_M/α if s_M is given by (2) and to $\tau_M \alpha$ if s_M is given by (3).

A possible way to derive a formula for τ_M is to balance the influence of the two terms depending on τ_M in the definition of C_M. Let us denote $\xi_M = \tau_M \, \gamma_{M,0}(\mathbf{b})$ and $\eta_M = \|\mathbf{b}\|_{0,\infty,M} \, h_M/\mu_k$. Then $C_M = \lambda_M + h_M^2 \, \nu_k^{-2} + \eta_M^2 \, \lambda_M^{-1}$ with $\lambda_M = \max\{\varepsilon, \xi_M\}$. If $\varepsilon \le \xi_M$, we have $C_M = \xi_M + h_M^2 \, \nu_k^{-2} + \eta_M^2/\xi_M$, which suggests to set $\xi_M \sim \eta_M$. If $\varepsilon \ge \xi_M$, we have $C_M = \varepsilon + h_M^2 \, \nu_k^{-2} + \eta_M^2/\varepsilon \le \varepsilon + h_M^2 \, \nu_k^{-2} + \eta_M^2/\xi_M$, which suggests to set $\xi_M \sim \eta_M^2/\varepsilon$. Thus, we may set $\xi_M = \min\{\eta_M, \eta_M^2/\varepsilon\}$. Then really $\varepsilon \le \xi_M$ if and only if $\xi_M = \eta_M$, and $\varepsilon \ge \xi_M$ if and only if $\xi_M = \eta_M^2/\varepsilon$. Therefore, $C_M \le \varepsilon + h_M^2 \, \nu_k^{-2} + 2\,\eta_M$. Returning to the previous notation, we come to the formula

$$\tau_M \sim \min\left\{ \frac{h_M}{\mu_k \, \|\mathbf{b}\|_{0,\infty,M}}, \frac{h_M^2}{\varepsilon \, \mu_k^2} \right\} \frac{\|\mathbf{b}\|_{0,\infty,M}^2}{\gamma_{M,0}(\mathbf{b})} . \tag{8}$$

If s_M are given by (3), Theorem 1 and relation (8) imply that

$$|||u - u_h|||_{LP} \lesssim \frac{h^{l+1/2}}{k^l} \|u\|_{l+1,\Omega} + \frac{h^l}{k^l} \frac{\mu_k}{k} \left(1 + \frac{1}{\beta}\right) \left(\varepsilon^{1/2} + h + \frac{h^{1/2}}{\mu_k^{1/2}} \right) \|u\|_{l+1,\Omega} .$$

Thus, if $\varepsilon < h$, we have the optimal convergence order $l + \frac{1}{2}$ with respect to h. On the other hand, if s_M are given by (2), the consistency error with τ_M from (8) may significantly deteriorate the convergence order. More precisely, we obtain

$$|||u - u_h|||_{LP} \lesssim \frac{h^{l+1/2}}{k^l} \|u\|_{l+1,\Omega} + \frac{h^l}{k^l} \frac{\mu_k}{k} \left(1 + \frac{1}{\beta}\right) \left(\varepsilon^{1/2} + \frac{h^{1/2}}{\mu_k^{1/2}}\right) \|u\|_{l+1,\Omega}$$
$$+ \left(\sum_{M \in \mathcal{M}_h} \min \left\{ \frac{h_M \, \|\mathbf{b}\|^2_{l,\infty,M}}{\mu_k \, \|\mathbf{b}\|_{0,\infty,M}}, \frac{\|\mathbf{b}\|^2_{l,\infty,M}}{\sigma_0} \right\} \frac{h_M^{2l}}{k^{2l}} \|u\|^2_{l+1,M} \right)^{1/2} . \quad (9)$$

The convergence order with respect to h of the last term on the right-hand side of (9) may be significantly smaller than for the other two terms. Let us demonstrate this for $\Omega = (0,1)^d$, $\sigma_0 = 0$, $b_2 = x_1^l$ and $b_i = 0$ for $i \neq 2$. We shall assume that $u \in C^{l+1}(\overline{\Omega})$ and $\sum_{|\alpha| \le l+1} |D^\alpha u| > 0$ in $\overline{\Omega}$. Moreover, we confine ourselves to meshes $\mathcal{M}_h$ consisting of N^d equal d-cubes, set $H = 1/N$ and assume that $H \gtrsim h$. Then the last term on the right-hand side of (9) can be estimated *from below* by

$$\left(\sum_{M \in \mathcal{M}_h} \frac{h_M^{2l+1} \, |\mathbf{b}|^2_{l,\infty,M}}{\mu_k \, k^{2l} \, \|\mathbf{b}\|_{0,\infty,M}} \|u\|^2_{l+1,M} \right)^{1/2} \gtrsim \left(\sum_{M \in \mathcal{M}_h} \frac{h^{2l+1} \, |M|}{\mu_k \, k^{2l} \, \|b_2\|_{0,\infty,M}} \right)^{1/2}$$
$$\geq \frac{h^{l+1/2}}{\mu_k^{1/2} \, k^l} \left(\int_H^1 \frac{1}{x_1^l} \, \mathrm{d}x_1 \right)^{1/2} \gtrsim \frac{h^{1+l/2}}{\mu_k^{1/2} \, k^l} \qquad \text{for } l \geq 2 . \quad (10)$$

A general estimate *from above* has not been established yet so that, for suitable data, the convergence might be even slower. Note, however, that (9) implies an optimal error estimate with respect to h if $\mathbf{b} \neq \mathbf{0}$ in $\overline{\Omega}$. If this is not the case but $\sigma_0 > 0$, then in general we obtain only the suboptimal convergence order l with respect to h. The discussed theoretical results have been also confirmed by numerical experiments.

The derivation of the formula (8) for τ_M was based on balancing the terms in the definition of C_M. If we take into account the consistency error as well and s_M are given by (2), we may come to the formula

$$\tau_M \sim \min \left\{ \frac{h_M}{\mu_k \, \|\mathbf{b}\|_{l,\infty,M}}, \frac{h_M^2}{\varepsilon \, \mu_k^2} \right\} . \quad (11)$$

This leads to the error estimate

$$|||u - u_h|||_{LP} \lesssim \frac{h^{l+1/2}}{k^l} \|u\|_{l+1,\Omega} + \frac{h^l}{k^l} \frac{\mu_k}{k} \left(1 + \frac{1}{\beta}\right) \left(\varepsilon^{1/2} + \frac{h^{1/2}}{\mu_k^{1/2}}\right) \|u\|_{l+1,\Omega} .$$

Although this estimate is optimal with respect to h, we do not think that the formula (11) is a good choice. First, the norm $\|\mathbf{b}\|_{l,\infty,M}$ is not convenient from the implementational point of view and second, which is more important, the various derivatives in the definition of this norm scale in different ways if the size of the computational domain is changed. Consequently, if the problem (1) is transformed into dimensionless variables before assembling the discrete problem, the discrete solution transformed back to Ω depends on the definition of the characteristic length. In other words, if a definition of τ_M based on (11) is optimal for a given problem, a rescaling of the space variable will generally cause a loss of this optimality. This unacceptable behaviour does not occur if the formula (8) is considered.

5 Numerical Results

Let us first consider the following example illustrating the deterioration of the convergence order discussed at the end of the previous section.

Example 1. We consider the problem (1) with $\Omega = (0,1)^2$, $\Gamma^D = \partial\Omega$, $\Gamma^N = \emptyset$, $\varepsilon = 10^{-12}$, $\mathbf{b}(x,y) = (0,x^2)$ and $c = 0$. The functions f and u_b are such that the solution of (1) is $u(x,y) = \sin(x+y)$.

The triangulations $\mathcal{T}_h$ are constructed by dividing Ω into equal squares and by cutting each square along the diagonals into four triangles. The space D_h consists of discontinuous piecewise linear functions on $\mathcal{M}_h = \mathcal{T}_h$ and the space W_h of continuous piecewise quadratic functions enriched elementwise by three quartic bubble functions, see [MST07] for details. The projection operator π_h defining the fluctuation operator in (2) and (3) is the orthogonal L^2 projection. The stabilization parameter is defined simply by the right-hand side of (8) with $\mu_k = 1$.

Tables 1 and 2 show errors of the discrete solutions computed using the local projection method defined by (2) and (3), respectively, for various values of h. The errors are measured in the (semi)norms $|||\cdot|||_{LP}$, $\|\cdot\|_{0,\Omega}$, $|\cdot|_{1,\Omega}$ and $\|\cdot\|_{0,\infty,h}$ where the discrete L^∞ norm $\|\cdot\|_{0,\infty,h}$ is defined as the maximum absolute value at vertices of the triangulation. The convergence orders are computed from the errors for the two finest meshes. The notation $r{-}n$ used in the tables means $r \cdot 10^{-n}$. We observe that, if s_M are given by (2), all convergence orders are suboptimal and the convergence order in the local projection norm is in agreement with the estimate (10). If s_M are given by (3), the accuracy is much higher and the convergence orders are nearly optimal although the assumption $\sigma_0 > 0$ is not satisfied. Nevertheless, our numerical tests show that, in most cases, both variants (2) and (3) of the local projection stabilization lead to comparable results, see [KL08].

Example 2. We consider the problem (1) with $\Omega = (0,1)^2$, $\Gamma^D = \partial\Omega$, $\Gamma^N = \emptyset$, $\varepsilon = 10^{-8}$, $\mathbf{b} = (1,0)$, $c = 0$, $f = 1$ and $u_b = 0$.

The solution $u(x,y)$ possesses an exponential boundary layer at $x = 1$ and parabolic boundary layers at $y = 0$ and $y = 1$. Away from the layers, the solution $u(x,y)$ is very close to x. The triangulation $\mathcal{T}_h$ consists of 32×32 equal

Table 1 Example 1, errors for the local projection method defined by (2)

h	$\|\|\|\cdot\|\|\|_{LP}$	$\|\|\cdot\|\|_{0,\Omega}$	$\|\cdot\|_{1,\Omega}$	$\|\|\cdot\|\|_{0,\infty,h}$
6.25−2	6.68−5	7.08−4	8.57−2	6.06−3
3.13−2	1.65−5	2.57−4	6.27−2	3.11−3
1.56−2	4.12−6	9.18−5	4.49−2	1.57−3
7.81−3	1.04−6	3.24−5	3.18−2	7.85−4
conv. order	1.99	1.50	0.50	1.00

Table 2 Example 1, errors for the local projection method defined by (3)

h	$\|\|\|\cdot\|\|\|_{LP}$	$\|\|\cdot\|\|_{0,\Omega}$	$\|\cdot\|_{1,\Omega}$	$\|\|\cdot\|\|_{0,\infty,h}$
6.25−2	1.19−5	5.59−5	6.73−3	3.27−4
3.13−2	2.06−6	6.20−6	1.45−3	4.28−5
1.56−2	3.63−7	7.42−7	3.33−4	5.59−6
7.81−3	6.47−8	9.36−8	8.09−5	7.21−7
conv. order	2.49	2.99	2.04	2.96

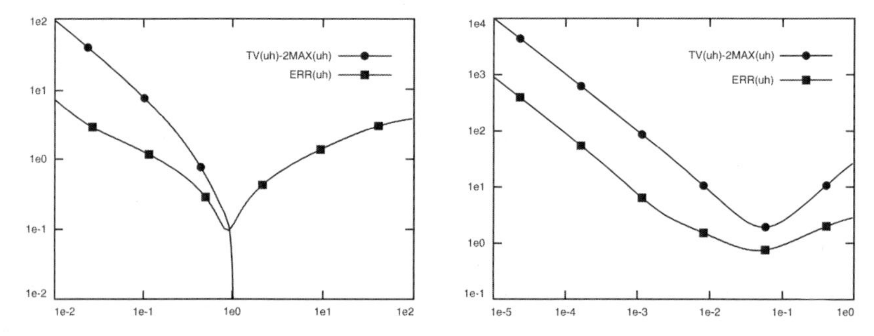

Fig. 1 Dependence of the total variation and error on the stabilization parameter for the SUPG method (*left*) and a local projection method (*right*)

squares and we set $\mathcal{M}_h = \mathcal{T}_h$. The space D_h consists of discontinuous piecewise linear functions and the space W_h of continuous piecewise biquadratic functions enriched elementwise by two bicubic bubble functions, see [MST07] for details. We shall also consider the SUPG method [BH82] with a space of continuous piecewise biquadratic functions.

We shall investigate the capability of the methods to remove spurious oscillations along the exponential boundary layer. For this we shall compute the discrete total variation $TV(u_h)$ of u_h along $y = 0.5$ using the values at vertices and midpoints of edges. Further, we compute the discrete maximum value $MAX(u_h)$ of u_h using the same values of u_h as before. Then u_h possesses no spurious oscillations along $y = 0.5$ if and only if $TV(u_h) = 2\,MAX(u_h)$. We shall also compute the error $ERR(u_h)$ of u_h as the l_2 norm of errors $u - u_h$ at vertices and midpoints of edges along $y = 0.5$.

In the left part of Fig. 1, we see the dependence of $TV(u_h) - 2\,MAX(u_h)$ and $ERR(u_h)$ on the stabilization parameter for the SUPG method. The values on the horizontal axis represent the ratio of the parameter and a theoretical 'optimal' value which is 1/128 in the present case. We observe that the spurious oscillations are not present for sufficiently large parameters. In the region without oscillations the error increases with increasing parameter since the discrete solution is more smeared. The left part of Fig. 1 is typical for SUPG approximations of exponential boundary layers whereas, for characteristic layers, large values of the stabilization parameter generally do not lead to a suppression of spurious oscillations in SUPG solutions.

In the right part of Fig. 1, results for the local projection method with s_M defined by (3) are depicted. The horizontal axis shows the ratio τ_M / h_M. Very similar results are obtained also for s_M defined by (2) and for any finite element spaces W_h, D_h of second order accuracy discussed in [MST07]. We see that the local projection method is not able to suppress the spurious oscillations sufficiently. Moreover, it is not easy to assure the highest possible suppression of spurious oscillations since the oscillations increase for τ_M both smaller and larger than the optimal value.

Thus, we can conclude that the local projection stabilization (which acts only on the fine scales of the discrete solution) assures stability and (optimal) convergence, even with respect to the SUPG norm [KT08], but the stabilization is too weak to suppress spurious oscillations sufficiently. Nevertheless, the oscillations are much smaller than in the Galerkin solution and they are localized in layer regions, which is a common feature with the SUPG method. A possible improvement of the local projection method could be achieved by introducing additional stabilization of the coarse scales of the discrete solution.

Acknowledgement This work is a part of the research project MSM 0021620839 financed by MSMT and it was partly supported by the Grant Agency of the Academy of Sciences of the Czech Republic under the grant No. IAA100190804.

References

[BB01] R. Becker and M. Braack. A finite element pressure gradient stabilization for the Stokes equations based on local projections. *Calcolo*, 38:173–199, 2001.

[BB06] M. Braack and E. Burman. Local projection stabilization for the Oseen problem and its interpretation as a variational multiscale method. *SIAM J. Numer. Anal.*, 43:2544–2566, 2006.

[BH82] A.N. Brooks and T.J.R. Hughes. Streamline upwind/Petrov-Galerkin formulations for convection dominated flows with particular emphasis on the incompressible Navier-Stokes equations. *Comput. Methods Appl. Mech. Eng.*, 32:199–259, 1982.

[BR06] M. Braack and T. Richter. Solutions of 3D Navier-Stokes benchmark problems with adaptive finite elements. *Comput. Fluids*, 35:372–392, 2006.

[Cia91] P.G. Ciarlet. Basic error estimates for elliptic problems. In P.G. Ciarlet and J.L. Lions, editors, *Handbook of Numerical Analysis, vol. 2 – Finite Element Methods (pt. 1)*, pages 17–351. North-Holland, Amsterdam, 1991.

[Geo08] E.H. Georgoulis. Inverse-type estimates on hp-finite element spaces and applications. *Math. Comput.*, 77:201–219, 2008.

[KL08] P. Knobloch and G. Lube. Local projection stabilization for advection–diffusion–reaction problems: one–level vs. two–level approach. 2008. Submitted.

[KT08] P. Knobloch and L. Tobiska. On the stability of finite element discretizations of convection–diffusion–reaction equations. Preprint 08–11, Otto von Guericke University, Magdeburg, 2008.

[MST07] G. Matthies, P. Skrzypacz, and L. Tobiska. A unified convergence analysis for local projection stabilizations applied to the Oseen problem. M^2AN, 41:713–742, 2007.

[MST08] G. Matthies, P. Skrzypacz, and L. Tobiska. Stabilization of local projection type applied to convection-diffusion problems with mixed boundary conditions. *Electron. Trans. Numer. Anal.*, 32:90–105, 2008.

[RLL08] G. Rapin, G. Lube, and J. Löwe. Applying local projection stabilization to inf-sup stable elements. In K. Kunisch, G. Of, and O. Steinbach, editors, *Numerical Mathematics and Advanced Applications*, pages 521–528. Springer, Berlin, 2008.

A Locally Adapting Parameter Design for the Divergence Stabilization of FEM Discretizations of the Navier–Stokes Equations

J. Löwe

Abstract We will first briefly summarize the previous efforts in constructing a parameter design for local projection and grad-div stabilization based on a-priori convergence analysis for the linearized problem given in [LRL08] and [MT07]. Especially for Taylor-Hood type elements this leads to a grad-div stabilization parameter $\mu \sim 1$. While this design works well for some academic testproblems it does not give satisfactory results for others. A review of the convergence estimate suggests an a-posteriori parameter design including local norms of velocity and pressure. Some first numerical results based on this parameter design will be presented.

1 Introduction

Consider the non-dimensional, unsteady, incompressible Navier–Stokes equations:

$$\begin{aligned} \partial_t \mathbf{u} - Re^{-1}\Delta\mathbf{u} + (\mathbf{u}\cdot\nabla)\mathbf{u} + \nabla p &= \tilde{\mathbf{f}} \qquad \text{in } \Omega\times(0,T) \\ \nabla\cdot\mathbf{u} &= 0 \qquad \text{in } \Omega\times(0,T) \end{aligned} \tag{1}$$

in the primitive variables velocity $\mathbf{u}$ and pressure p in a bounded, polyhedral domain $\Omega \subset \mathbb{R}^d$, $d = 2, 3$ and with given source term $\tilde{\mathbf{f}}$. The dimensionless Reynolds number is given by $Re = \frac{UL}{\nu}$ with U and L being a characteristic velocity and length, respectively, and ν the kinematic viscosity.

A standard approach for solving (1) is to apply a semi-discretization in time with an implicit A-stable scheme first and then to linearize the problem with a fixed point

J. Löwe
Institute for Numerical and Applied Mathematics, Georg-August-University of Göttingen, D-37083 Göttingen, Germany, E-mail: loewe@math.uni-goettingen.de

A.F. Hegarty et al. (eds.), *BAIL 2008 - Boundary and Interior Layers.* Lecture Notes in Computational Science and Engineering,
DOI: 10.1007/978-3-642-00605-0,

or Newton-type method. The fixed point iteration leads to a series of Oseen-type problems:

$$-Re^{-1}\Delta\mathbf{u} + (\mathbf{b}\cdot\nabla)\mathbf{u} + \sigma\mathbf{u} + \nabla p = \mathbf{f} \qquad \text{in } \Omega$$
$$\nabla\cdot\mathbf{u} = 0 \qquad \text{in } \Omega.$$

We consider σ to be constant and proportional to the inverse of the chosen timestep size and $\mathbf{b} \in \mathbf{H}_{div}(\Omega) \cap \mathbf{L}^\infty(\Omega)$ with $\sigma - \frac{1}{2}\nabla\cdot\mathbf{b} \geq \sigma_0 \geq 0$ almost everywhere. For simplicity we impose homogeneous Dirichlet boundary conditions $\mathbf{u} = \mathbf{0}$ on $\partial\Omega$.

The appropriate solution space for the continuous problem is

$$(\mathbf{u}, p) \in \mathbf{V}\times Q := \left[H_0^1(\Omega)\right]^d \times L_0^2(\Omega).$$

The weak formulation for the Oseen problem then reads

$$\text{Find } U = (\mathbf{u}, p) \in \mathbf{V}\times Q \text{ s.t.}$$
$$A(U,V) = (\mathbf{f},\mathbf{v}) \qquad \forall V = (\mathbf{v}, q) \in \mathbf{V}\times Q$$

with the bilinear form

$$A(U,V) := Re^{-1}(\nabla\mathbf{u},\nabla\mathbf{v}) + ((\mathbf{b}\cdot\nabla)\mathbf{u} + \sigma\mathbf{u},\mathbf{v}) - (\nabla\cdot\mathbf{v}, p) + (\nabla\cdot\mathbf{u}, q),$$

where $(\,,\,)$ denotes the inner product on $L^2(\Omega)$ or $[L^2(\Omega)]^d$.

As a spatial discretization we consider quadrilateral ($d = 2$) and hexahedral elements ($d = 3$) and require a shape-regular triangulation $\mathcal{T}_h$. Let F_K be the mapping from the reference cell $\hat{K}$ to real cell K and let $\mathbb{Q}_r$ be the space of tensor polynomials, i.e. polynomials of maximum degree r in each coordinate direction. Then we can define the mapped finite element space

$$Y_{r,h} = \{v \in C(\overline{\Omega}) \mid v|_K \circ F_K \in \mathbb{Q}_r(\hat{K}) \quad \forall K \in \mathcal{T}_h\}\,.$$

We choose the discrete ansatz spaces $\mathbf{V}_h = \left[Y_{s,h}\right]^d \cap \mathbf{V}$ and $Q_h = Q_{t,h} \cap Q$ for velocity and pressure with polynomial degrees s and t, respectively.

2 The Local Projection Stabilization Framework

The standard Galerkin approximation with finite elements suffers from two problems. On the one hand the case $Re \gg 1$ gives raise to spurious oscillations in the velocity component of the solution due to dominating advection and poor mass conservation; on the other hand, a pressure instability occurs for spaces that do not satisfy the discrete inf-sup condition.

A widespread framework to deal with all these problems is the residual based stabilization. Especially the combination of *Streamline-Upwind/Petrov-Galerkin (SUPG)* and *Pressure-Stabilization/Petrov-Galerkin (PSPG)* is often used, sometimes supplemented with *Grad-Div stabilization*, see [BBJL07] and references therein.

The class of residual based methods has several drawbacks. For example the SUPG and PSPG methods are non-symmetric and introduce additional coupling terms between velocity and pressure. These create some difficulties in the analysis and lead to upper bounds on the stabilization parameters in order to prove the stability of the method.

As a remedy for the drawbacks of the class of residual based methods several symmetric stabilization methods have been proposed. They all have in common that they add a symmetric, positive semi-definite bilinear form S_h to the original weak formulation of the problem.

The stabilized variational formulation is then given by:

$$\text{Find } U_h = (\mathbf{u}_h, p_h) \in \mathbf{V}_h \times Q_h \text{ s.t.}$$
$$(A + S_h)(U_h, V_h) = (\mathbf{f}, \mathbf{v}_h) \qquad \forall V_h = (\mathbf{v}_h, q_h) \in \mathbf{V}_h \times Q_h.$$

There are several ways to define the penalty term S_h, see [BBJL07]. Here we will focus on the local projection stabilization (LPS) following the framework introduced in [MST07]. The idea of LPS is to penalize only the small scales of the quantities of interest defined by some fluctuation operator.

Let V_H/Q_H be a pair of scalar and discontinuous coarse spaces on a suitable macro triangulation $\mathcal{M}_h$ and let $\pi^{v/q} : L^2(\Omega) \to V_H/Q_H$ be the local L^2-projections into the coarse spaces. Then we can define the fluctuation operators

$$\kappa^{v/q} := id - \pi^{v/q} : L^2(\Omega) \to L^2(\Omega).$$

We will use boldface notation $\boldsymbol{\kappa}^v$ if we apply the operator component-wise. The stabilizing bilinear form S_h can then be defined as

$$S_h(U, V) := \sum_{M \in \mathcal{M}_h} \tau_M \big(\boldsymbol{\kappa}^v((\nabla \cdot \mathbf{b})\mathbf{u}), (\nabla \cdot \mathbf{b})\mathbf{v}\big)_M$$
$$+ \sum_{M \in \mathcal{M}_h} \mu_M \big(\kappa^q(\nabla \cdot \mathbf{u}), \nabla \cdot \mathbf{v}\big)_M + \sum_{M \in \mathcal{M}_h} \alpha_M \big(\boldsymbol{\kappa}^v(\nabla p), \nabla q\big)_M.$$

It contains penalty terms for the fluctuations of the streamline derivative and divergence of the velocity and the pressure gradients, weighted element-wise by user chosen parameters τ_M, μ_M and α_M. Other variants that stabilize fluctuations of the full gradient of the velocity are possible.

3 Parameter Design

Two typically used conforming spatial discretizations are the family of Taylor-Hood elements (TH, $s = t + 1$) and approximations with equal order for velocity and pressure (EO, $s = t$). The coarse spaces are chosen in a so called two-level manner, where $\mathcal{T}_h$ is a suitable global refinement of $\mathcal{M}_h := \mathcal{T}_{2h}$. The full a-priori analysis on stability and error estimates for this method can be found in [LRL08] and [MT07].

Under the assumptions given there one can derive the following estimate for the error between the continuous solution $U = (\mathbf{u}, p)$ and the discrete solution $U_h = (\mathbf{u}_h, p_h)$ in the stabilized energy norm:

$$|||U - U_h|||_{LP}^2 \leq C \sum_{M \in \mathcal{M}_h} \Big(\tau_M h_M^{2s} |(\mathbf{b} \cdot \nabla)\mathbf{u}|_{s,\omega_M}^2 + C_M^u h_M^{2s} |\mathbf{u}|_{s+1,\omega_M}^2 + C_M^p h_M^{2t} |p|_{t+1,\omega_M}^2 \Big) \tag{2}$$

where ω_M denotes a certain neighborhood of the macro element and

$$C_M^u := Re^{-1} + h_M^2(\sigma + \tau_M^{-1} + \alpha_M^{-1}) + \mu_M + \tau_M \|\mathbf{b}\|_{\infty,M}^2,$$
$$C_M^p := \alpha_M + \mu_M^{-1} h_M^2.$$

The energy norm itself is given by:

$$|||(\mathbf{v}, q)|||_{LP}^2 = Re^{-1}|\mathbf{v}|_1^2 + \sigma_0 \|\mathbf{v}\|_0^2 + \delta \|q\|_0^2 + S_h(\mathbf{v}, q; \mathbf{v}, q).$$

In order to get asymptotically optimal rates of convergence, the stabilization parameters must satisfy a certain scaling with respect to h_M given in Table 1. These parameter designs are based on the assumption $|\mathbf{u}|_{k+1,M} \sim |p|_{k,M}$ and obtained by balancing the parameter dependent terms in the a-priori error estimate (2) in order to minimize the upper bound on the error.

For the Taylor-Hood element the divergence parameter μ_M is notably conspicuous because it is of order 1 and might dominate the whole PDE. In [OR04], where the grad-div stabilization for the Stokes problem is analyzed, it is remarked, that the larger the norm of the pressure is compared to the norm of the velocity, the more important the divergence stabilization is. We propose that balancing the μ_M–dependent terms should include the local norms of $\mathbf{u}$ and p because there may be large differences in the scaling of both. Following this approach gives:

Table 1 Selected space combinations with parameter scaling ($Re^{-1} < h_M$)

	V_h	Q_h	V_H	Q_H	τ_M	μ_M	α_M	error
TH	$Y_{k,h}$	$Y_{k-1,h}$	$Y_{k-1,2h}^{\text{disc}}$	$\{0\}$	$\sim h_M$	~ 1	0	$\mathcal{O}(h_M^k)$
EO	$Y_{k,h}$	$Y_{k,h}$	$Y_{k-1,2h}^{\text{disc}}$	$\{0\}$	$\sim h_M$	$\sim h_M$	$\sim h_M$	$\mathcal{O}\left(h_M^{k+1/2}\right)$

$$\mu_M \, |\mathbf{u}|_{k+1,\omega_M}^2 \sim \mu_M^{-1} \, |p|_{k,\omega_M}^2 \quad \Longrightarrow \quad \mu_M \sim \frac{|p|_{k,\omega_M}}{|\mathbf{u}|_{k+1,\omega_M}}.$$

Since the solution $(\mathbf{u}, p)$ is generally unknown, these norms must be replaced by norms of the discrete solution $(\mathbf{u}_h, p_h)$. This leads to a local and nonlinear parameter design. We should further note, that it may be difficult to recover approximations of the high order derivatives from the discrete solution to evaluate the norms for large k.

4 Numerical Results

As test cases we considered two stationary Navier–Stokes problems with special properties.

Problem 1. On the unit square $\Omega = (0,1)^2$ we define

$$\mathbf{u}(x,y) = \begin{pmatrix} \cos(2x-1)e^{2y-1} \\ \sin(2x-1)e^{2y-1} \end{pmatrix}, \qquad p(x,y) = \frac{e^2 - e^{-2}}{8} - \frac{e^{4y-2}}{2}$$

and right hand side $\mathbf{f} = 0$. Then the Laplacian vanishes, $\Delta \mathbf{u} = 0$. The sole contribution from the velocity field to the PDE is the nonlinear term that cancels out with the pressure gradient.

Problem 2. Again on the unit square $\Omega = (0,1)^2$ we prescribe a fixed velocity profile and a channel-like linear pressure

$$\mathbf{u}(x,y) = \begin{pmatrix} \sin(\pi y) \\ 0 \end{pmatrix}, \qquad p(x,y) = Re^{-1}\pi(x - \tfrac{1}{2})$$

and get a non-vanishing right-hand side. This time the convective term $(\mathbf{u} \cdot \nabla)\mathbf{u}$ is zero and the pressure is scaled with the inverse of the Reynolds number. A vector plot of the velocity field for both examples is given in Fig. 1.

Remark. We did not use the quadratic Poiseuille profile for the second example because it is contained in the ansatz spaces for $k \geq 2$.

The following numerical tests were carried out on an unstructured, quasi uniform mesh with $h \approx \frac{1}{32}$ and the Taylor-Hood element with $k = 2$. The nonlinearity was resolved by a damped defect correction iteration and the norm of the residual was reduced below 10^{-12}.

Figure 2 shows how the various errors of the discrete solution depend on the Reynolds number without stabilization. For the first problem we see almost a linear increase of the errors in the velocity with the Reynolds number, while the pressure

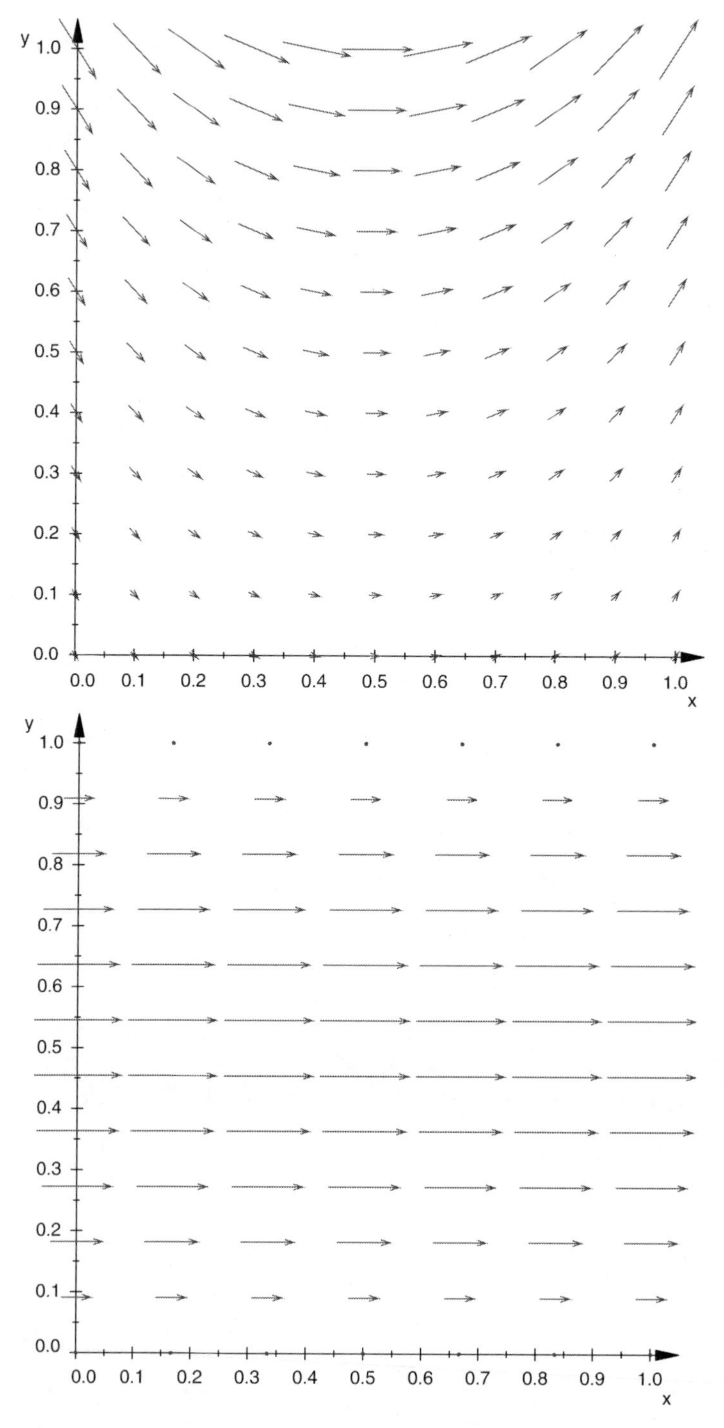

Fig. 1 Vector plot of velocity for problems 1 (*top*) and 2 (*bottom*)

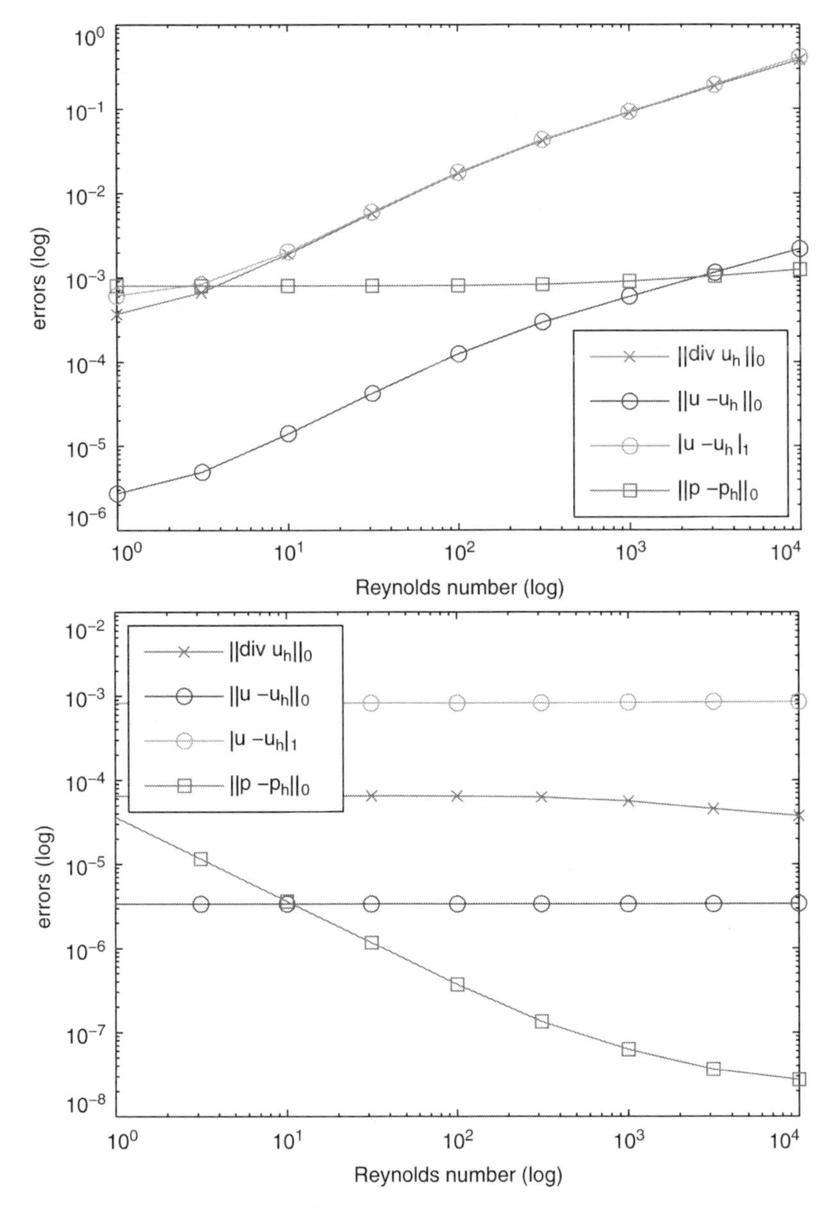

Fig. 2 Errors vs. Reynolds number Re for problems 1 (*top*) and 2 (*bottom*)

error remains constant. The error of the velocity in the H^1-seminorm is dominated by the divergence error. For the second problem we can observe a linear decrease of the pressure error that is caused by the scaling of the pressure with Re^{-1}. The velocity errors are not affected by the Reynolds number and the divergence error is smaller than the H^1-seminorm error.

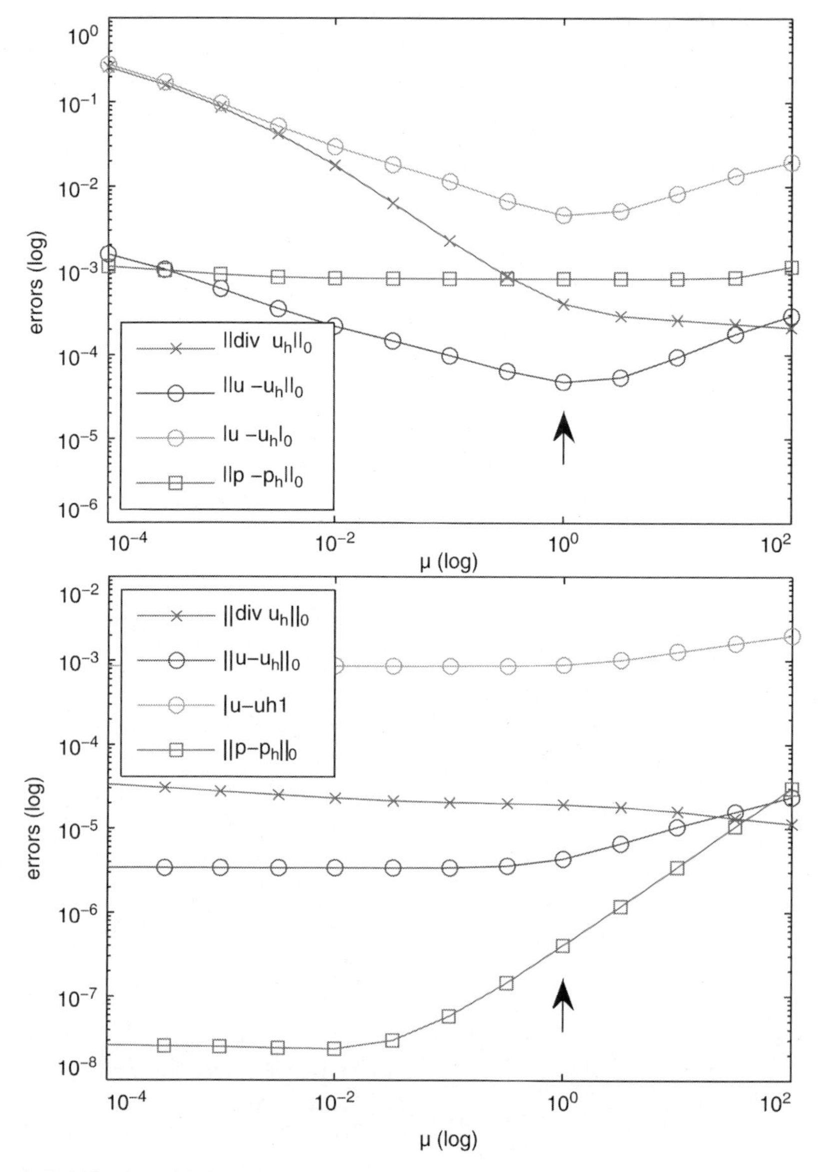

Fig. 3 Stabilization with the old parameter design, problems 1 (*top*) and 2 (*bottom*)

The effect of divergence stabilization on the errors for the original parameter design and both examples with $Re = 10^4$ is shown in Fig. 3. For the first problem the divergence stabilization improves the velocity errors by several orders of magnitude and decouples the divergence error from the H^1-seminorm error. The optimal parameter $\mu_M \approx 1$ reduces the divergence error to the level it had for $Re = 1$. However, the behavior is different for the second problem. At some point

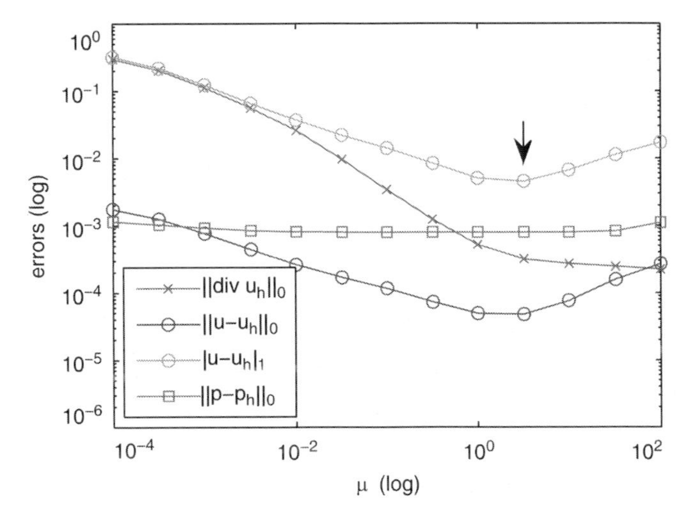

Fig. 4 Stabilization with the new parameter design, Example 1

the pressure error starts to increase linearly with the stabilization parameter. The previously optimal value now increases the pressure error by more than one order of magnitude. Over the whole range of tested parameters only a marginal improvement of the error can be observed. The errors without stabilization are almost optimal.

To get some first results for the new parameter design we used the reference solution and inserted it into the new parameter design. Due to vanishing second derivatives of the pressure for the second problem the parameter design reduces to $\mu_M = 0$ and reproduces what we could see in the previous numerical result: for this problem the divergence stabilization is superfluous. For the first problem the original assumption on the norms is valid and the new parameter design gives results (shown in Fig. 4) comparable to the old parameter design.

More realistic flows, like the flow around a cylinder used in benchmark computations [TS96], show locally varying properties. Close to the cylinder nonlinear effects are stronger, while far behind the cylinder channel like flow can be observed. The proposed parameter is an indicator for the flow type and varies by two orders of magnitude for the flow around the cylinder.

5 Conclusion

Parameter designs for the divergence stabilization did not take into account the local norms of velocity and pressure so far. This leads to parameters far from being optimal for some types of flow (e.g. channel type flow) that actually increase the errors. By a careful look into existing a-priori analysis and error estimates we were able to derive a new parameter design for the divergence stabilization that includes local norms of velocity and pressure in order to minimize the upper bound of the error. The rate of convergence is not affected by the new choice. Unfortunately the new parameter design has several drawbacks that are an obstacle to an efficient implementation.

We should note, that similar observations can be made for the pressure stabilization parameter, because it appears in front of velocity and pressure norms in the error estimate. In practice the effect of badly chosen parameters is less visible there, because the parameter typically is proportional to h_M or h_M^2 for pressure stabilization.

We have not yet implemented the proposed nonlinear parameter design, because we belive that balancing the parameter using the asymptotic a-priori error estimate is still not optimal. What we finally want to do is to determine the load on the divergence constraint, for example by using a Helmholtz-decomposition of the convective and external forcing terms in the momentum equation.

The question whether it is possible to construct a reliable and robust parameter design, that works over a broad range of problems without case by case parameter tuning and can be efficiently implemented, is still open.

References

[BBJL07] M. Braack, E. Burman, V. John, and G. Lube. Stabilized finite element methods for the generalized Oseen problem. *Computer Methods in Applied Mechanics and Engineering*, 196(4–6):853–866, 2007.

[LRL08] G. Lube, G. Rapin, and J. Löwe. Local projection stabilization for incompressible flows: equal-order vs. inf-sup stable interpolation. *ETNA*, 32:106–122, 2008.

[MST07] G. Matthies, P. Skrzypacz, and L. Tobiska. A unified convergence analysis for local projection stabilisation applied to the oseen problem. *Mathematical Modelling and Numerical Analysis*, 41(4):713–742, 2007.

[MT07] G. Matthies and L. Tobiska. Local projection type stabilisation applied to inf-sup stable discretisations of the Oseen problem. *preprint*, 2007.

[OR04] M. A. Olshanskii and A. Reusken. Grad-div stablilization for Stokes equations. *Mathematics of Computation*, 73(248):1699–1718, 2004.

[TS96] S. Turek and M. Schäfer. Benchmark computations of laminar flow around cylinder. *Flow Simulation with High-Performance Computers II*, 52:547–566, 1996.

Distributed and Boundary Control of Singularly Perturbed Advection–Diffusion–Reaction Problems

G. Lube and B. Tews

Abstract We consider the numerical analysis of quadratic optimal control problems with distributed and Robin boundary control governed by an elliptic problem. The Galerkin discretization is stabilized via the local projection approach which leads to a symmetric discrete optimality system. In the singularly perturbed case, the Robin control at parts of the boundary can be seen as regularized Dirichlet control.

1 Introduction

Let $\Omega \subset \mathbb{R}^d, d \in \{2,3\}$ be a bounded polyhedral domain with Lipschitz boundary $\partial\Omega = \overline{\Gamma_R \cup \Gamma_D}, \Gamma_D \cap \Gamma_R = \emptyset$ and outer normal unit vector $\mathbf{n}$. We address some aspects of the numerical analysis of the quadratic optimal control problem

$$\begin{aligned}\text{Minimize } J(u, q_\Omega, q_\Gamma) := &\frac{1}{2}\lambda_\Omega \|u - \tilde{u}_\Omega\|^2_{L^2(\Omega)} + \frac{1}{2}\lambda_\Gamma \|u - \tilde{u}_\Gamma\|^2_{L^2(\Gamma_R)} \\ &+ \frac{1}{2}\alpha_\Omega \|q_\Omega\|^2_{L^2(\Omega)} + \frac{1}{2}\alpha_\Gamma \|q_\Gamma\|^2_{L^2(\Gamma_R)} \qquad (1)\end{aligned}$$

where $(u, q_\Omega, q_\Gamma) \in V \times Q_\Omega \times Q_\Gamma := \{v \in H^1(\Omega) : u|_{\Gamma_D} = 0\} \times L^2(\Omega) \times L^2(\Gamma_R)$ solves the mixed boundary value problem of advection-diffusion-reaction type

$$\begin{aligned}-\varepsilon\Delta u + \mathbf{b}\cdot\nabla u + \sigma u &= f + q_\Omega \quad \text{in } \Omega, \qquad (2)\\ u = 0 \quad \text{on } \Gamma_D, \qquad \varepsilon\nabla u\cdot\mathbf{n} + \beta u &= g + q_\Gamma \quad \text{on } \Gamma_R.\end{aligned}$$

G. Lube (✉)
Department of Mathematics, NAM, Georg-August-University Göttingen, Lotzestrasse 16–18, D-37083 Göttingen, Germany, E-mail: lube@math.uni-goettingen.de

A.F. Hegarty et al. (eds.), *BAIL 2008 - Boundary and Interior Layers.* Lecture Notes in Computational Science and Engineering,
DOI: 10.1007/978-3-642-00605-0,

We assume that $\varepsilon > 0$ and $\sigma \geq 0$ are constants and that the advective field $\mathbf{b}$ is divergence-free. In (1), the desired states are $\tilde{u}_\Omega$ and $\tilde{u}_\Gamma$. The constants $\lambda_\Omega, \lambda_\Gamma \geq 0$ with $\lambda_\Omega^2 + \lambda_\Gamma^2 > 0$ describe the weights of the distributed and boundary control in (1) whereas $\alpha_\Omega, \alpha_\Gamma \geq 0$ with $\alpha_\Omega^2 + \alpha_\Gamma^2 > 0$ serve as regularisation parameters. The state equation (2) describes the dependence of the state u on the control (q_Ω, q_Γ).

Problem (1)–(2) with $\Gamma_R = \emptyset$ has been considered in [3, 10] for the singularly perturbed case $0 < \epsilon \ll 1$, see also the references therein. Here one goal is to consider problem (1)–(2) simultaneously for distributed and (Robin) boundary control. Notably, for $0 < \epsilon \ll 1$, the Robin control can be seen as regularized Dirichlet control.

The Galerkin discretization is stabilized as in [3] via the local projection approach (LPS for short below) which leads to a symmetric optimality system. This implies that discretization and optimization commute as opposed to residual-based stabilization techniques. Another aim of the present paper is a more general LPS approach, including a two-level variant (as in [3]) and a one-level variant introduced in [9]. Let us emphasize two aspects of the analysis: (1) The regularity of the solution of problem (2) is taken into account by using Sobolev–Slobodeckij spaces and adapting the analysis of the LPS method. (2) The analysis is performed for shape regular meshes (as opposed to quasi-uniform meshes in [3]) which allows for (isotropic) mesh refinement at corners or edges of the domain and in boundary layers.

An outline of the paper is as follows: In Sect. 2, we address the solvability of problem (1)–(2). Then, in Sect. 3, we consider the finite element (FE) discretization of the optimality system whereas Sect. 4 presents its convergence properties. In Sects. 5 and 6, we address a numerical experiment and the interpretation of Robin control as regularized Dirichlet control. For full proofs we refer to [8].

Standard notations for Lebesgue and Sobolev spaces are used, e.g., the L^2-inner product and the L^2-norm in $G \subseteq \Omega$ are denoted by $(\cdot,\cdot)_G$ and $\|\cdot\|_{0,G}$.

2 Continuous Optimal Control Problem

Here we consider the optimality system for the continuous optimal control problem (1)–(2). To this goal, we first consider the solvability of the state equation (2) with $\tilde{f} := f + q_\Omega$ and $\tilde{g} := g + q_\Gamma$. The variational form of problem (2) reads:

$$\text{Find } u \in V \text{ such that } a(u,v) = f(v) \; \forall v \in V, \tag{3}$$

$$a(u,v) := \varepsilon(\nabla u, \nabla v)_\Omega + (\mathbf{b}\cdot\nabla u + \sigma u, v)_\Omega + (\beta u, v)_{\Gamma_R},$$

$$f(v) := (\tilde{f}, v)_\Omega + (\tilde{g}, v)_{\Gamma_R}.$$

Lemma 1. *There exists a unique solution $u \in H^1(\Omega)$ of problem (3) under the assumptions:*

i) $\mathbf{b} \in [L^\infty(\Omega)]^d, \quad \tilde{f} \in L^2(\Omega), \quad \tilde{g} \in L^2(\Gamma_R), \quad \beta \in L^\infty(\Gamma_R),$
ii) $\varepsilon > 0, \; \sigma \geq 0$ *and* $\nabla \cdot \mathbf{b} = 0$ *a.e. in* Ω,

iii) $\tilde{\beta} := \beta + \frac{1}{2}(\mathbf{b}\cdot\mathbf{n}) \geq \beta_0 \geq 0, \quad \beta \geq 0$ *a.e. on* Γ_R,
iv) There holds: $(iv)_1 \quad \mu_{d-1}(\Gamma_D) > 0,$ *and/or* $(iv)_2 \quad \sigma > 0$ *or* $\beta_0 > 0$.

Moreover, the optimal control problem (1)–(2) has a unique solution $(\overline{u}, \overline{q}_\Omega, \overline{q}_\Gamma)$.

The proof can be found in [8], Lemma 2.1. Please note that the assumption $\beta \geq 0$ is not needed for this result, but it will be used later on in the analysis in Sect. 4.

In general, the solution of (3) is not in $W^{2,2}(\Omega)$. Let $\mathcal{S}$ be the set of points (for $d = 2$) or edges (for $d = 3$) which subdivide the polyhedral boundary $\partial\Omega$ into smooth disjoint connected components. The weighted Sobolev space $V_\delta^{k,2}(\Omega)$ denotes the closure of $C^\infty(\Omega)$ w.r.t.

$$\|v\|_{V_\delta^{k,2}(\Omega)} = \Big(\sum_{|\alpha|\leq k}\int_\Omega r^{2(\delta-k+|\alpha|)}|D^\alpha u|^2\,dx\Big)^{\frac{1}{2}}$$

where $r = r(x) = \operatorname{dist}(x, \mathcal{S})$, $\delta \in \mathbb{R}$, and $k \in \mathbb{N}$. The parameter δ is defined via eigenvalues of eigenvalue problems (in local coordinate systems at parts of the set $\mathcal{S}$) associated with problem (3). As it is not the goal here to give sufficient conditions for the solution of problem (3) to belong to $V_\delta^{k,2}(\Omega)$, we refer to [6]. Moreover, we do not intend to consider graded FE meshes in the neighborhood of the set $\mathcal{S}$ although the forthcoming numerical analysis allows such kind of refinement. For such approach to optimal control problems, see [1].

Here we consider on a subdomain $G \subseteq \Omega$ the Sobolev–Slobodeckij spaces

$$W^{k+\lambda,2}(G) := \Big\{v \in W^{k,2}(G) : \|u\|_{k+\lambda,2,G} < \infty\Big\}, \qquad k \in \mathbb{N}_0,\ \lambda \in [0,1)$$

$$\|u\|_{k+\lambda,2,G} := \Big(\|u\|_{k,2,G}^2 + \sum_{|\alpha|=k}\int_G\int_G \frac{|D^\alpha u(x) - D^\alpha u(y)|^2}{|x-y|^{d+2\lambda}}dx\,dy\Big)^{\frac{1}{2}}.$$

The spaces $W^{k+\lambda,2}(\Gamma_R)$ are defined in a similar way.

Remark 1. The embeddings $V_\delta^{2,2}(\Omega) \subset W^{\frac{d}{2}+\kappa,2}(\Omega) \subset C(\overline{\Omega})$ are valid for $\delta < 2 - \frac{d}{2} + \kappa$ with $\kappa > 0$, cf. [6]. In particular, for the case $\partial\Omega = \Gamma_D$ in polyhedral domains, the conditions $\delta \leq \frac{1}{2} + \kappa, \kappa > 0$ are sufficient.

As problem (3) is uniquely solvable, we define the affine linear solution operator $S : L^2(\Omega) \times L^2(\Gamma_R) \to V,\ u = S(q_\Omega + f, q_\Gamma + g)$. Due to the linearity of (2) we can split S in its linear and affine linear part. Inserting $u = S(q_\Omega + f, q_\Gamma + g) = S(q_\Omega, q_\Gamma) + S(f, g)$ in (1), we obtain (with trace operator γ) and the definitions $u_\Omega := \tilde{u}_\Omega - S(f,g)$ and $u_\Gamma := \tilde{u}_\Gamma - \gamma \circ S(f,g)$ the reduced cost functional:

$$\begin{aligned} j(q_\Omega, q_\Gamma) = J\,(q_\Omega, q_\Gamma, S(q_\Omega, q_\Gamma)) &= \frac{1}{2}\lambda_\Omega\|S(q_\Omega, q_\Gamma) - u_\Omega\|_{0,\Omega}^2 \\ &+ \frac{1}{2}\lambda_\Gamma\|\gamma \circ S(q_\Omega, q_\Gamma) - u_\Gamma\|_{0,\Gamma_R}^2 + \frac{1}{2}\alpha_\Omega\|q_\Omega\|_{0,\Omega}^2 \\ &+ \frac{1}{2}\alpha_\Gamma\|q_\Gamma\|_{0,\Gamma_R}^2. \end{aligned} \tag{4}$$

Now the reduced optimization problem reads

$$\text{Minimize} \quad j(q_\Omega, q_\Gamma), \quad (q_\Omega, q_\Gamma) \in Q_\Omega \times Q_\Gamma. \tag{5}$$

The reduced cost functional j is continuously differentiable. In order to formulate the optimality conditions for problem (5), we define the associated adjoint state $\overline{p} \in V$ to $(\overline{q}_\Omega, \overline{q}_\Gamma)$ as the solution of

$$\text{Find } p \in V: \quad a_{adj}(p, v) = \lambda_\Omega(\overline{u} - u_\Omega, v)_\Omega + \lambda_\Gamma(\overline{u} - u_\Gamma, v)_{\Gamma_R} \quad \forall v \in V, \tag{6}$$
$$a_{adj}(p, v) := \epsilon(\nabla p, \nabla v)_\Omega - (\mathbf{b} \cdot \nabla p, v)_\Omega + \sigma(p, v)_\Omega + ((\beta + \mathbf{b} \cdot \mathbf{n})p, v)_{\Gamma_R}.$$

The necessary (and sufficient) optimality conditions read

$$D_{q_\Omega} j(\overline{q}_\Omega, \overline{q}_\Gamma) \cdot (k_\Omega - \overline{q}_\Omega) = (\alpha_\Omega \overline{q}_\Omega + \overline{p}, k_\Omega - \overline{q}_\Omega)_\Omega = 0, \quad \forall k_\Omega \in Q_\Omega, \tag{7}$$
$$D_{q_\Gamma} j(\overline{q}_\Omega, \overline{q}_\Gamma) \cdot (k_\Gamma - \overline{q}_\Gamma) = (\alpha_\Gamma \overline{q}_\Gamma + \gamma \circ \overline{p}, k_\Gamma - \overline{q}_\Gamma)_{\Gamma_R} = 0, \quad \forall k_\Gamma \in Q_\Gamma, \tag{8}$$

leading to

$$\alpha_\Omega \overline{q}_\Omega + \overline{p} = 0, \text{ in } \Omega \qquad \alpha_\Gamma \overline{q}_\Gamma + \gamma \circ \overline{p} = 0 \text{ on } \Gamma_R. \tag{9}$$

The optimality system (KKT-system) for problem (1)–(2) is formed by (9) together with the state problem (3) and the adjoint state problem (6). The second order derivatives of $j(q_\Omega, q_\Gamma)$ do not depend on (q_Ω, q_Γ) and are positive definite.

As already said, the solution of (1)–(2) is in general not arbitrarily smooth.

Assumption 1: The optimal solution $(\overline{u}, \overline{p}, \overline{q}_\Omega, \overline{q}_\Gamma)$ of the optimal control problem (1)–(2) belongs to $[W^{1+\lambda,2}(\Omega)]^3 \times W^{\frac{1}{2}+\lambda,2}(\Gamma_R)$ with $1 + \lambda > \frac{d}{2}$.

Assume that $\alpha_\Omega, \alpha_\Gamma > 0$. Then Assumption 1 is valid if the solution u of (3) belongs to $W^{1+\lambda,2}(\Omega)$, $1 + \lambda > d/2$, eventually for sufficiently smooth data $\tilde{f}, \tilde{g}, \beta$. For sufficient conditions, see Remark 1. Then the same statement is valid for the solution p of (6) for sufficiently smooth data u_Ω, u_Γ. Moreover, the regularity of $\overline{q}_\Omega$ and $\overline{q}_\Gamma$ follows via (9). Finally, we remark that Assumption 1 allows later on Lagrangian interpolation of the solution.

3 Stabilized Discrete Optimality System

Here we introduce the discretized optimal control problem to (1)–(2). A more general approach to the discretization as in [3] is applied by considering shape-regular FE meshes and a more flexible stabilization concept.

Consider a family of shape-regular, admissible decompositions $\mathcal{T}_h$ of Ω into d-dimensional simplices, quadrilaterals ($d = 2$) or hexahedra ($d = 3$). Let h_T be the diameter of a cell $T \in \mathcal{T}_h$ and $h = \max_{T \in \mathcal{T}_h} h_T$. Assume that, for each $T \in \mathcal{T}_h$, there exists an affine mapping $F_T : \hat{T} \to T$ which maps the reference element $\hat{T}$ onto T. This quite restrictive assumption for quadrilaterals/ hexahedra can be weakened to asymptotically affine linear mappings [2]. Let e_h denote the set of element faces (for $d = 3$) or element edges (for $d = 2$) induced by $\mathcal{T}_h$ on $\partial\Omega$. Moreover, we assume that the Robin part Γ_R of the boundary is exactly triangulated by e_h.

Set $P_{\mathcal{T}_h} = \{v_h \in L^2(\Omega) : v_h \circ F_T \in \mathbb{P}_1(\hat{T}), T \in \mathcal{T}_h\}$ within $\mathbb{P}_1(\hat{T})$, the space of complete linear polynomials on $\hat{T}$, and $R_{\mathcal{T}_h} = \{v_h \in L^2(\Omega) : v_h \circ F_T \in \mathbb{Q}_1(\hat{T}), T \in \mathcal{T}_h\}$ within $\mathbb{Q}_1(\hat{T})$, the space of all polynomials on $\hat{T}$ with maximal first degree in each coordinate direction. The state space V is approximated by a FE space $V_h \supset P_{\mathcal{T}_h} \cap V$ or $V_h \supset R_{\mathcal{T}_h} \cap V$. Similarly, let $Q_{h,\Omega} \subset H^1(\Omega)$ be a FE space for the control variable and $Q_{h,\Gamma} = Q_{h,\Omega}|_{\Gamma_R}$ its restriction to Γ_R.

The basic Galerkin discretization of the state problem (3) reads:

$$\text{find } u_h \in V_h \text{ such that } \quad a(u_h, v_h) = f(v_h), \qquad \forall v_h \in V_h. \tag{10}$$

The solution u_h of (10) may suffer from spurious oscillations. As a remedy, we consider the local projection stabilization (LPS) approach which results in a symmetric discrete optimality system. LPS methods split the discrete function spaces into small and large scales and add stabilization terms of diffusion-type acting only on the small scales. There are basically a two- and a one-level variant (indicated by $\mathcal{M}_h = \mathcal{T}_{2h}$ and $\mathcal{M}_h = \mathcal{T}_h$, respectively).

The *two-level variant* starts from the given space $V_h = P_{\mathcal{T}_h} \cap V$ or $V_h = R_{\mathcal{T}_h} \cap V$ for simplicial or hexahedral elements. The large scales are determined by means of a coarse, non-overlapping and shape-regular mesh $\mathcal{M}_h = \{M_i\}_{i \in I}$ which is constructed by coarsening $\mathcal{T}_h$ s.t. each $M \in \mathcal{M}_h$ with diameter h_M is the union of neighboring cells $T \in \mathcal{T}_h$. (A more practical approach is to start from the coarse grid $\mathcal{M}_h$ and to construct $\mathcal{T}_h$ by an appropriate refinement, see [4], Sect. 4.) Moreover, we assume:

$$\exists\, C > 0 : \quad h_M \leq C h_T, \quad \forall T \in \mathcal{T}_h,\ M \in \mathcal{M}_h \text{ with } T \subset M. \tag{11}$$

We introduce a discontinuous FE space $D_h \subset L^2(\Omega)$ of piecewise constant functions on $\mathcal{M}_h$ and its restriction $D_h(M) := \{v_h|_M \ ;\ v_h \in D_h\}$ to $M \in \mathcal{M}_h$. The next ingredient is the local L^2-projection $\pi_M : L^2(M) \to D_h(M)$ which defines the global projection $\pi_h : L^2(\Omega) \to D_h$ by $(\pi_h v)|_M := \pi_M(v|_M)$ for all $M \in \mathcal{M}_h$. The fluctuation operator $\kappa_h : L^2(\Omega) \to L^2(\Omega)$ is defined by $\kappa_h := id - \pi_h$.

The *one-level variant* starts from the given discontinuous FE space D_h of piecewise constant functions on $\mathcal{M}_h = \mathcal{T}_h$ and uses an appropriate FE space V_h on $\mathcal{T}_h$. For simplicial elements, define

$$P_1^{bub}(\hat{T}) = P_1(\hat{T}) + \hat{b} \cdot P_0(\hat{T}), \quad \hat{b}(\hat{x}) := (d+1)^{d+1} \hat{\lambda}_1(\hat{x}) \cdot \ldots \cdot \hat{\lambda}_{d+1}(\hat{x})$$

with the barycentric coordinates $\hat{\lambda}_1, \ldots, \hat{\lambda}_{d+1}$. The enriched space is defined as

$$V_h = \{v \in H^1(\Omega) \cap V : v|_T \circ F_T \in P_1^{bub}(\hat{T})\ \forall T \in \mathcal{T}_h\}.$$

A similar construction is given in Sect. 4 of [9] for hexahedral elements. Then the same framework as in the two-level approach can be used by setting $\mathcal{M}_h = \mathcal{T}_h$.

For both variants, the stabilized discrete formulation reads: find $u_h \in V_h$ such that

$$a_{lps}(u_h, v_h) := a(u_h, v_h) + s_h(u_h, v_h) = f(v_h), \qquad \forall v_h \in V_h, \tag{12}$$

$$s_h(u_h, v_h) := \sum_{M \in \mathcal{M}_h} \tau_M (\kappa_h (\mathbf{b} \cdot \nabla u_h), \kappa_h (\mathbf{b} \cdot \nabla v_h))_M. \tag{13}$$

The stabilization s_h with parameters $\tau_M \geq 0$ acts solely on the small scales. Another variant uses $\tilde{s}_h(u_h, v_h) = \sum_M \tilde{\tau}_M (\tilde{\kappa}_h(\nabla u_h), \tilde{\kappa}_h(\nabla v_h))_M$ instead of $s_h(\cdot, \cdot)$. Here $\tilde{\kappa}_h$ denotes a vector-valued version of the fluctuation operator κ_h.

For a discussion of "pro's and con's" of the two variants, we refer to [4].

The discretized control problem associated with (1)–(2) reads as follows:

$$\min \; J(u_h, q_{h,\Omega}, q_{h,\Gamma}), (u_h, q_{h,\Omega}, q_{h,\Gamma}) \in V_h \times Q_{h,\Omega} \times Q_{h,\Gamma}, \tag{14}$$

$$a_{lps}(u_h, v_h) = (f + q_{h,\Omega}, v_h)_\Omega + (g + q_{h,\Gamma}, v_h)_{\Gamma_R}, \forall v_h \in V_h. \tag{15}$$

Problem (14)–(15) has a unique solution $(\overline{u}_h, \overline{q}_{h,\Omega}, \overline{q}_{h,\Gamma})$ which allows us to define the discrete solution operator $S_h : Q_\Omega \times Q_\Gamma \to V_h$ by

$$a_{lps}(S_h(q_{h,\Omega}, q_{h,\Gamma}), v_h) = (f + q_{h,\Omega}, v_h)_\Omega + (g + q_{h,\Gamma}, v_h)_{\Gamma_R} \quad \forall v_h \in V_h$$

and the discrete reduced cost functional as $j_h(q_{h,\Omega}, q_{h,\Gamma}) = J(S_h(q_{h,\Omega}, q_{h,\Gamma}), q_{h,\Omega}, q_{h,\Gamma})$. The necessary (and here also sufficient) optimality conditions read

$$\alpha_\Omega \overline{q}_{h,\Omega} + \overline{p}_h = 0, \qquad \alpha_\Gamma \overline{q}_{h,\Gamma} + \gamma \circ \overline{p}_h = 0.$$

Here the discrete adjoint state $p_h \in V_h$ solves the discrete adjoint state problem

$$a_{lps}(v_h, p_h) = \lambda_\Omega (\overline{u}_h - u_\Omega, v_h)_\Omega + \lambda_\Gamma (\overline{u}_h - u_\Gamma, v_h)_{\Gamma_R}. \tag{16}$$

where $\overline{u}_h = S_h(q_\Omega, q_\Gamma)$ is the discrete state according to (15).

Remark 2. The symmetry of the LPS term implies that the operations "optimize" and "discretize" commute, see [3].

4 A-Priori Error Analysis

Here we provide the error analysis for the optimal control problem (1)–(2). It turns out that additional assumptions for the LPS method are required.

Assumption 2: The fluctuation operator $\kappa_h = id - \pi_h$ has the property:

$$\exists\, C_\kappa > 0 : \; \|\kappa_h q\|_{0,M} \leq C_\kappa h_M^s |q|_{s,M}, \quad \forall q \in W^{s,2}(M),\ s \in [0, 1],\ \forall M \in \mathcal{M}_h. \tag{17}$$

Remark 3. The original version of (17) in [9] only considers $s \in \{0, 1\}$.

Following [9], we construct an interpolation $j_O : V \to V_h$ such that the error $v - I_h v$ is L^2-orthogonal to D_h for all $v \in V$. The following assumption is valid for the discrete spaces discussed in the previous section and allows us to conserve standard approximation properties.

Assumption 3: There exists a constant $\beta_S > 0$ such that, for any $M \in \mathcal{M}_h$,

$$\inf_{q_h \in D_h(M)} \sup_{v_h \in Y_h(M)} \frac{(v_h, q_h)_M}{\|v_h\|_{0,M} \|q_h\|_{0,M}} \geq \beta_S > 0. \tag{18}$$

where $Y_h(M) := \{v_h|_M \ : \ v_h \in V_h, v_h = 0 \text{ on } \Omega \setminus M\}$.

Condition (18) implies that D_h must not be too rich. On the other hand, D_h must be rich enough to fulfil (17) .

The following result extends the proof in [9] to $\lambda \in \{0, 1\}$, see [8], Lemma 4.1.

Lemma 2. *Under Assumption 3 there exists an operator $j_O : V \to V_h$ such that*

$$(v - j_O v, q_h)_\Omega = 0, \quad \forall q_h \in D_h, \forall v \in V, \tag{19}$$

and for all $M \in \mathcal{M}_h$, for all $E \in \mathrm{e}_h$, and for $v \in V \cap W^{1+\lambda,2}(\Omega)$ with $1 + \lambda > \frac{d}{2}$

$$\|v - j_O v\|_{0,M} + h_M |v - j_O v|_{1,M} + h_M^{\frac{1}{2}} \|v - j_O v\|_{0,E} \lesssim h_M^{1+\lambda} \|v\|_{1+\lambda,2,\omega(M)}. \tag{20}$$

The next goal is to derive error estimates for the state problems (15) and (16). First, the stability of the scheme will be given in the mesh-dependent norm

$$|||v||| := \left(\varepsilon |v|_{1,\Omega}^2 + \sigma \|v\|_{0,\Omega}^2 + \|\tilde{\beta}^{\frac{1}{2}} v\|_{0,\Gamma_R}^2 + s_h(v, v) \right)^{\frac{1}{2}}, \qquad \forall v \in V.$$

Lemma 3. *The LPS schemes (15) and (16) have unique solutions.*

Proof. We consider, e.g., problem (15) with $v_h = u_h$. The application of the Cauchy–Schwarz inequality and the definition of the triple norm yields the a priori estimate

$$|||u_h||| \leq C_\Omega \|f + q_{h,\Omega}\|_{0,\Omega} + C_\Gamma \|g + q_{h,\Gamma}\|_{0,\Gamma_R}$$

with $C_\Omega := \min\{\sigma^{-\frac{1}{2}}; C_P \epsilon^{-\frac{1}{2}}\}$, $C_\Gamma := \min\{\beta_0^{-\frac{1}{2}}; C_P \epsilon^{-\frac{1}{2}}\}$ and Poincare constant C_P.

The following a priori estimates are based on the standard technique of combining stability and consistency results based on the previous auxiliary results. Here, and in the following Lemma, we fix some controls $(p_\Omega, p_\Gamma) \in Q_\Omega \times Q_\Gamma$ which will be later on, for the main theorem, chosen as the Lagrangian interpolants of the optimal controls $(\overline{q}_\Omega, \overline{q}_\Gamma)$.

Lemma 4. *For $(q_\Omega, q_\Gamma) \in Q_\Omega \times Q_\Gamma$, let $u = S(q_\Omega, q_\Gamma) \in V$ be the solution of (2). For some $(p_\Omega, p_\Gamma) \in Q_\Omega \times Q_\Gamma$, let $w_h = S_h(p_\Omega, p_\Gamma) \in V_h$ be the solution of*

$$a_{lps}(w_h, v_h) = (f + p_\Omega, v_h)_\Omega + (g + p_\Gamma, v_h)_{\Gamma_R} \qquad \forall v_h \in V_h \tag{21}$$

with

$$\tau_M \sim h_M / \|\mathbf{b}\|_{[L^\infty(M)]^d}. \tag{22}$$

Then, under the assumptions of Lemma 1, there holds the a-priori error estimate

$$\begin{aligned} |||u - w_h||| &\le C_\Omega \|q_\Omega - p_\Omega\|_{0,\Omega} + C_\Gamma \|q_\Gamma - p_\Gamma\|_{0,\Gamma_R} \\ &\quad + C\Big(\sum_{M\in\mathcal{M}_h} h_M^{2\lambda+1}\Big\{\frac{|\mathbf{b}\cdot\nabla u|^2_{\lambda,2,M}}{\|\mathbf{b}\|_{[L^\infty(M)]^d}} + C_M \|u\|^2_{1+\lambda,2,M}\Big\}\Big)^{\frac{1}{2}} \end{aligned} \tag{23}$$

with constants C_M and C_Γ as in the proof of Lemma 3 and

$$C_M := \varepsilon h_M^{-1} + \sigma h_M + \|\mathbf{b}\|_{[L^\infty(M)]^d} + \|\beta\|_{L^\infty(\partial M\cap\Gamma_R)} + \|\mathbf{b}\cdot\mathbf{n}\|_{L^\infty(\partial M\cap\Gamma_R)}.$$

For a full proof of Lemma 4, see [8], Lemma 4.3. Similarly, we obtain an a-priori error estimate for the adjoint problem (16) where $|||u - w_h|||$ in (23) can be further estimated via Lemma 4. A full proof of Lemma 5 is given in [8], Lemma 4.4.

Lemma 5. *For $(q_\Omega, q_\Gamma) \in Q_\Omega \times Q_\Gamma$, let $p \in V$ be the solution of the adjoint state problem (6) and for some $(p_\Omega, p_\Gamma) \in Q_\Omega \times Q_\Gamma$, let $y_h \in V_h$ be the adjoint discrete solution. Then, there holds the a-priori error estimate*

$$\begin{aligned} |||p - y_h||| &\le (C_\Omega^2 \lambda_\Omega + C_\Gamma^2 \lambda_\Gamma)|||u - w_h||| \\ &\quad + C\Big(\sum_M h_M^{2\lambda+1}\Big\{\frac{|\mathbf{b}\cdot\nabla p|^2_{\lambda,2,M}}{\|\mathbf{b}\|_{[L^\infty(M)]^d}} + C_M \|p\|^2_{1+\lambda,2,M}\Big\}\Big)^{\frac{1}{2}} \end{aligned}$$

with τ_M as in (22) and constants C_M, C_Ω and C_Γ as in the previous Lemma.

We can now give the main result for the optimal control problem. For a full proof of Theorem 1, we refer to [8], Theorem 4.5.

Theorem 1. *Let the assumptions of Lemma 1 and Assumption 2 be valid. Moreover, let $(\overline{u}, \overline{q}_\Omega, \overline{q}_\Gamma)$ be the solution of the optimal control problem (1)–(2) and $(\overline{u}_h, \overline{q}_{h,\Omega}, \overline{q}_{h,\Gamma})$ the solution of the discretized problem (14)–(15). Finally, let α_Ω, $\alpha_\Gamma > 0$. Then there exists a constant $C > 0$ depending on $\lambda_\Omega, \lambda_\Gamma, \alpha_\Omega, \alpha_\Gamma, C_\Omega, C_\Gamma$ such that the following error estimate holds:*

$$\begin{aligned} &\|\overline{q}_\Omega - \overline{q}_{h,\Omega}\|_{0,\Omega} + \|\overline{q}_\Gamma - \overline{q}_{h,\Gamma}\|_{0,\Gamma_R} \\ &\le C\Big\{\Big(\sum_{M\in\mathcal{M}_h} h_M^{1+2\lambda}|\overline{q}_\Omega|^2_{1+\lambda,2,M}\Big)^{\frac{1}{2}} + \Big(\sum_{E\in\mathrm{e}_h\cap\Gamma_R} h_E^{1+2\lambda}|\overline{q}_\Gamma|^2_{1+\lambda,2,E}\Big)^{\frac{1}{2}} \\ &\quad + \Big(\sum_M h_M^{1+2\lambda}\Big(\frac{|\mathbf{b}\cdot\nabla\overline{u}|^2_{\lambda,2,M}}{\|\mathbf{b}\|_{[L^\infty(M)]^d}} + \frac{|\mathbf{b}\cdot\nabla\overline{p}|^2_{\lambda,2,M}}{\|\mathbf{b}\|_{[L^\infty(M)]^d}} \\ &\quad + C_M\big(\|\overline{u}\|^2_{1+\lambda,2,M} + \|\overline{p}\|^2_{1+\lambda,2,M}\big)\Big)\Big)^{\frac{1}{2}}\Big\} \end{aligned}$$

with τ_M as in (22), $h_E = diam(E)$, $E \in \mathrm{e}_h$ and C_M, C_Ω, C_Γ as in Lemma 4.

Remark 4. In the limit case $\lambda = 1$, we obtain the optimal convergence rate $\mathcal{O}(h_M^{\frac{3}{2}})$.

5 Numerical Experiment

Consider the following numerical example:

$$\begin{aligned} \min J(q_\Omega, u) &:= \frac{1}{2}\|u - \tilde{u}_\Omega\|^2_{L^2(\Omega)} + \frac{1}{2}\alpha_\Omega \|q_\Omega\|^2_{L^2(\Omega)}, \\ -\epsilon \Delta u + (\mathbf{b} \cdot \nabla) u + \sigma u &= f + q_\Omega \quad \text{in } \Omega, \quad u = 0 \quad \text{on } \partial\Omega \end{aligned}$$

with $q_\Omega \in L^2(\Omega)$, $\epsilon = 10^{-5}$, $\mathbf{b} = (-1, -2)^t$, $\sigma = 1$, $f = 1$, $\tilde{u}_\Omega = 1$ and $\alpha_\Omega = 0.1$. The numerical solution in [3] (for box-constraints of control) with the two-level LPS method and $\varepsilon = 10^{-3}$ gave strong oscillations in the boundary layer regions.

Table 1 gives the convergence history and the numerical convergence rate of the cost functional J. Figure 1 shows the discrete control and state on the coarse grid for the two-level approach with Q_1-elements and $h = \frac{1}{128}$. Spurious oscillations in the boundary layer regions are significantly reduced as compared to the results in [3].

There is an ongoing scientific discussion on the strength of the LPS-method vs. classical residual-based stabilization techniques (like the streamline diffusion method). In [5] it is shown for the one-level LPS method that the LPS-norm gives additional control of the streamline derivative, i.e., on $(\sum_M \delta_M \|\mathbf{b} \cdot \nabla(\cdot)\|^2_{0,M})^{\frac{1}{2}}$ with $\delta_M \sim \min(h_M/\|\mathbf{b}\|_{0,\infty,M}; h_M^2/\varepsilon)$. A further reduction of remaining spurious oscillations in boundary layers is possible with adaptive mesh refinement based on a posteriori error estimators. For the streamline diffusion method applied to optimization problems for advection-diffusion problems, we refer to [10].

Table 1 h-convergence of the cost functional

$h = 2^{-l}$	$J(\overline{q}_h, \overline{u}_h)$	$J(\overline{q}_h, \overline{u}_h) - J(\overline{q}_{2h}, \overline{u}_{2h})$	num. conv. rate
2	3.082E−01	–	–
3	2.767E−01	3.152E−02	–
4	2.639E−01	1.277E−02	1.303
5	2.602E−01	3.748E−03	1.769
6	2.592E−01	9.138E−04	2.036
7	2.591E−01	1.743E−04	2.390

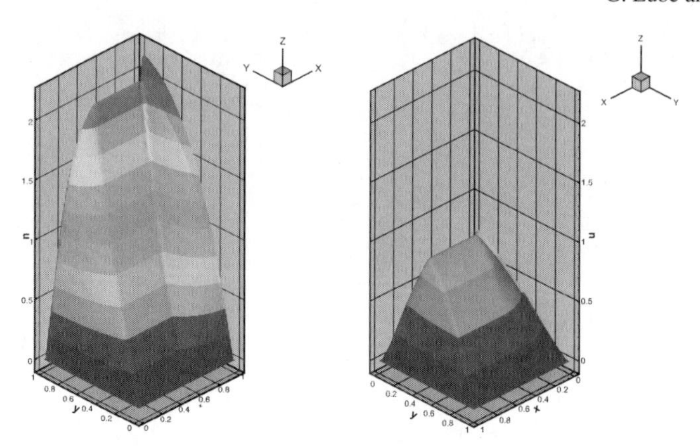

Fig. 1 Optimal discrete control and state for Example 2 with $\varepsilon = 10^{-5}$ and $\tau = 0.1\,h$

6 Further Application: Regularized Dirichlet Control

In applications, a Dirichlet boundary control $u = q$ is desirable. A review of some variants is given in [7]. One possibility is to approximate the Dirichlet control by a Robin control

$$\hat{\varepsilon}\nabla u \cdot \mathbf{n} + \beta(u - q) = 0, \quad \beta = \mathcal{O}(1) \tag{24}$$

for $\hat{\varepsilon} \to +0$, but the choice of $\hat{\varepsilon}$ is delicate. For the singularly perturbed problem (2) with $\hat{\varepsilon} = \varepsilon$, one can interpret the Robin control as regularized Dirichlet control.

Define the subsets Γ_-, Γ_0 and Γ_+ of the boundary $\partial\Omega$, depending on the sign of $(\mathbf{b} \cdot \mathbf{n})(x)$. The solution u of problem (2) has boundary layers at the outflow part Γ_+ with gradient $|\varepsilon\nabla u \cdot \mathbf{n}| \sim 1$ and at characteristic boundaries Γ_0 with (at most) $|\varepsilon\nabla u \cdot \mathbf{n}| \sim \sqrt{\varepsilon}$. At the inflow part Γ_-, one has only $|\varepsilon\nabla u \cdot \mathbf{n}| \sim \varepsilon$. This motivates us to exclude a Dirichlet control at the outflow boundary Γ_+. On $\Gamma_- \cup \Gamma_0$, the Robin regularization (24) with $\hat{\varepsilon} = \varepsilon$ and $\beta + \frac{1}{2}\mathbf{b} \cdot \mathbf{n} \geq \beta_0 > 0$ is a good approximation of the Dirichlet control $u = q$.

A typical situation is the flow in a domain of channel type $\Omega = (0, L) \times (-\frac{H}{2}, \frac{H}{2})$ with the flow field $\mathbf{b}(x) = ((\frac{H}{2} - |x_2|)^\kappa, 0)^T$ with $\kappa \geq 0$. The solution u of (2) can be seen as a temperature field or as the density of some chemical reactant. Let us describe potential applications of Dirichlet control: A Dirichlet condition $u = q$ is given at $\Sigma \subset \Gamma_- = \{0\} \times (-\frac{H}{2}, \frac{H}{2})$ whereas a Robin condition $\varepsilon\frac{\partial u}{\partial x_1} + \beta(u - g) = 0$ with $\beta + \frac{1}{2}\mathbf{b} \cdot \mathbf{n} \geq \beta_0 > 0$ is prescribed on $\Gamma_- \setminus \Sigma$. A Neumann condition $\varepsilon\frac{\partial u}{\partial x_1} = 0$ might be prescribed on $\Gamma_+ = \{1\} \times (-\frac{H}{2}, \frac{H}{2})$. An "insulation" condition $\varepsilon\frac{\partial u}{\partial x_2} = 0$ is given at the channel walls $\Gamma_0 = (0, L) \times \{-\frac{H}{2}, \frac{H}{2}\}$. Similarly, one can assume a Dirichlet condition $u = q$ at $\Sigma \subset \Gamma_0$ of the channel walls. Finally, replacing the Dirichlet control on $\Sigma \subset \Gamma_- \cup \Gamma_0$ by Robin boundary control leads to the problem considered within this report. An analytical justification of this approach and numerical results will be given elsewhere.

References

1. Th. Apel, A. Rösch, and G. Winkler. Optimal control in nonconvex domains: A priori discretization error estimate. *Calcolo*, 44:137–158, 2007.
2. D.N. Arnold, D. Boffi, and R.S. Falk. Approximation by quadrilateral finite elements. *Mathematics of Computation*, 71:909–922, 2002.
3. R. Becker and B. Vexler. Optimal control of the convection-diffusion equation using stabilized finite element methods. *Numerische Mathematik*, 106(3):349–367, 2007.
4. P. Knobloch and G. Lube. Local projection stabilization for advection-diffusion-reaction problems: One-level vs. two-level approach, 2008. submitted.
5. P. Knobloch and L. Tobiska. On the stability of finite element discretizations of convection-diffusion-reaction equations, 2008. submitted.
6. A. Kufner and A.-M. Sändig. *Some Applications of Weighted Sobolev Spaces*. Teubner Verlagsgesellschaft, 1987.
7. K. Kunisch and B. Vexler. Constrained Dirichlet boundary control in L^2 for a class of evolution equations. *SIAM Journal of Control Optimization*, 46 (5):1726–1753, 2007.
8. G. Lube and B. Tews. Optimal control of singularly perturbed advection-diffusion-reaction problems. Technical report, Georg-August University of Göttingen, NAM, Preprint 2008.15, 2008.
9. G. Matthies, P. Skrzypacz, and L. Tobiska. A unified convergence analysis for local projection stabilizations applied to the Oseen problem. M^2AN, 41(4):713–742, 2007.
10. N. Yan and Z. Zhou. A priori and a posteriori error estimates of streamline diffusion finite element method for optimal control problem governed by convection dominated diffusion equation. *Numerical Mathematics: Theory, Methods and Application*, 1:297–320, 2008.

Antisymmetric Aspects of a Perturbed Channel Flow

J. Mauss, P. Cathalifaud, and J. Cousteix

Abstract This paper aims at studying steady laminar flows of incompressible newtonian fluids in channels at high Reynolds numbers when wall deformations can lead to separation. Thanks to the use of generalized asymptotic expansions, cases are examined for which linearized Euler equations are a good approximation in the core flow. The extraction of the antisymmetric part of the problem leads to a new and promising approach of the flow structure understanding. Comparisons with Navier–Stokes solutions demonstrate the relevance of the proposed approach.

1 Introduction

We consider a steady, two-dimensional, incompressible, laminar flow in a channel at high Reynolds numbers. When the walls are parallel the fully developed flow, Poiseuille's flow, constitutes the reference flow. The channel geometry is perturbed by wall deformations, troughs or bumps, which can be sufficiently severe to induce flow separation.

Here, the flow is analyzed by using the Successive Complementary Expansion Method [1], SCEM, in which we seek a Uniformly Valid Approximation, UVA, based on generalized asymptotic expansions.

In the study of high Reynolds number flows, the first idea is to consider Euler equations formally obtained from Navier–Stokes equations when the Reynolds number tends to infinity. Then, an asymptotic analysis can be applied and it is tempting to call for a hierarchical process. The first step is to solve the Euler equations. In the vicinity of singular zones, near the walls or in the wakes, the second step consists in trying to correct the first approximation by a boundary layer analysis.

J. Mauss (✉)
Institut de Mécanique des Fluides de Toulouse UMR-CNRS and Université Paul Sabatier, 118 route de Narbonne, 31062 Toulouse Cedex, France, E-mail: mauss@cict.fr

A.F. Hegarty et al. (eds.), *BAIL 2008 - Boundary and Interior Layers.*
Lecture Notes in Computational Science and Engineering,
DOI: 10.1007/978-3-642-00605-0,

However, in many problems involving a strong coupling, this type of hierarchical approach is known not to be possible. Excluding a multi-layer approach of triple deck type [4, 5], which introduces very restrictive hypotheses on the scales, a possibility is to use generalized asymptotic expansions. According to this method, the small parameters of the problem can be included in the functions which form the expansions. This idea is very different because the small parameters are not considered as tending towards zero but are only small. Thanks to the generalized expansions, the effects of the eulerian region on the boundary layer region and the reciprocal effects are considered simultaneously and not hierarchically. Moreover, the construction of a UVA does not require any matching principle, only the boundary conditions of the problem are used.

After the formulation of the problem (Sect. 2), a direct analysis (Sect. 3) with small wall deformations shows that the Navier–Stokes equations reduce to a coupled system consisting of generalized boundary layer equations *uniformly valid in the whole flow* – the so-called field equations – and linearized Euler equations – the so-called core equations. A deeper study is performed by separating geometrically the symmetric and antisymmetric parts (Sect. 4). The analysis of the flow enables us to improve the usual asymptotic hypotheses and to consider original configurations. Comparisons of the evolution of the skin-friction coefficient with Navier–Stokes solutions show the relevance of the proposed approach (Sect. 5).

2 Formulation of the Problem

The Navier–Stokes equations are written in nondimensional form in an orthogonal axis-system (x, y)

$$\frac{\partial \mathcal{U}}{\partial x} + \frac{\partial \mathcal{V}}{\partial y} = 0, \tag{1a}$$

$$\mathcal{U}\frac{\partial \mathcal{U}}{\partial x} + \mathcal{V}\frac{\partial \mathcal{U}}{\partial y} = -\frac{\partial \mathcal{P}}{\partial x} + \frac{1}{\mathcal{R}}\left(\frac{\partial^2 \mathcal{U}}{\partial x^2} + \frac{\partial^2 \mathcal{U}}{\partial y^2}\right), \tag{1b}$$

$$\mathcal{U}\frac{\partial \mathcal{V}}{\partial x} + \mathcal{V}\frac{\partial \mathcal{V}}{\partial y} = -\frac{\partial \mathcal{P}}{\partial y} + \frac{1}{\mathcal{R}}\left(\frac{\partial^2 \mathcal{V}}{\partial x^2} + \frac{\partial^2 \mathcal{V}}{\partial y^2}\right). \tag{1c}$$

with $\mathcal{R}$ denoting the Reynolds number based on the width of the non-perturbed channel and a reference velocity such that the basic plane Poiseuille flow is

$$u_0 = \frac{1}{4} - y^2,\ v_0 = 0,\ p_0 = -\frac{2x}{\mathcal{R}} + p_c. \tag{2}$$

where p_c is an arbitrary constant. The channel is perturbed by indentations of the lower and upper walls

$$y_l = -\frac{1}{2} + F(x, \varepsilon),\ y_u = \frac{1}{2} - G(x, \varepsilon), \tag{3}$$

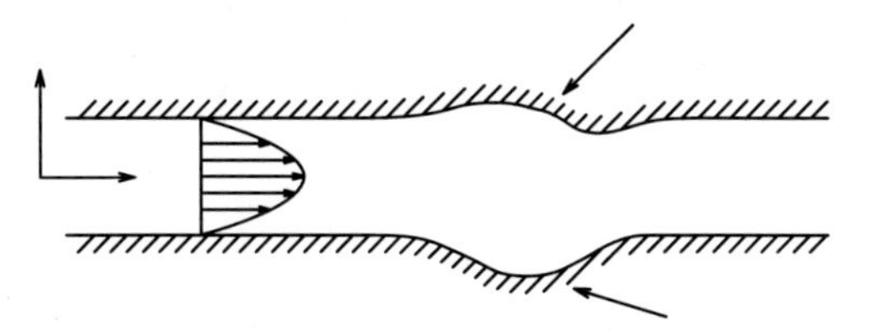

Fig. 1 Flow in a two-dimensional channel with deformed walls. In this figure, all quantities are dimensionless

where ε is a parameter (Fig. 1). At high Reynolds number, the reduced equations obtained formally by taking their limit when the Reynolds number goes to infinity are of first order. A singular perturbation problem arises.

3 Direct Analysis

To go further, it is usual [4, 5] to consider small wall perturbations leading to assumption $(H1)$

$$(H1): F = \varepsilon f,\ G = \varepsilon g. \tag{4}$$

A perturbation is said to be significant when flow separation is possible. To translate this, it is required that, in boundary layers of thickness ε, the perturbation of the longitudinal velocity is of the same order as u_0, i.e. of order $\mathrm{O}(\varepsilon)$. Thus, according to SCEM, we are seeking a UVA of the form

$$\mathcal{U} = u_0(y) + \varepsilon \hat{u}(x, y, \varepsilon) + \cdots = u(x, y, \varepsilon) + \cdots, \tag{5a}$$

$$\mathcal{V} = \varepsilon \hat{v}(x, y, \varepsilon) + \cdots = v(x, y, \varepsilon) + \cdots, \tag{5b}$$

$$\mathcal{P} = p_0(x) + \varepsilon \hat{p}(x, y, \varepsilon) + \cdots = p(x, y, \varepsilon) + \cdots. \tag{5c}$$

It must be noted that $\hat{u}$, $\hat{v}$, $\hat{p}$ are functions not only of x and y but also of ε. Expansions (5a–5c) are said to be generalized to underline the difference with regular expansions in which $\hat{u}$, $\hat{v}$, $\hat{p}$ would not depend on ε. An asymptotic expansion is not necessarily based on regular expansions and it has been shown that generalized expansions are more powerful for certain boundary layer problems [1].

In the *whole flow field*, Navier–Stokes equations reduce to [1]

$$\frac{\partial \hat{u}}{\partial x} + \frac{\partial \hat{v}}{\partial y} = 0, \tag{6a}$$

$$u_0 \frac{\partial \hat{u}}{\partial x} + \hat{v}\frac{\mathrm{d}u_0}{\mathrm{d}y} + \varepsilon\left(\hat{u}\frac{\partial \hat{u}}{\partial x} + \hat{v}\frac{\partial \hat{u}}{\partial y}\right) = -\frac{\partial \hat{p}_1}{\partial x} + \frac{1}{\mathcal{R}}\frac{\partial^2 \hat{u}}{\partial y^2}, \tag{6b}$$

where the index "1" denotes the characteristics of the flow perturbation in the core. As shown in [1], it must be noted that in the streamwise momentum equation, $\frac{\partial \hat{p}}{\partial x}$ is replaced by $\frac{\partial \hat{p}_1}{\partial x}$. Equations (6a–6b) have the same form as the standard boundary layer equations but $\hat{p}_1(x, y)$ is a solution of core flow equations given below. In the core, (6a–6b) reduce to the core flow equations up to negligible terms and therefore (6a–6b) *are valid in the whole field.*

The global interactive boundary layer model described by (6a–6b) and the core flow equations is the best approximation of Navier–Stokes model we can propose but it is not easy to solve. Fortunately, it can be shown that the core flow (Euler) equations can be linearized and the solution of the resulting model is much easier [2]. Thus, the field equations are structurally non-linear whereas the core flow equations are linear. With notations defined by (5a–5c), the field and core flow equations can now be written

$$\frac{\partial u}{\partial x} + \frac{\partial v}{\partial y} = 0, \tag{7a}$$

$$u\frac{\partial u}{\partial x} + v\frac{\partial u}{\partial y} = -\frac{\partial p_1}{\partial x} + \frac{1}{\mathcal{R}}\frac{\partial^2 u}{\partial y^2}, \tag{7b}$$

$$\frac{\partial u_1}{\partial x} + \frac{\partial v_1}{\partial y} = 0, \tag{8a}$$

$$u_0\frac{\partial u_1}{\partial x} + v_1\frac{\mathrm{d}u_0}{\mathrm{d}y} = -\frac{\partial}{\partial x}\left(p_1 - p_0\right), \tag{8b}$$

$$u_0\frac{\partial v_1}{\partial y} = -\frac{\partial}{\partial y}\left(p_1 - p_0\right). \tag{8c}$$

In the above equations index "1" refers to quantities satisfying the core flow equations. From (8a–8c), it is found that v_1 is solution of Poisson's equation

$$u_0\left(\frac{\partial^2 v_1}{\partial x^2} + \frac{\partial^2 v_1}{\partial y^2}\right) = v_1\frac{\mathrm{d}^2 u_0}{\mathrm{d}y^2}, \tag{9}$$

and the x-component of the pressure gradient required to solve the generalized boundary layer equations is given by (8b) in which the continuity equation (8a) is taken into account

$$-u_0\frac{\partial v_1}{\partial y} + v_1\frac{\mathrm{d}u_0}{\mathrm{d}y} = -\frac{\partial}{\partial x}\left(p_1 - p_0\right). \tag{10}$$

It can be shown that (9) associated to (10) gives the y-momentum equation (8c) if the perturbations vanish at upstream infinity. This establishes the equivalence between (8a–8c) and (9–10).

To sum up, the problem to solve comprises (7a–7b), (9) and (10). At the walls, the boundary conditions are

$$y = y_\ell \text{ and } y = y_u : u = 0, \ v = 0, \tag{11}$$

and the coupling between the core flow equations and the generalized boundary layer equations is expressed by identifying u, v and u_1, v_1 in the core

$$(u, v) \to (u_1, v_1). \tag{12}$$

The model presented above belongs to a class of strong coupling method since there is no hierarchy between the boundary layer equations and the core flow equations. The triple deck theory, or more precisely its equivalent for channel flows as developed by Smith [4, 5], belongs also to this class of strong coupling models. In fact, Smith's model is included in the present model since the expansions are regular whereas in the present model the expansions are generalized. It is interesting to note that the first approximation of Smith's model for v_1 is symmetric with respect to y and corresponds to a geometrically antisymmetric problem. In the core, Smith's model gives

$$v_1 = -u_0(y)\frac{\mathrm{d}A(x)}{\mathrm{d}x}, \tag{13}$$

where A is defined as the displacement function. It must be noted that (13) is an eigensolution of (10) but not of (9). This remark leads us to try to separate as far as possible the symmetric and antisymmetric problems which leads, as we will see, to a new approach of the asymptotic problem. The issue of asymmetry has been approached earlier by Lagrée et al. [3].

4 Influence of Asymmetry

The analysis starts from (1a–1c) in which we introduce the transformation

$$X = x, \ Y = y - \nu H(\mu x, \varepsilon), \ U = \mathcal{U}, \ V = \mathcal{V} - \mathcal{U}\nu\frac{\mathrm{d}H}{\mathrm{d}x}, \ P = \mathcal{P}, \tag{14}$$

where ν and μ are order functions such that $\nu \preceq 1$ and $\mu \prec 1$. We have

$$H = \mathrm{O_S}(1) \text{ and } \frac{\mathrm{d}^n H}{\mathrm{d}x^n} = \mathrm{O}(\mu^n). \tag{15}$$

where $\mathrm{O_S}$ means "is of strict order of" whereas O means "is at most of order of" [1]. With these hypotheses, Navier–Stokes equations become

$$\frac{\partial U}{\partial X} + \frac{\partial V}{\partial Y} = 0, \tag{16a}$$

$$U\frac{\partial U}{\partial X} + V\frac{\partial U}{\partial Y} = -\frac{\partial P}{\partial X} + \frac{1}{\mathcal{R}}\left(\frac{\partial^2 U}{\partial X^2} + \frac{\partial^2 U}{\partial Y^2}\right) + \mathrm{O}(\nu\mu), \tag{16b}$$

$$U\frac{\partial V}{\partial X} + V\frac{\partial V}{\partial Y} = -\frac{\partial P}{\partial Y} + \frac{1}{\mathcal{R}}\left(\frac{\partial^2 V}{\partial X^2} + \frac{\partial^2 V}{\partial Y^2}\right) + \mathrm{O}(\nu\mu). \tag{16c}$$

If we set

$$E = \frac{F+G}{2}, \quad H = \frac{F-G}{2}, \tag{17}$$

the problem is geometrically symmetrized. Note that the channel is deformed symmetrically when $H = 0$. The wall conditions then become

$$Y = Y_\ell = -\frac{1}{2} + E \text{ and } Y = Y_u = \frac{1}{2} - E : U = 0, \ V = 0. \tag{18}$$

Moreover, for small μ, the basic flow corresponding to $E = 0$ is

$$U_0 = \frac{1}{4} - Y^2, \ V_0 = 0, \ P_0 = -2\frac{X}{\mathcal{R}} + P_c, \tag{19}$$

where P_c is an arbitrary constant. We introduce assumption $(H2)$

$$(H2) : E = \varepsilon e, \ \nu\mu^2 \preceq \varepsilon. \tag{20}$$

With $(H2)$, the complete system to solve comprises the field equations

$$\frac{\partial U}{\partial X} + \frac{\partial V}{\partial Y} = 0, \tag{21a}$$

$$U\frac{\partial U}{\partial X} + V\frac{\partial U}{\partial Y} = -\frac{\partial P_1}{\partial X} + \frac{1}{\mathcal{R}}\frac{\partial^2 U}{\partial Y^2}, \tag{21b}$$

where P is replaced by P_1 and the core flow equations which can be linearized

$$\frac{\partial U_1}{\partial X} + \frac{\partial V_1}{\partial Y} = 0, \tag{22a}$$

$$U_0\frac{\partial U_1}{\partial X} + V_1\frac{\mathrm{d}U_0}{\partial Y} = -\frac{\partial}{\partial X}\left(P_1 - P_0\right), \tag{22b}$$

$$U_0\frac{\partial V_1}{\partial X} + \nu\frac{\mathrm{d}^2 H}{\mathrm{d}X^2}U_0^2 = -\frac{\partial}{\partial Y}\left(P_1 - P_0\right). \tag{22c}$$

This system is solved with (18) and the coupling condition in the core

$$V \to V_1. \tag{23}$$

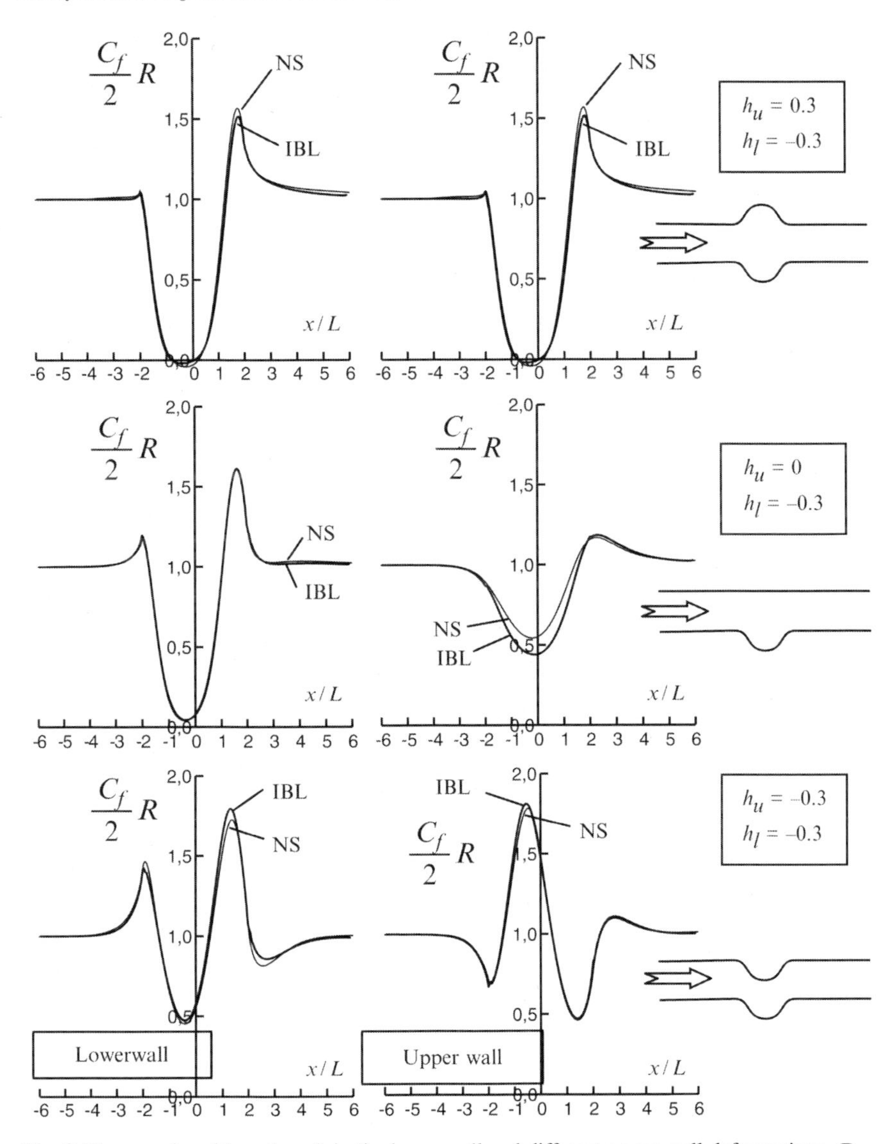

Fig. 2 Flow produced by a trough in the lower wall and different upper wall deformations. $\mathcal{R} = 1{,}000$. NS-Navier–Stokes results, IBL-Interactive Boundary Layer results

5 Results and Conclusions

To assess the validity of the Interactive Boundary Layer method, IBL, it is chosen to examine the evolution of the skin-friction coefficient which is a very sensitive flow feature, $C_f = \dfrac{2}{\mathcal{R}}\tau_w$ where τ_w is the reduced wall shear stress. Details on the

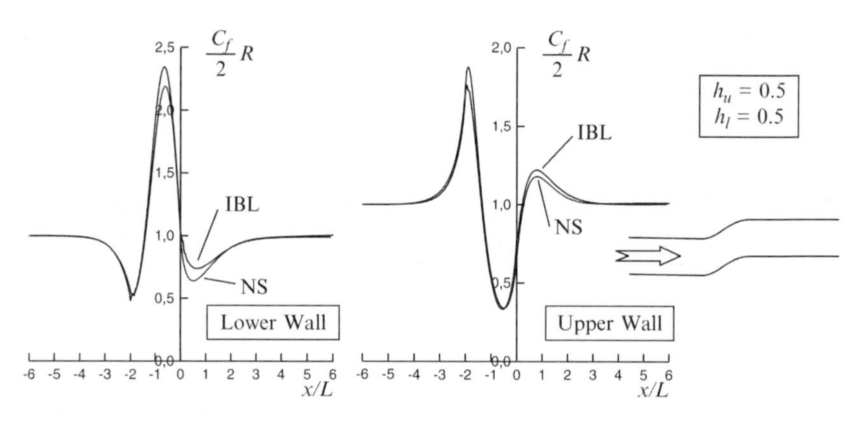

Fig. 3 Flow produced by a channel bend. $\mathcal{R} = $ **1,000**. NS-Navier–Stokes results; IBL = Interactive Boundary Layer results

numerical procedure can be found elsewhere [2]. The Navier–Stokes equations are solved with the commercial code FLUENT. Different cases are calculated in which the walls are deformed in a domain $x_1 \leq x \leq x_2$

$$F = \frac{h_l}{2}\left(1 + \cos\frac{2\pi x}{L}\right); G = -\frac{h_u}{2}\left(1 + \cos\frac{2\pi x}{L}\right); L = 4. \qquad (24)$$

For all cases, the Reynolds number is $\mathcal{R} = 1{,}000$.

At first, comparisons between IBL and Navier–Stokes results are given in Fig. 2. The lower wall is deformed by a trough located in the domain $-2 \leq x \leq 2$ with $h_l = -0.3$. The upper wall is deformed in the same domain $-2 \leq x \leq 2$ but different upper wall shapes have been investigated between the symmetric case ($h_u = 0.3$) and the antisymmetric case ($h_u = -0.3$). Even though the amplitude of the wall deformation is not really small as required by the theory, an excellent agreement with Navier–Stokes results is observed.

The IBL model enables us also to treat original problems. In the case of a bend, when the channel does not recover its initial position at the downstream end, the usual techniques of small perturbations do not work any longer. As an example, the walls are deformed in the domain $-2 \leq x \leq 0$ with $h_l = 0.5$ and $h_u = 0.5$; for $x > 0$, we have $y_l = 0$, $y_u = 1$ so that the channel axis is displaced from $y = 0$ upstream to $y = 0.5$ downstream. In this case again, a good agreement with Navier–Stokes results is observed (Fig. 3). This shows that $\frac{\mathrm{d}H}{\mathrm{d}x}$ which characterizes the influence of the antisymmetric part of the wall deformation plays an important role in the definition of what could be the small parameter of the problem.

Other non usual cases can be treated by this method, for example dilated or constricted channels, . . . The IBL calculations are much faster than Navier–Stokes calculations and, in addition, the new asymptotic analysis helps us to understand the flow structure. Moreover, this step is necessary to approach the important problem of separation control.

References

1. J. Cousteix and J. Mauss. *Asymptotic analysis and boundary layers*, volume XVIII, Scientific Computation. Springer, Berlin, 2007.
2. J. Cousteix and J. Mauss. Interactive boundary layer models for channel flow. *European Journal of Mechanics B: Fluids*, 2008. doi:1016/j.euromechflu.2008.01.003.
3. P.Y. Lagrée, A. van Hirtum, and X. Pelorson. Asymmetrical effects in a 2d stenosis. *European Journal of Mechanics B: Fluids*, 26:83–92, 2007.
4. F.T. Smith. Flow through constricted or dilated pipes and channels: part 1. *Quarterly Journal of Mechanics and Applied Mathematics*, XXIX(Pt 3):343–364, 1976.
5. F.T. Smith. Flow through constricted or dilated pipes and channels: part 2. *Quarterly Journal of Mechanics and Applied Mathematics*, XXIX(Pt 3):365–376, 1976.

Turbulence Receptivity of Longitudinal Vortex-Dominated Flows

C. Moldoveanu, A. Giovannini, and H.C. Boisson

Abstract The transient effect of turbulence forcing on an isolated vortex has been investigated by means of Large Eddy Simulation. A dynamic mixed model is used to relate resolved and subgrid scale fluctuations in an incompressible CFD code. The initial condition is an Oseen vortex developing in a periodic box and subsequently forced by a superimposed turbulence. Turbulence is imposed in three different situations, separately outside or inside the initial vortex core and in the whole computational domain. The instantaneous flow field is observed for an outside forcing and it exhibits a wandering motion of the vortex as a whole. The velocity fluctuations grow near the vortex axis while they decrease for an inside forcing. The whole domain forcing is close to the external forcing, implying that most of the turbulence interaction is induced by the outer zone. An analytical investigation of the wandering effect shows that the near-axis fluctuations are mainly produced by this effect.

1 Introduction

Longitudinal vortices are known as permanent low decay structures that preserve vorticity in a quasi-2D situation. In aircraft traffic, trailing vortices are known to create a serious hazard and this produces a severe limitation in distance and time spacing for aeroplanes following each other in take-off and landing. During the last decade, an important research effort has been conducted at the European level on the problem of vortex wake decay [1]. However this is not the only situation in which longitudinal vortices are involved. Many atmospheric phenomena or many industrial situations are due to vortices in isolated, counter-rotating or co-rotating

H.C. Boisson (✉)
Université de Toulouse, Institut de Mécanique des Fluides de Toulouse, UMR CNRS/INPT/UPS, Allée Camille Soula, 31400 Toulouse, France, E-mail: Henri.Boisson@imft.fr

A.F. Hegarty et al. (eds.), *BAIL 2008 - Boundary and Interior Layers.* Lecture Notes in Computational Science and Engineering, DOI: 10.1007/978-3-642-00605-0,

vortex systems. Melander and Hussain [2] consider these as the basic form of a coherent structure in turbulent flow including wall elongated structures.

The present work is devoted to the influence of turbulence on a single isolated vortex in order to determine its intrinsic behaviour while excluding mutual interactions that exist in vortex systems.

Turbulence can be entrained by rolling up during the vortex core formation or transferred during the exchanges between the vortex core fluid and the free stream. In experiments, vortices are generated by wings [3,4] and grid turbulence is imposed. The point measurements made lead to the conclusion that the fluctuating motion can penetrate the vortex core and that an annular distribution is observed for the longitudinal velocity fluctuation intensity. All these measurements are made at fixed positions and it should be pointed out that they cannot directly distinguish between instantaneous block motion of the vortex and local turbulence.

However some observations lead one to propose that the vortex core can remain unperturbed and that the apparent turbulence level measured by a fixed probe is mainly due to the effect of wandering – that is to say, the displacement of the whole vortex in space. This effect was pointed out by Devenport et al. [5], who have proposed a correction to Laser Doppler Anemometer measurements in order to obtain the actual turbulence level inside the core. Numerical simulation seems to confirm this view and some authors have noticed the appearance of bending modes of the centreline of the longitudinal vortex, but they [2, 6, 7] attribute the enhanced decay of the vortex structure to an increased dissipation rate of turbulence by secondary coherent structures in the outer layer of the vortex.

The objective of the present numerical contribution is to understand the underlying mechanisms and to address the following questions: (1) what mechanism excites the perturbations inside the vortex core? (2) what contribution comes from the wandering oscillation of the vortex tubes centreline? To answer both these questions, Large Eddy Simulation is used. The subgrid scale model, which was proved to respect the physics of boundary layer transition in the work of Péneau et al. [8] and was adapted for wake trailing vortices by Moldoveanu et al. [9], has been used to study the present case of an isolated vortex. Forcing is applied separately first in the domain external to the core then inside the core and finally in the whole domain. Simulation allows such experiments.

The paper is divided into three main sections: Mathematical model and strategy (Sect. 2), Results of the forcing (Sect. 3), Analytical development of wandering (Sect. 4).

2 Mathematical Model and Strategy

The numerical simulations are carried out using the JADIM code. The Navier–Stokes equations (1) are discretized using a second-order centred scheme on a staggered grid. The resulting terms are integrated in space on finite volumes and the solution is advanced in time by means of a three-step Runge–Kutta procedure.

The nonlinear terms are computed explicitly while the diffusive terms are calculated using the semi-implicit Crank–Nicholson algorithm. To satisfy the incompressibility condition, a Poisson equation is solved by combining a direct inversion in the (x, y) plane with a spectral Fourier method in the z direction.

The Dynamic Mix Model (DMM) was proposed by Zang et al. [10] and modified by Calmet and Magnaudet [11]. The implicit spatial filtering of the Navier–Stokes equations is imposed by the finite volume numerical method, namely the box filter on $\overline{\Delta_i}$, the local mesh spacing in the ith direction. Decomposing the velocity field into a resolved part $\overline{V_i}$ (field computed by the code), and a subgrid part V'_i (field to model) leads to $V_i(x,t) = \overline{V_i}(x,t) + V'_i(x,t)$ and to

$$\frac{\partial \overline{V_i}}{\partial x_i} = 0, \qquad \frac{\partial \overline{V_i}}{\partial t} + \frac{\partial \overline{V_i} \ \ \overline{V_j}}{\partial x_j} = -\frac{1}{\rho}\frac{\partial \overline{P}}{\partial x_i} + \frac{\partial (2\nu \bar{S}_{ij} + \tau_{ij}^{sm})}{\partial x_j}, \tag{1}$$

where $\bar{S}_{ij} = \frac{1}{2}(\frac{\partial \overline{V_i}}{\partial x_j} + \frac{\partial \overline{V_j}}{\partial x_i})$ denotes the resolved strain rate tensor, ρ the density and ν the kinematic viscosity.

In the DMM used for this study, the Leonard term (resolved velocity correlations) is obtained using a double filtering operation and calculated explicitly while the cross term and the Reynolds stress term are modelled. The final expression for the double filtered stress tensor T_{ij}^{sm} is

$$T_{ij}^{sm} = (\nu + C_S^M \overline{\Delta})(|S| \cdot |S_{ij}|) - C_S^M \overline{\Delta}^F (|S^F| \cdot |S_{ij}^F|). \tag{2}$$

This stress tensor is deduced from the subgrid scale tensor τ_{ij}^{sm} by expressing the explicit space filtering operation. All tensors denoted by $(\cdot)^F$ are double filtered contributions. The first right-hand side term represents viscous and subgrid contributions, and the second term the exchanges between the resolved and unresolved parts of the flow. The coefficient C_S^M is calculated by an optimizing process for each grid point. See Calmet and Magnaudet [11] for more elaborate details.

The computational domain comprises a cubic box with a periodic boundary condition in two directions (including the axis Oz of the vortex tube) and a symmetry condition in the last direction. The box dimensions and grid size are given in Table 1. The grid is refined in the plane xOy perpendicular to the vortex tube in order to have a good resolution (37 grid points) of the vortex core.

The vortex profile is defined as the so-called Oseen vortex with the following distributions of vorticity and tangential velocity:

Table 1 Dimensions and grid size of the computational domain

Direction	L_x	L_y	L_z
Size	3	3	3
Grid points	128	128	64

$$\omega(r) = \frac{\Gamma_0}{\pi r_0^2} e^{-\frac{r^2}{r_0^2}}, \qquad v_\theta(r) = \frac{\Gamma_0}{2\pi r}\left(1 - e^{-\frac{r^2}{r_0^2}}\right), \tag{3}$$

where $\Gamma_0 = -1$ is the circulation of the vortex, (r,θ) are the radial and azimuthal coordinates, r_0 is the radius ($r_0 = 0.1$), and the Reynolds number of the flow is $Re_v = \Gamma_0/\nu = 20.000$. The reference velocity $V_0 = \Gamma_0/(2\pi r_0)$ is equal to 1.59.

The first objective is to run an extensive LES simulation in order to calibrate the ability of the code to handle vortex dynamics without any excessive numerical dissipation. Introducing the viscous reference time $t_v = r_0^2/(4\nu)$ and the reduced time $T = t/t_v$, in Fig. 1 we compare the numerical evolution in time of the kinetic energy E_c (a) and the enstrophy Z (b) with their theoretical expressions. The conclusion is that the agreement is acceptable and the numerical diffusivity low.

In order to investigate the influence of the turbulence, perturbations are supplemented to the vortex system. This is achieved by superposition of a uniform random perturbation on each velocity component with, in this case, the amplitude maximum of 0.03 V_0. After a few time steps the turbulence field is established.

3 Results of the Forcing

3.1 External Forcing

In the first stage of the numerical experiment, white noise perturbation is introduced only outside the vortex core. In Fig. 2 the effect of rotation is visible as it stretches the large scale structures in the outer mixing layer. At $T = 0.04$ the flow is just becoming established after the initial perturbation but at $T = 0.24$ the system has undergone revolutions and vorticity has diffused considerably in the wrapped shear layer.

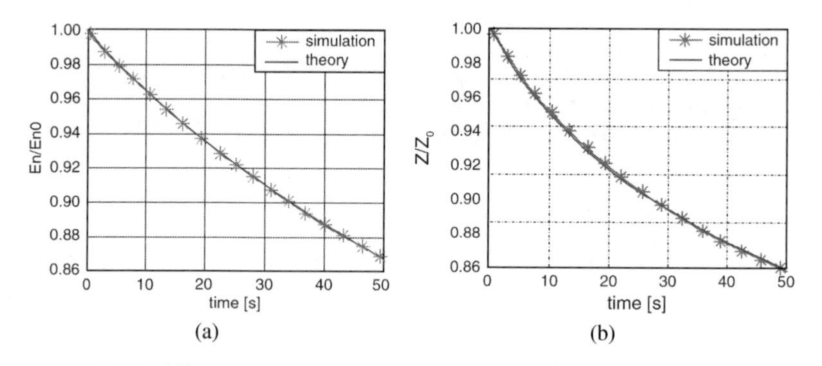

Fig. 1 Numerical evolutions: (**a**) kinetic energy (**b**) enstrophy

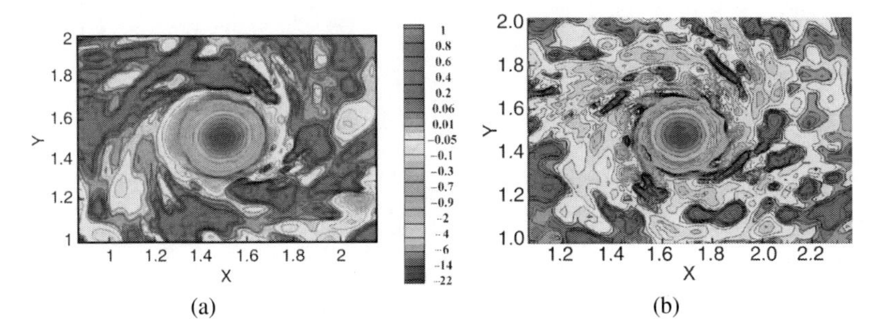

Fig. 2 Isovorticity contours: (**a**) $T = \mathbf{0.04}$ (**b**) $T = \mathbf{0.24}$

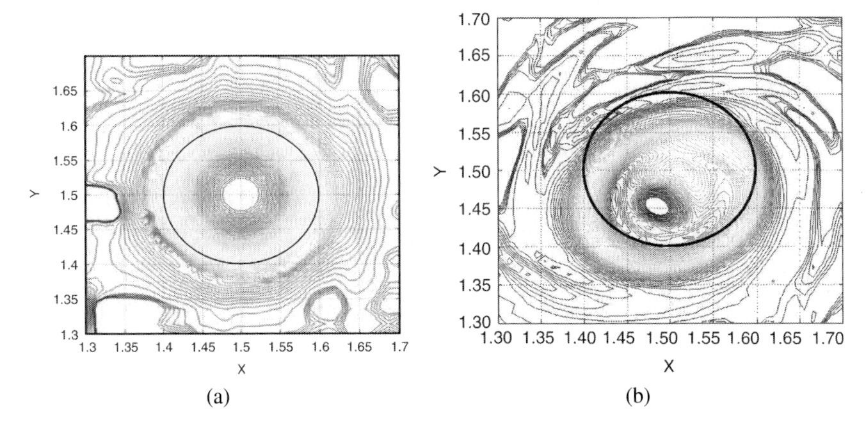

Fig. 3 Close-up of isovorticity contours: (**a**) $T = \mathbf{0.00}$ (**b**) $T = \mathbf{0.24}$

A close-up of the vorticity lines is provided in Fig. 3. This shows a very significant pattern of the processing motion of the vortex centre. It is obvious that a wandering effect is observed and that one must consider the block displacement of the vortex around its initial position.

The apparent turbulent velocity components have been recorded after a space averaging operation along the Z-direction on the line $x = 1.5$ (which is the position of the axis crossing the centre of the vortex) before introducing the forcing effect of imposed turbulence in the outer region; (see Fig. 4). No attempt has been made to correct for any wandering effect. A large amplification of the fluctuations of the u and v components is observed. This leads to a quasi-isotropic level for both directions in the plane, a situation which is also observed in physical experiments. The case of the longitudinal w-fluctuation is different: the level obtained is not generally of the same order as the other components. The guess is that the observed inner fluctuation level is mainly due to the wandering effect and it seems obvious that this effect is not the same in the in-plane components as in the off-plane component. This latter exhibits a double peak in the shear layer that develops around the initial vortex core.

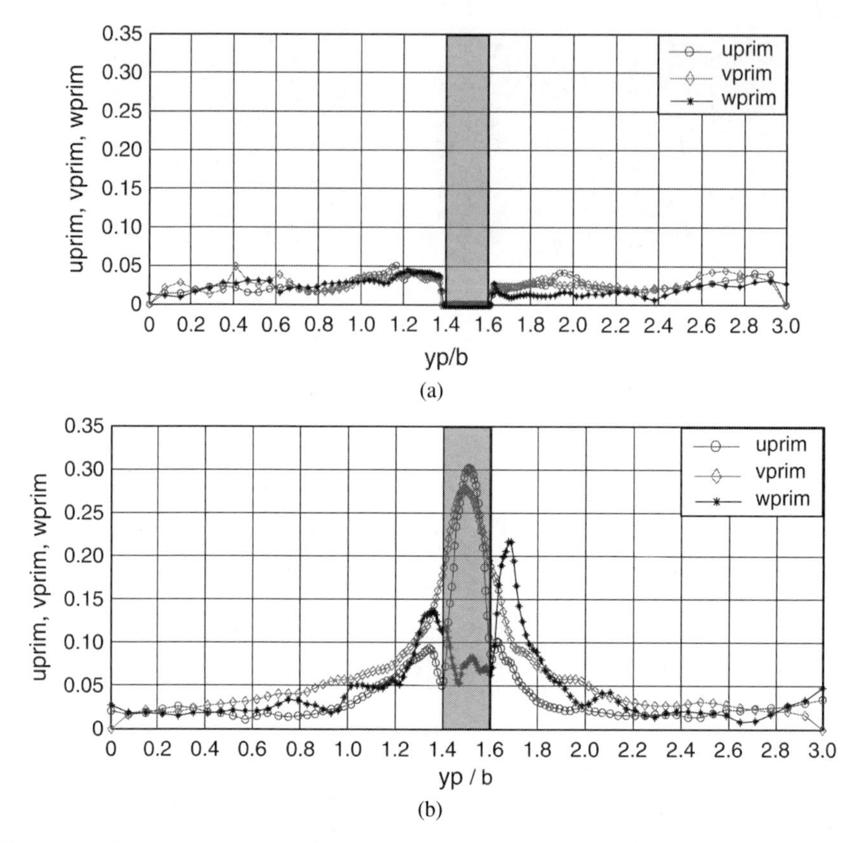

Fig. 4 Eulerian space-averaged velocity fluctuations at $x = 1.5$: (**a**) $T = \mathbf{0.00}$ (**b**) $T = \mathbf{0.07}$ (external forcing)

3.2 Internal Forcing

Under the same conditions as the previous case, the perturbations were introduced only in the core of the vortex and the fluctuation level (rms) is displayed in Fig. 5. A quasi-constant fluctuation level is obtained in the inner part of the core and this level rapidly decreases in the outer flow and remains at zero in almost all the external domain. The fluctuation in the inner part decreases also as shown for $T = 0.084$. This inner level continues to decrease up to the time $T = 0.18$. At this value of T, the u' and v' levels are still more than twice the w' levels. However it does not seems that this difference can be attributed to wandering as for the case of forcing the outer layer of the vortex.

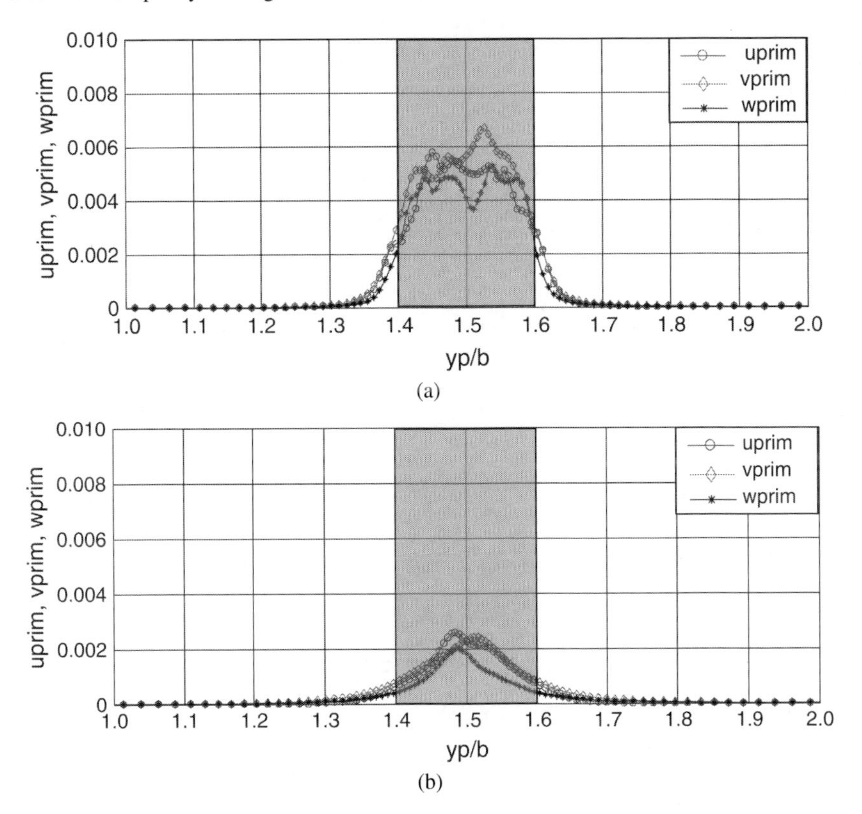

Fig. 5 Eulerian space averaged velocity fluctuations at $x = 1.5$: (**a**) $T = \mathbf{0.00}$ (**b**) $T = \mathbf{0.084}$ (internal forcing)

3.3 *Whole Domain Forcing*

The case that should probably be the more usual is where turbulence is introduced both in the core by rolling up of an initially turbulent layer and outside the core by external turbulence. Of course both turbulent layers would probably have different properties. In this simulation a uniform level of fluctuation is introduced inside and outside the vortex core without any discontinuity in the initial turbulence levels or length-scales.

In Fig. 6 the results of the forcing are presented in the axis of the vortex on the line $x = 1.5$. Roughly speaking, the pattern observed is closer to the case with outer forcing than the one with inner forcing. The levels obtained in the core for the u' and v' levels are almost identical to the ones observed for the first situation. However some differences can be noticed. First, the u' level in the outer zone shows an amplification with respect to the initially-imposed turbulence level. Second, the peaks for the w' component seems to develop inside the initial vortex core. One can infer that the shear layers are modified by turbulence in the inner core.

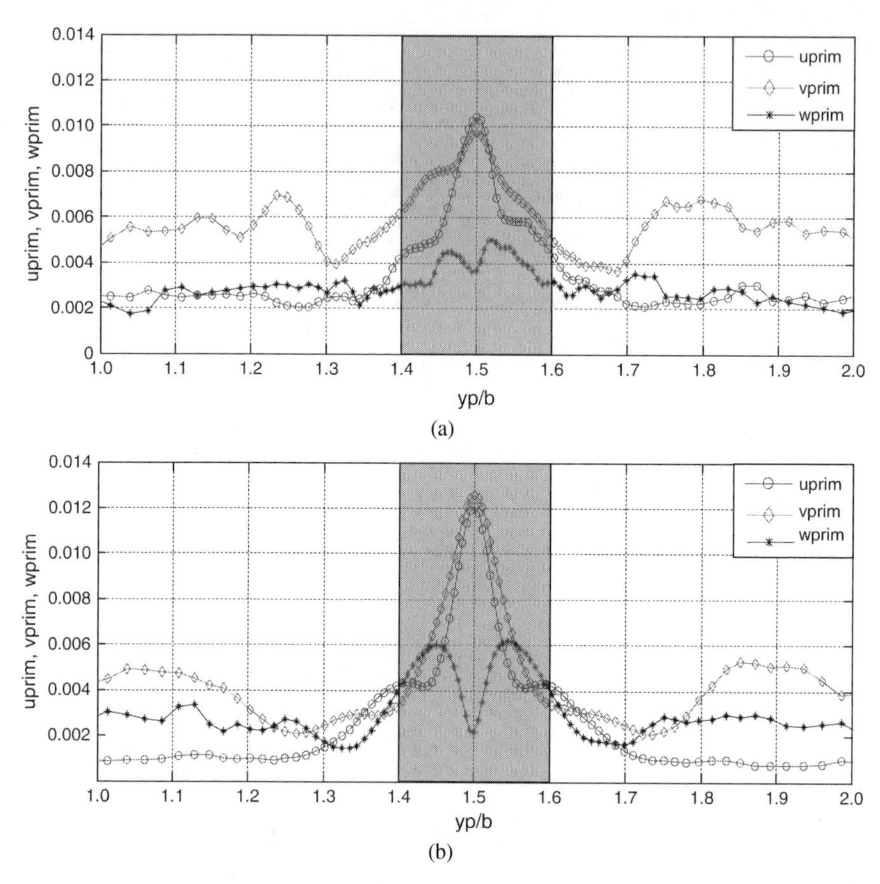

Fig. 6 Eulerian space averaged velocity fluctuations at $x = 1.5$: (**a**) $T = \mathbf{0.03}$ (**b**) $T = \mathbf{0.16}$ (whole domain forcing)

A comparison of the three different cases is provided in Fig. 7, which displays the u' velocity fluctuations with respect to time at the axis position of the initial vortex. It can be noticed that the external forcing effect is close to the one of the whole domain forcing on a large part of the development but is amplified in a pseudo-periodic oscillating motion and reaches larger amplitudes.

4 Vortex Wandering: A Simple Analytical Development

In the theoretical frame of Devenport [5] with Gaussian random fluctuations of the centre of the vortex tube for Lamb-Oseen vortex distributions (U_{L-O} and V_{L-O}), and an amplitude $\sigma << r_0$, the statistics upon U and V components on the plane $x0y$ are written as

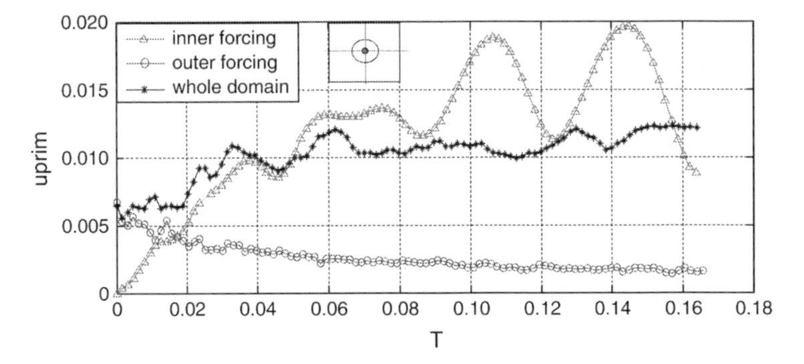

Fig. 7 Velocity fluctuations on the vortex axis (u′): inner forcing (*circle*), outer forcing (*triangle*), whole domain forcing (*plus*)

$$\overline{<U(x,y)>} = U_{L-O}, \qquad \overline{<V(x,y)>} = V_{L-O}$$

for the averaged values with $\overline{<.>}$ denoting a statistical average. Then

$$\begin{aligned} \overline{<u^2>} &= \sigma^2\left(\left(\frac{\partial U_{L-O}}{\partial x}\right)^2 + \left(\frac{\partial U_{L-O}}{\partial y}\right)^2\right), \\ \overline{<v^2>} &= \sigma^2\left(\left(\frac{\partial V_{L-O}}{\partial x}\right)^2 + \left(\frac{\partial V_{L-O}}{\partial y}\right)^2\right) \end{aligned} \tag{4}$$

$$\overline{<uv>} = \sigma^2\left(\frac{\partial U_{L-O}}{\partial x}\frac{\partial V_{L-O}}{\partial x} + \frac{\partial U_{L-O}}{\partial y}\frac{\partial V_{L-O}}{\partial y}\right) \tag{5}$$

After some analytical development, this leads to:

$$\overline{<u^2>} = V_0^2\frac{\sigma^2}{r_0^2}\frac{1}{\xi^4}\Big(1 - e^{-\xi^2}(2 + 4y^2/r_0^2) + e^{-2\xi^2}(1 + 4\frac{y^2r^2}{r_0^4} + 4\frac{y^2}{r_0^2})\Big) \tag{6}$$

$$\overline{<v^2>} = V_0^2\frac{\sigma^2}{r_0^2}\frac{1}{\xi^4}\Big(1 - e^{-\xi^2}(2 + 4x^2/r_0^2) + e^{-2\xi^2}(1 + 4\frac{x^2r^2}{r_0^4} + 4\frac{x^2}{r_0^2})\Big) \tag{7}$$

$$\overline{<uv>} = 4V_0^2\frac{\sigma^2}{r_0^2}\frac{1}{\xi^4}\Big(e^{-\xi^2}xy\big(1 + \xi^2e^{-\xi^2}\big) - 1\Big) \tag{8}$$

for the fluctuations expressed with $\xi = r/r_0$, and $r = \sqrt{(x-x_c)^2 + (y-y_c)^2}$. (x_c, y_c) is the center of the intitial vortex.

This analytical distribution defines peaks for the rms of the fluctuations at the centre of the initial vortex similar to the ones observed in the profiles at a value $u'_{rms} = v'_{rms} = V_0\sigma/r_0$. It can be noticed that the $\overline{<u^2>}$ and $\overline{<v^2>}$ decrease with a power law ξ^{-4}. The cross correlation $\overline{<uv>}$ has the shape of a quadripole with alternated maxima away from the central position for which it is equal to zero. These distributions are also similar to the one of the simulation confirming the necessity of taking this effect into account.

5 Conclusions

We have run numerical simulations in order to separate the effect of external and internal turbulence. The external turbulence organizes itself by wrapping and stretching around the vortex tube in such a way that both organized azimuthal structures around the core location and bending of the centreline of the tube occur. The external turbulence produces fluctuations inside the core but the internal turbulence does not propagate outside and is damped. These effects compete in the real life of a vortex tube during the roll-up process. These results are compared qualitatively to an analytical model for wandering that shows the same behaviour as the simulations.

References

1. Far wake Program final report (2008), http://www.far-wake.org/
2. Melander, M.V., Hussain, F.: Coupling between a coherent structure and fine-scale turbulence, Physical Review E **48**, 2669–2689 (1993)
3. Beninati, M.L., Marshall, J.S.: An experimental study of the effect of free-stream turbulence on a trailing vortex, Experiments in Fluids **38**, 244–257 (2005)
4. Bailey, S.C.C., Tavoularis, S.: Measurements of the velocity field of a wing-tip vortex, wandering in grid turbulence, Journal of Fluid Mechanics **601**, 281–315 (2008)
5. Devenport, W.J., Rife, M.C., Liapis, S.I., Follin, G.S.: The structure and development of a wing-tip vortex, Journal of Fluid Mechanics, **312**, 67–106 (1996)
6. Holzpfel, F., Hofbauer, T., Darracq, D., Moet, H., et al.: Analysis of wake vortex decay mechanisms in the atmosphere, Aerospace Science and Technology **7**, 263–275 (2003)
7. Takahashi, N., Ishii, H., Miyazaki, T.: The influence of turbulence on a columnar vortex, Physics of Fluids **17**, 035105 (2005); DOI:10.1063/1.1858532
8. Péneau, F., Boisson, H.C., Kondjoyan, A., Uranga, A.: Bypass transition of a boundary layer subjected to free-stream turbulence, In: International Conference on Boundary and Internal Layers BAIL 2004, Toulouse, July 5–9, (2004)
9. Moldoveanu, C., Boisson, H.C., Giovannini, A.: Receptivity of a longitudinal contra-rotating vortex pair in an external flow: a numerical experimentation, In: International Conference on Boundary and Internal Layers BAIL 2004, Toulouse, July 5–9, (2004)
10. Zang, Y., Streer, R.L., Koseff, J.R.: A dynamic mixed subgrid-scale model and its application to turbulent recirculating flows, Physics of Fluids A **5**(12), 3186–3196 (1993)
11. Calmet, I., Magnaudet, J.: Large eddy simulation of high-Schmidt number mass transfer in a turbulent channel flow, Physics of Fluids **9**(2), 438–455 (1996)

A Thermally Induced Singularity in a Wake

Herbert Steinrück and Bernhard Kotesovec

Abstract The two-dimensional laminar flow past a heated horizontal plate is studied in a distinguished limit of large Reynolds and Grashof numbers. The limiting problem constitutes an interaction between the potential flow and the wake flow. It turns out that solutions exist only if the interaction parameter $\kappa = Gr/Re^{9/4}$ is below a critical value. When approaching this critical value a singularity in the wake forms. The nature of this singularity will be analyzed.

1 Introduction

We consider the wake behind a heated horizontal plate of length L and temperature T_p in a parallel free stream of velocity U_∞, temperature T_∞ and angle ϕ with respect to the horizontal direction in a distinguished limit of large Reynolds number $Re = U_\infty L/\nu$ and Grashof number $Gr = g\beta\Delta T L^3/\nu^2$, where ν, β, g and $\Delta T = T_p - T_\infty$ are the kinematic viscosity, the isobaric expansion coefficient of the fluid, which is assumed to be positive, the gravity acceleration, and the difference of the plate temperature and the free stream temperature, respectively. We consider a two-dimensional incompressible flow where buoyancy effects are taken into account using the Boussinesq approximation, see Fig. 1.

Due to the large Reynolds number the flow field can be decomposed into the outer inviscid potential flow, the boundary-layer flow along the plate and the wake behind the plate. However, temperature perturbations are limited to boundary-layers and the wake. The temperature perturbation in the wake causes a vertical hydro-static pressure gradient in the wake. Thus there is a pressure difference across the wake which induces a perturbation of the potential flow [1].

H. Steinrück (✉)
Vienna University of Technology, Institute of Fluid Mechanics and Heat Transfer, Resselgasse 3, 1040 Vienna, Austria, E-mail: herbert.steinrueck@tuwien.ac.at

A.F. Hegarty et al. (eds.), *BAIL 2008 - Boundary and Interior Layers.*
Lecture Notes in Computational Science and Engineering,
DOI: 10.1007/978-3-642-00605-0,

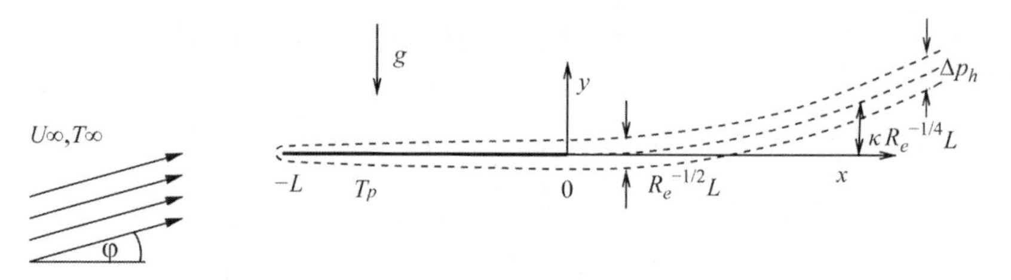

Fig. 1 Mixed convection flow past a horizontal plate

From the viewpoint of the outer (potential) flow the wake is located around the stream line starting from the trailing edge of the plate. Due to the (small) inclination of the wake there is a component of the hydro-static pressure gradient in the main flow direction of the wake flow. Thus the velocity profile and the temperature profile depend on the potential flow, namely on the inclination of the streamline emanating from the trailing edge.

As a consequence the wake flow problem and the potential flow problem form an interaction problem and thus both problems have to be solved simultaneously. This problem has been first formulated in [2] and solved numerically. It turned out that solutions exist only when the interaction parameter κ, which will be introduced in Sect. 2, is below a critical value κ_c. However, this critical value and the behavior of the wake when approaching the critical value has not been discussed. In this paper we focus on the nature of the singularity in the wake when κ approaches its critical value κ_c.

In Sect. 2 we introduce the governing equations in dimensionless form and introduce a suitable interaction parameter κ. Numerical solutions of the interaction problem are discussed in Sect. 3. The nature of this singularity will be discussed in Sect. 4.

2 Governing Equations

To identify the interaction parameter we estimate the magnitudes of the physical quantities involved in the interaction mechanism, see Table 1.

Here ρ denotes the density of the fluid. The interaction parameter κ^2 can be interpreted as the hydro-static pressure gradient acting on the wake flow referred to the double stagnation pressure of the oncoming parallel flow. The parameter K which is here a scale of the inclination of the wake centerline has been introduced first by Schneider and Wasel [3] as buoyancy parameter which describes the influence of buoyancy effects onto the mixed convection boundary-layer flow past a horizontal plate.

An inclination of the wake can be induced by an additional obstacle or an angle of inclination of the parallel oncoming flow. Let us assume that ϕ is a measure for corresponding contribution to the inclination of the wake. Thus the scaled

Table 1 Magnitudes of physical quantities involved in the interaction mechanism

Thickness of the wake	$L\,Re^{-1/2}$
Hydro-static pressure gradient in the wake	$\rho g \beta \Delta T$
Hydro-static pressure jump across the wake	$\Delta p_h = \rho g \beta \Delta T L Re^{-1/2}$
Velocity perturbation induced by Δp_h	$\dfrac{\Delta p_h}{\rho U_\infty}$
Inclination of the wake centerline induced by Δp_h	$K = \dfrac{\Delta p_h}{\rho U_\infty^2} = Gr\,Re^{-5/2}$
Hydro-static pressure gradient in the direction of the wake centerline	$\rho g \beta \Delta T K$
Interaction parameter	$\kappa^2 = \dfrac{g\beta\Delta T L}{U_\infty^2} K = Gr^2 Re^{-9/2}$

corresponding pressure gradient along the centerline of the wake is of the order $\lambda = \phi K Re^{1/2}$.

2.1 The Potential Flow

We introduce a Cartesian coordinate system such that its x-axis is horizontal and its origin is at the trailing edge of the plate. In the following we will use dimensionless variables. All lengths (if not stated otherwise) are scaled with the plate length L. Scaling the velocity with the velocity U_∞ of the parallel free stream and using the notation of complex valued functions of a complex variable $z = x + iy$ the potential flow can be written as

$$u - iv = 1 - i\phi\sqrt{\frac{z}{z+1}} + K(u_1 - i\,v_1). \tag{1}$$

The first two terms on the right side of (1) describe the potential flow past a horizontal plate of a free stream with an angle ϕ to the horizontal axis. The third term on the right side of (1) takes the buoyancy effects into account. Along the plate the vertical velocity component v_1 has to vanish.

From the viewpoint of the potential flow the scaled pressure has a jump discontinuity of size γ_w across the wake. Using the linearized Bernoulli equation we have

$$-u_1(x, 0+) + u_1(x, 0-) = \gamma_w(x), \quad x > 0, \tag{2}$$

where γ_w is the dimensionless pressure jump across the wake. If $\gamma(x)$ is given, following [2], we obtain for the dimensionless inclination of the wake

$$y_w'(x) = \phi\sqrt{\frac{x}{x+1}} + K v_1(x, 0), \tag{3}$$

where

$$v_1(x,0) = \frac{1}{2\pi} \int_0^\infty \sqrt{\frac{x}{\xi} \frac{\xi+1}{x+1}} \frac{\gamma_w(\xi)\,d\xi}{x-\xi}. \tag{4}$$

is the vertical velocity component $v_1(x,0)$ at the center line of the wake.

2.2 The Wake Flow

If the wake is inclined the hydro-static pressure gradient has a non-vanishing component in the main flow direction. Thus the fluid in the wake is accelerated in case of positive inclination and decelerated in case of negative inclination. Thus the equations for the flow and temperature field are

$$u_x + v_{\bar{y}} = 0, \tag{5}$$

$$u\,u_x + v\,u_{\bar{y}} = Y_w'\theta + u_{\bar{y}\bar{y}}, \quad u_{\bar{y}}(x,0) = 0, \quad u(x,\infty) = 1, \tag{6}$$

$$u\theta_x + v\theta_{\bar{y}} = \frac{1}{Pr}\theta_{\bar{y}\bar{y}}, \quad \theta_{\bar{y}}(x,0) = 0, \quad \theta(x,\infty) = 0, \tag{7}$$

where $\theta = (T - T_\infty)/(T_p - T_\infty)$ denotes the dimensionless temperature perturbation and $Pr = \nu/a$ is the Prandtl number, where a denotes the thermal diffusivity of the fluid. Note that in the wake equations $\bar{y} = (y - y_w(x))\sqrt{Re}$ denotes the vertical wake coordinate referred to the centerline of the wake. With $Y_w = y_w K \sqrt{Re}$ we denote the appropriately scaled centerline of the wake.

At the trailing edge, $x = 0$, the velocity and temperature profiles are given by the Blasius velocity profile and the corresponding temperature profile for the case of forced convection. The hydro-static pressure difference across the wake is given by

$$\gamma_w(x) = 2\int_0^\infty \theta(x,\bar{y})\,d\bar{y}. \tag{8}$$

From the viewpoint of the potential flow γ_w can be interpreted as a vortex distribution along the wake centerline, see [1].

3 Numerical Solution

A necessary condition for the existence of the integral in (4), such that v_1 exists, is that $\gamma_w(x)$ decays to zero for $x \to \infty$. Since the total enthalpy flux $\int_0^\infty u\theta\,d\bar{y}$ in the wake is constant, γ_w can only vanish, if the velocity u in the wake tends to infinity. This is case when $\lambda > 0$. Then in the far field similarity solutions of the form

$$u \sim \lambda^{2/5} x^{1/5} F'(\eta), \quad \theta \sim \frac{1}{\lambda^{1/5} x^{3/5}} D(\eta), \quad \eta = \lambda^{1/5} \frac{\bar{y}}{x^{2/5}} \tag{9}$$

exist, see [2], where F and D are the solutions of the similarity equations

$$F''' + \frac{3}{5} F\, F'' - \frac{1}{5} F' F' + D = 0, \quad \frac{1}{Pr} D' + \frac{3}{5} F D = 0, \tag{10}$$

with the boundary conditions

$$F(0) = F''(0) = F'(\infty) = 0, \quad \int_0^\infty F' D \, \mathrm{d}\eta = \int_0^\infty u_B \theta_B \, \mathrm{d}\bar{y}. \tag{11}$$

Thus in the far field the velocity and temperature profiles of the wake flow tend to the velocity and temperature profiles of a two-dimensional laminar plume. Since the flow and temperature profile of the wake flow is symmetric with respect to the centerline, it is sufficient to integrate the enthalpy flux only over one half of the wake. For the numerical solution an iterative method is proposed.

(1) First a suitable wake centerline $\frac{\mathrm{d}Y_w^{(0)}}{\mathrm{d}x} = \lambda \sqrt{\frac{x}{x+1}}$ is chosen
(2) The wake equations are integrated for a velocity $u^{(i)}$ and temperature field $\theta^{(i)}$ by a marching technique for a prescribed inclination $\frac{\mathrm{d}Y_w^{(i-1)}}{\mathrm{d}x}$ of the wake.
(3) Then the pressure jump $\gamma_w^{(i)} = 2 \int_0^\infty \theta^{(i)} \, \mathrm{d}\bar{y}$ across the wake is determined
(4) Evaluating (4) a new centerline $Y_w^{(i)}$ of the wake is determined and steps (2)–(4) are repeated until convergence is obtained.

We note that for $\kappa = 0$ no iterations are necessary. In the following we keep the inclination parameter $\lambda = 1$ and the Prandtl number $Pr = 0.71$ fixed. The interaction parameter κ will be increased starting from zero.

In Figs. 2–4 the velocity at the centerline of the wake, the pressure jump across the wake and the vertical velocity at the centerline of the wake are shown. For $\kappa = 0$ the shape of the wake is given by the well known 2D potential flow solution of the flow past an inclined plate [4]. The centerline velocity increases from $u = 0$ at the trailing edge due to viscosity. Then buoyancy leads to further acceleration and a velocity overshoot forms. Accordingly the vortex distribution $\gamma_w(x)$ (or the pressure jump across the wake) decreases.

Evaluating the integral (4) shows that the induced vertical velocity component v_1 is negative. Thus for κ sufficiently large the wake turns downwards about a plate length behind the trailing edge. After attaining a minimum the wake turns upwards again. Accordingly the graph of centerline velocity first becomes flat. Increasing κ further a minimum forms. When κ attains a critical value $\kappa = \kappa_c$ this minimum becomes zero. Since this solution is singular at the zero of the centerline velocity a further increase of κ is not possible. The physical mechanism which causes the singularity is the following: In the parts of the wake with downward inclination the wake flow is decelerated. The deceleration of the wake causes the wake to broaden there. The increase of the wake thickness causes finally an increase of the hydro-static pressure jump across the wake. In the limiting case, $\kappa = \kappa_c$, the wake thickness becomes infinite in wake coordinates and thus γ_w also tends to infinity.

In order to compute solutions with κ close to the critical value κ_c a different strategy has to be employed. First a value $u_{\min}$ for the minimum of the centerline velocity is prescribed. We chose a suitable vertical velocity perturbation $v_1^{(n)}$ and determine κ such that the minimum of centerline velocity has the prescribed value. This has to be done iteratively. Then a new vorticity distribution is computed and a new vertical velocity $v_1^{(n+1)}$ is determined. The process is repeated until convergence is obtained.

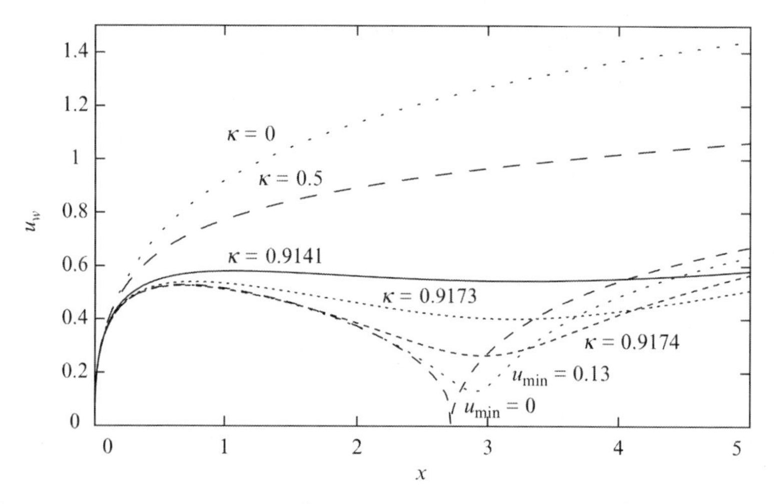

Fig. 2 Centerline velocity in the wake, $\lambda = 1$, $Pr = 0.71$

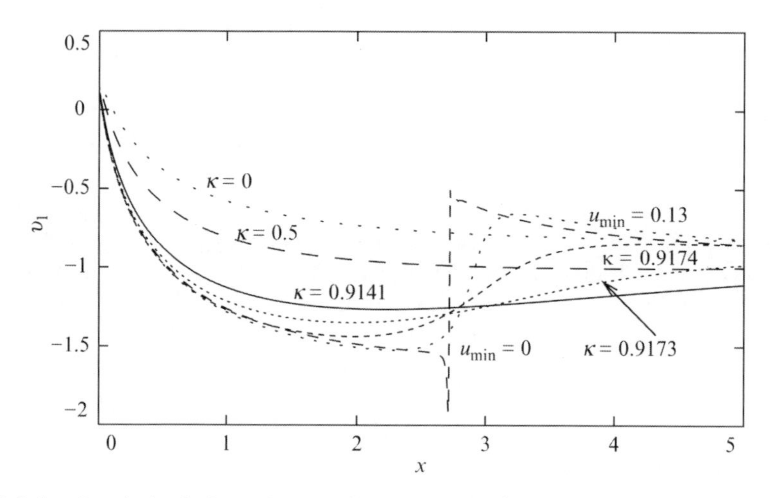

Fig. 3 Induced vertical velocity at the centerline of the wake, $\lambda = 1$, $Pr = 0.71$

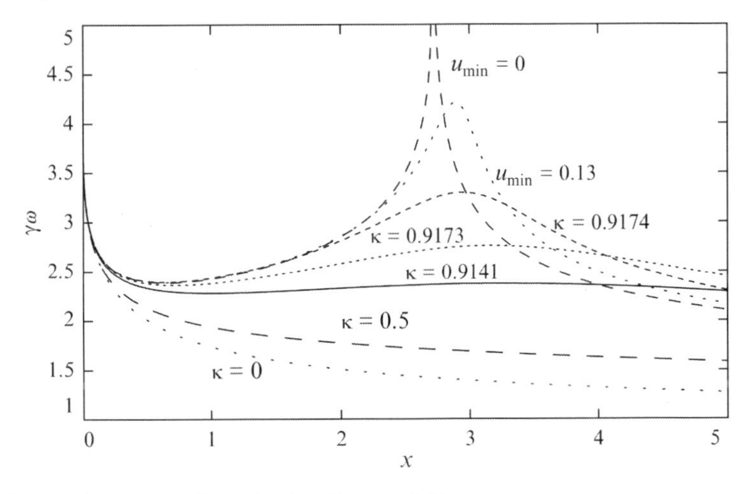

Fig. 4 Pressure jump across the wake, $\lambda = 1$, $Pr = 0.71$

4 Analysis of the Singularity

In order to study the singularity it is convenient to transform the wake equations to the von Mises coordinates and use $W = u^2$ as dependent variable. We define

$$u(x, \bar{y}) = \sqrt{W(s, \psi(x, \bar{y}))}, \quad \theta(x, \bar{y}) = \Theta(s, \psi(x, \bar{y})) \tag{12}$$

with

$$s = x - x_0, \quad \psi(x, \bar{y}) = \int_0^{\bar{y}} u(x, y')\,\mathrm{d}y', \tag{13}$$

where x_0 is the location of the singularity for $\kappa = \kappa_c$ and ψ is the stream function of the wake flow. Thus we obtain

$$W_s = 2Y_w'\Theta + \sqrt{W}W_{\psi\psi}, \quad \Theta_s = \left(\sqrt{W}\Theta_\psi\right)_\psi \tag{14}$$

with the boundary conditions $W_\psi(s, 0) = \Theta_\psi(s, 0) = 0$ and $W(s, \infty) = 1$, $\Theta(s, \infty) = 0$. The pressure jump across the wake is given by

$$\gamma_w = 2\int_0^\infty \frac{\Theta(s, \psi)}{\sqrt{W(s, \psi)}}\,\mathrm{d}\psi. \tag{15}$$

At $s = 0$ for $|\kappa - \kappa_c| \ll 1$ we have $W(0, \psi) \sim \varepsilon + W_0(\psi)$ with $W_0(0) = W_0'(0) = 0$. The parameter ε represents the value of the minimum of W. In the limit $\kappa = \kappa_c$ it vanishes.

We expand W asymptotically for $|s| \ll 1$, $\varepsilon \ll 1$:

$$W \sim \varepsilon + W_0 + \hat{Y}_w(s)W_1 + sW_2 + ..., \quad \Theta = \Theta_0 + s\Theta_1 + ... \tag{16}$$

with $\hat{Y}_w(s) = Y_w(x) - Y_w(x_0)$. The local behavior of the centerline $\hat{Y}_w$ is not known a priorily and thus a corresponding term in the expansion of W is added. Inserting into the differential equation (14) we obtain

$$W_1 = 2\Theta_0, \quad W_2 = \sqrt{W_0} W_0'', \quad \Theta_1 = (\sqrt{W_0}\Theta_0')'. \tag{17}$$

However, W_1, W_2 and Θ_1 do not satisfy the boundary condition at $\psi = 0$. Thus a sub-layer has to be introduced. It turns out that the sub-layer does not influence the leading order equations and thus it will not be discussed here. Note that $W_2(0) = 0$.

Using the local asymptotic expansion we can determine the local behavior of γ_w. We choose some value $\psi^* > 0$ and approximate $W \sim \varepsilon + W_0''\psi/2 + \hat{Y}_w(s)W_1(0)$ and $\Theta \sim \Theta_0(0)$ for $\psi < \psi^*$ and obtain

$$\begin{aligned} \gamma_w(s) = 2\int_0^\infty \frac{\Theta}{\sqrt{W}}\, d\psi \sim 2\int_0^{\psi^*} \frac{\Theta_0(0)}{\sqrt{\varepsilon + W_0''\psi^2/2 + \hat{Y}_w(s)W_1(0)}}\, d\psi \\ \sim -\frac{\sqrt{2}\Theta_0(0)}{\sqrt{W_0''(0)}} \ln|\varepsilon + \hat{Y}_w(s)W_1|. \end{aligned} \tag{18}$$

It can be shown that the singular part of the $\gamma_w(s)$ is independent of the choice of ψ^*.

Considering that $u_1 - i v_1$ is a potential flow, using the complex valued function theory and $u_1 = -\gamma_w/2$ we conclude

$$u_1 - i\, v_1 = \frac{\Theta_0(0)}{\sqrt{2W_0''(0)}} \ln F(z, \varepsilon), \tag{19}$$

where $F(z; \varepsilon)$ is a complex valued function of z with

$$|F(s)| = |\varepsilon + \hat{Y}_w(s)\, W_1|, \quad -\frac{\Theta_0(0)}{\kappa^2\sqrt{2W_0''(0)}} \arg \mathrm{F(s)} = \hat{Y}_w' \quad \text{for } s \text{ real}, \tag{20}$$

and $\hat{Y}_w(0, \varepsilon) = 0$, $\hat{Y}_w'(0) = 0$. This constitutes a problem for finding $F(z, \varepsilon)$ and $\hat{Y}_w(s, \varepsilon)$ simultaneously. We can express the solution $F(z, \varepsilon) = \varepsilon\tilde{F}(z/\varepsilon)$, $\hat{Y}_W(s, \varepsilon) = \varepsilon\tilde{Y}_w(s/\varepsilon)$ of (20) for arbitrary values of ε by the solution $\tilde{F}$ and $\tilde{Y}_w$ of (20) for $\varepsilon = 1$. In the limiting case $\varepsilon = 0$ we can guess the solution $F(z, 0) = z$. Thus in that case the centerline of the wake has a corner of size

$$[y_w'] = K\frac{\Theta_0\pi}{\sqrt{2W_0''(0)}}. \tag{21}$$

As a consequence the centerline velocity behaves in the limiting case $u_{\min} = 0$ locally like $u \sim \sqrt{|s|}$ (Fig. 2) and the hydro-static pressure difference γ_w has a logarithmic singularity (Fig. 4). For $\varepsilon > 0$ the corner is smoothed, see Fig. 3.

5 Conclusions

The laminar two-dimensional wake behind a horizontal wake has been discussed. It turned out that increasing the scaled buoyancy parameter a dip in the near wake forms. Due to the negative inclination of the wake the fluid in the wake decelerates and the wake thickness increases. A limiting case, where the centerline velocity of the wake vanishes at one point has been identified. Due to the interaction with the outer flow field in that case the centerline of the wake has a corner. If the buoyancy parameter is increased beyond this limiting value no stationary two dimensional laminar flows exist.

References

1. W. Schneider. Lift, thrust and heat transfer due to mixed convection flow past a horizontal plate. *J. Fluid Mech.*, 529:51–69, 2005.
2. Lj. Savić and H. Steinrück. Mixed convection flow past a horizontal plate. *Theor. Appl. Mech.*, 32:1–19, 2005.
3. W. Schneider and M. G. Wasel. Breakdown of the boundary-layer approximation for mixed convection above a horizontal plate. *Int. J. Heat Mass Transfer*, 28:2307–2313, 1985.
4. W. Schneider. *Mathematische Methoden der Strömungslehre*. Vieweg, 1978.

A Schwarz Technique for a System of Reaction Diffusion Equations with Differing Parameters

Meghan Stephens and Niall Madden

Abstract We describe an overlapping Schwarz method for a coupled system of two singularly perturbed reaction-diffusion equations which may have differing perturbation parameters. We give an outline of the analysis that shows that the algorithm is parameter-uniform. Supporting numerical results are presented.

1 Introduction

We consider the implementation and analysis of a standard finite difference method applied using an overlapping Schwarz domain decomposition algorithm for a singularly perturbed problem. In particular we are interested in a system of two linear, one-dimensional, singularly perturbed reaction-diffusion equations which may have distinct singular perturbation parameters. Our model problem is: *Find* $\mathbf{u} \in [C^4(0,1)]^2$ *such that*

$$\mathbf{L}\mathbf{u} := -\begin{pmatrix} \varepsilon_1^2 & 0 \\ 0 & \varepsilon_2^2 \end{pmatrix} \mathbf{u}'' + A\mathbf{u} = \mathbf{f} \qquad in \quad (0,1), \tag{1}$$

$$\mathbf{u}(0) = \mathbf{b_0}, \qquad \mathbf{u}(1) = \mathbf{b_1}. \tag{2}$$

The perturbation parameters may be small and of different magnitudes: $0 < \varepsilon_1 \leq \varepsilon_2 \leq 1$. We assume that the coefficients of A satisfy the conditions

$$a_{ij} \begin{cases} > 0 & \text{if} \quad i = j, \\ \leq 0 & \text{if} \quad i \neq j, \end{cases} \qquad \text{and} \qquad \sum_{j=1}^{2} a_{ij} > \alpha^2 > 0. \tag{3}$$

M. Stephens (✉)
Department of Mathematics, National University of Ireland, Galway, Ireland,
E-mail: Meghan.Stephens@NUIGalway.ie

A.F. Hegarty et al. (eds.), *BAIL 2008 - Boundary and Interior Layers.*
Lecture Notes in Computational Science and Engineering,
DOI: 10.1007/978-3-642-00605-0,

This ensures that the differential operator $\mathbf{L}$ satisfies a maximum principle, as does its associated finite difference analog.

The solution to (1)–(2) typically exhibits two overlapping boundary layers, see, e.g., [5]. That study shows that the standard finite difference method applied on a Shishkin mesh is almost first order accurate, improved in [2] to show the rate of convergence of the method is actually almost second order. In [3], analysis is provided for a system of $M \geq 2$ equations, and for more general meshes.

This study is concerned with the application of a Schwarz domain decomposition technique for coupled systems. Such methods are used to construct a sequence of discrete approximations $\{\mathbf{U}^{[0]}, \mathbf{U}^{[1]}, \dots\}$ to the true solution $\mathbf{u}$ for which it can be shown that there is a constant C such that

$$\|\mathbf{u} - \mathbf{U}^{[k]}\|_{\infty,\Omega^N} \leq C\big((N^{-1}\ln N)^2 + \rho^k\big),$$

for some $\rho \in (0,1)$. At each iteration, $\mathbf{U}^{[\mathbf{k}]}$ is formed by combining the solution to certain discrete sub-problems posed on different but overlapping subregions. These sub-problems are smaller and so, for example, demand less memory resources than the corresponding direct algorithm.

The algorithm we present here is based on that of [4] for a single (uncoupled) singularly perturbed reaction-diffusion equation, where it is proved that

$$\|u - U^{[k]}\|_{\infty,\Omega^N} \leq C\big((N^{-1}\ln N)^2 + 2^{-k}\big).$$

This result is analogous to that one would obtain for a classical problem. However, numerical results in that paper suggest that, for small values of the perturbation parameter, far less iterations are required than is suggested by this estimate.

In [8], the algorithm is extended to a coupled system of $M \geq 2$ equations with identical singular perturbation parameters and it is proved that, in practice, only one iteration is required. An analogous result for a single semilinear reaction-diffusion equation is given in [1].

The purpose of this report is to describe how the Schwarz algorithm can be extended to a system of two equations with differing parameters, and to investigate the number of iterations required in the case $0 < \varepsilon_1 \ll \varepsilon_2 \ll 1$.

This paper is structured as follows. We outline the algorithm in Sect. 2. An analysis of the algorithm is outlined in Sect. 3. The results of supporting numerical experiments are given in Sect. 4.

Notation

We denote C, with or without a subscript, to be a constant independent of ε_1, ε_2, N and k. Similarly $\mathbf{C} = C(1,1,\dots,1)^T$.

A mesh of N intervals on the domain $\bar{\Omega} = [a,b]$ is denoted as $\overline{\Omega}^N = \{a = x_0 < x_1 < \cdots < x_N = b\}$. The mesh on the open interval $\Omega = (a,b)$ is $\Omega^N = \{x_1 < \cdots < x_{N-1}\}$.

We define the usual max-norm on vectors $\mathbf{y} \in \mathbb{R}^m$, that is, $\|\mathbf{y}\| = \max_{p=1,\dots,m} |y_p|$. For a real-valued function $y \in C(\Omega)$, and vectors of real-valued functions $\mathbf{z} \in [C(\Omega)]^2$ we have

$$\|y\|_\Omega = \max_{x\in\Omega} |y(x)|, \quad \|\mathbf{z}\|_\Omega = \max\{\|z_0\|_\Omega, \|z_1\|_\Omega, \dots, \|z_m\|_\Omega\}.$$

For a vector of mesh functions $\mathbf{Z}(x_i) = (Z_0(x_i), Z_1(x_i), \dots Z_m(x_i))^T$ define

$$\|Z\|_{\Omega^N} = \max_j \left(\max_{x_i\in\Omega^N} |Z_j(x_i)| \right).$$

Given a mesh function Z, its piecewise linear interpolant is denoted $\overline{Z}$.

2 Schwarz Algorithm

The domain $\Omega = (0, 1)$ is split into the five overlapping subdomains, as shown in Fig. 1,

$$\Omega_{LL} = (0, 4\tau_1), \quad \Omega_L = (\tau_1, 4\tau_2), \quad \Omega_C = (\tau_2, 1 - \tau_2),$$
$$\Omega_R = (1 - 4\tau_2, 1 - \tau_1), \quad \Omega_{RR} = (1 - 4\tau_1, 1),$$

where we choose the *Shishkin transition points* as in [2],

$$\tau_2 = \min\left\{\frac{1}{8}, 2\frac{\varepsilon_2 \ln N}{\alpha}\right\}, \quad \tau_1 = \min\left\{\frac{\tau_2}{4}, 2\frac{\varepsilon_1 \ln N}{\alpha}\right\}.$$

On a given subdomain, $\Omega_d = (a, b)$, we construct a uniform mesh $\overline{\Omega}_d^N := \{a = x_0 < x_1 < \dots < x_N = b\}$, with $h_d = x_i - x_{i-1} = (b - a)/N$.

On each subdomain $\Omega_d = (a, b)$ the discrete problem is

$$-\begin{pmatrix} \varepsilon_1^2 & 0 \\ 0 & \varepsilon_2^2 \end{pmatrix} \delta^2 \mathbf{U_d}(x_i) + A(x_i)\mathbf{U_d}(x_i) = \mathbf{f}(x_i), \text{ for } x_i \in \Omega_d^N, \tag{4}$$

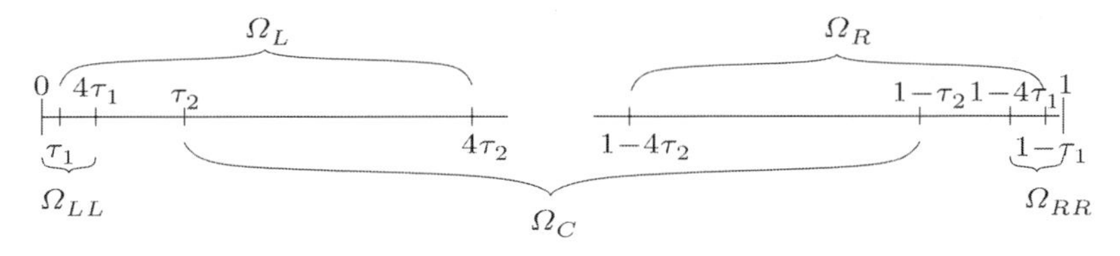

Fig. 1 Overlapping subdomains of Ω

where

$$\delta^2 z_d(x_i) := \frac{1}{h_d^2}\big(z_d(x_{i-1}) - 2z_d(x_i) + z_d(x_{i+1})\big).$$

We use the initial approximation

$$\mathbf{U}^{[0]}(x_i) = \mathbf{0} \text{ on } \Omega^S, \quad \mathbf{U}^{[0]}(0) = \mathbf{b_0}, \quad \mathbf{U}^{[0]}(1) = \mathbf{b_1}.$$

Then for $k \geq 1$, define $\mathbf{U_{LL}}^{[k]}$, $\mathbf{U_L}^{[k]}$, $\mathbf{U_C}^{[k]}$, $\mathbf{U_R}^{[k]}$ and $\mathbf{U_{RR}}^{[k]}$ to be the solutions to the appropriate version of (4) subject to the boundary conditions

$$\begin{aligned}
\mathbf{U_{LL}}^{[k]}(0) &= \mathbf{b_0}, & \mathbf{U_{LL}}^{[k]}(4\tau_1) &= \overline{\mathbf{U}}^{[k-1]}(4\tau_1);\\
\mathbf{U_{RR}}^{[k]}(1-4\tau_1) &= \overline{\mathbf{U}}^{[k-1]}(1-4\tau_1), & \mathbf{U_{RR}}^{[k]}(1) &= \mathbf{b_1};\\
\mathbf{U_L}^{[k]}(\tau_1) &= \overline{\mathbf{U}}_{\mathbf{LL}}^{[k]}(\tau_1), & \mathbf{U_L}^{[k]}(4\tau_2) &= \overline{\mathbf{U}}^{[k-1]}(4\tau_2);\\
\mathbf{U_R}^{[k]}(1-4\tau_2) &= \overline{\mathbf{U}}^{[k-1]}(1-4\tau_2), & \mathbf{U_R}^{[k]}(1-\tau_1) &= \overline{\mathbf{U}}_{\mathbf{RR}}^{[k]}(1-\tau_1);\\
\mathbf{U_C}^{[k]}(\tau_2) &= \overline{\mathbf{U}}_{\mathbf{L}}^{[k]}(\tau_2), & \mathbf{U_C}^{[k]}(1-\tau_2) &= \overline{\mathbf{U}}_{\mathbf{R}}^{[k]}(1-\tau_2).
\end{aligned}$$

The approximation $\mathbf{U}^{[k]}$ is taken to be

$$\mathbf{U}^{[k]} = \begin{cases}
\mathbf{U_{LL}}^{[k]}(x_i), & x_i \in \overline{\Omega}_{LL}^N \backslash \overline{\Omega}_L,\\
\mathbf{U_L}^{[k]}(x_i), & x_i \in \overline{\Omega}_{L}^N \backslash \overline{\Omega}_C,\\
\mathbf{U_C}^{[k]}(x_i), & x_i \in \overline{\Omega}_{C}^N,\\
\mathbf{U_R}^{[k]}(x_i), & x_i \in \overline{\Omega}_{R}^N \backslash \overline{\Omega}_C,\\
\mathbf{U_{RR}}^{[k]}(x_i), & x_i \in \overline{\Omega}_{RR}^N \backslash \overline{\Omega}_R.
\end{cases}$$

Remark 1. The iterate $\mathbf{U}^{[k]}$ is a mesh function defined on the mesh

$$\Omega^S := \left(\overline{\Omega}_{LL}^N \backslash \overline{\Omega}_L\right) \bigcup \left(\overline{\Omega}_{L}^N \backslash \overline{\Omega}_C\right) \bigcup \overline{\Omega}_{C}^N \bigcup \left(\overline{\Omega}_{R}^N \backslash \overline{\Omega}_C\right) \bigcup \left(\overline{\Omega}_{RR}^N \backslash \overline{\Omega}_R\right).$$

That is, it is the piecewise uniform mesh with $N/4$ intervals on each of $[0, \tau_1]$, $[\tau_1, \tau_2]$, $[1-\tau_2, 1-\tau_1]$, $[1-\tau_1, 1]$ and N intervals on the region $[\tau_2, 1-\tau_2]$.

The algorithm terminates when

$$\|\mathbf{U}^{[k]} - \mathbf{U}^{[k-1]}\|_{\Omega^S} \leq \lambda N^{-2}, \tag{5}$$

where λ is chosen so that the relative error is $\mathcal{O}(N^{-1} \ln N)^2$. Typically, one takes λ to be $\mathcal{O}(\|\mathbf{u}\|_{\overline{\Omega}})$, estimated by noting that,

$$\|\mathbf{u}\|_{\overline{\Omega}} \leq \max\{\|\mathbf{b_0}\|, \|\mathbf{b_1}\|\} + \alpha^{-2}\|\mathbf{f}\|_{\overline{\Omega}}.$$

3 Analysis of the Schwarz Algorithm

A detailed analysis of the iterative scheme is contained in [7]. Here we present the key result, and give a brief outline of how it is proved. The analysis makes use of standard maximum principle arguments, and bounds on derivatives of $\mathbf{u}$ similar to those in [3].

Theorem 3.1 *Let* $\mathbf{U}^{[k]}$ *be the* k^{th} *iterate of the discrete Schwarz method of Sect. 2. Then, there is a constant* C*, such that*

$$\|\mathbf{U}^{[k]} - \mathbf{u}\|_{\Omega^S} \leq C\left(2^{-k} + (N^{-1}\ln N)^2\right).$$

Outline of proof. Here we give only a flavour of the analysis, with emphasis of the region Ω_{LL}.

Clearly there exists C such that

$$\|\mathbf{U}^{[0]} - \mathbf{u}\|_{\Omega^S} = \|\mathbf{u}\|_{\Omega^S} \leq C\,(2^0 + (N^{-1}\ln N)^2).$$

Assume that for an arbitrary integer $k \geq 0$ there exists C such that

$$\|\mathbf{U}^{[k]} - \mathbf{u}\|_{\Omega^S} \leq C\left(2^{-k} + (N^{-1}\ln N)^2\right). \tag{6}$$

The proof proceeds by induction. Define $\mathbf{E_{LL}}^{[k+1]} := \mathbf{U_{LL}}^{[k+1]} - \mathbf{u}$. Using Taylor expansions we can show that $\left|\mathbf{L}^N\mathbf{E_{LL}}^{[k+1]}(x_i)\right| \leq \mathbf{C}(N^{-1}\ln N)^2$. Also $\mathbf{E_{LL}}^{[k+1]}(0) = \mathbf{0}$ and using standard interpolation error bounds along with (6) we can show that

$$\left|\mathbf{E_{LL}}^{[k+1]}(4\tau_1)\right| = \left|(\bar{\mathbf{U}}_\mathbf{L}{}^{[k+1]} - \mathbf{u})(4\tau_1)\right| \leq \mathbf{C}\left(2^{-k} + (N^{-1}\ln N)^2\right).$$

On the subdomain Ω_{LL} we can construct a barrier function $\Phi(x_i)$ which satisfies

$$\mathbf{L}^N\left(\Phi(x_i) \pm \mathbf{E_{LL}}^{[k+1]}(x_i)\right) \geq \mathbf{0},$$

$$\left(\Phi(0) \pm \mathbf{E_{LL}}^{[k+1]}(0)\right) \geq \mathbf{0}, \quad \left(\Phi(4\tau_1) \pm \mathbf{E_{LL}}^{[k+1]}(4\tau_1)\right) \geq \mathbf{0},$$

and $\Phi(x_i) \leq \mathbf{C}\left(2^{-(k+2)} + (N^{-1}\ln N)^2\right)$. Using a discrete maximum principle

$$\left|(\mathbf{U_{LL}}^{[k+1]} - \mathbf{u})(x_i)\right| \leq \mathbf{C}\left(2^{-(k+2)} + (N^{-1}\ln N)^2\right), \qquad \text{for } x_i \in \overline{\Omega}_{LL}^N \backslash \overline{\Omega}_L.$$

If follows directly that $|\mathbf{U_L}^{[k+1]} - \mathbf{u})(\tau_1)| \leq \mathbf{C}\left(2^{-(k+2)} + (N^{-1}\ln N)^2\right)$. Then a similar style of analysis (see [7] for details) gives that

$$\left|(\mathbf{U_L}^{[k+1]} - \mathbf{u})(x_i)\right| \leq \mathbf{C}\left(2^{-(k+1)} + (N^{-1}\ln N)^2\right), \qquad \text{for } x_i \in \overline{\Omega}_L^N \backslash \overline{\Omega}_C.$$

The analysis for ${\mathbf{U_{RR}}}^{[k+1]}$ and ${\mathbf{U_R}}^{[k+1]}$ is very similar, and can then be extended to ${\mathbf{U_C}}^{[k+1]}$ to complete the proof.

4 Numerical Results

In this section we present numerical results for the Schwarz algorithm, which support the analysis in Sect. 3. The exact solution to the test problem is unknown so we use the two-mesh difference approach [6, Theorem 8.6] to estimate the errors.

Let $\mathbf{U}^N = \mathbf{U}^{[k]}$ where k is such that condition (5) is satisfied, and similarly let $\widehat{\mathbf{U}}^{2N}$ be the computed solution obtained using the Schwarz algorithm where the subdomains are defined as in Sect. 2, but with $2N$ intervals on each subdomain. Then the estimate of the error, for a particular ε_1 and ε_2, is given by

$$D^N_{\varepsilon_1,\varepsilon_2} := \max_{i=1,\dots,N-1} |\mathbf{U}^N(x_i) - \widehat{\mathbf{U}}^{2N}(x_{2i})|.$$

For $\varepsilon_1 = 10^{-\ell}$, for $\ell = 0, 1, 2\dots$, we denote the ε_2-uniform error estimate, and corresponding rate of convergence as

$$D^N_{\varepsilon_1} = \max_{\kappa=0}^{\ell} D^N_{\varepsilon_1,10^{-\kappa}}, \qquad \rho^N_{\varepsilon_1} := \log_2\left(\frac{D^N_{\varepsilon_1}}{D^{2N}_{\varepsilon_1}}\right).$$

For our test problem we take

$$A = \begin{pmatrix} 2(x+1)^2 & -(1+x^3) \\ -2\cos\frac{\pi x}{4} & (1+\sqrt{2})e^{1-x} \end{pmatrix}, \qquad \mathbf{f} = \begin{pmatrix} 2e^x \\ 10x+1 \end{pmatrix}, \tag{7}$$

and $\mathbf{b_0} = \mathbf{0}$, $\mathbf{b_1} = \mathbf{6}$.

For these experiments, $\alpha = 1$ and the user-chosen parameter in (5) is taken to be $\lambda = 6$.

Table 1 lists $D^N_{\varepsilon_1}$, $\rho^N_{\varepsilon_1}$ for various values of N and ε_1. We can see that the errors are independent of the singular perturbation parameters ε_1 and ε_2 and are decreasing as N increases. The computed rates of convergence are second order, with the usual logarithmic factor which is expected when using the Shishkin transition points for Schwarz algorithms.

Table 2 displays k_{ε_1}, the maximum number of iterations recorded when computing $\mathbf{U}$ for $\varepsilon_2 = 1, 10^{-1}, 10^{-2}, \dots \varepsilon_1$, for various values of N and ε_1. In [7], analysis shows that the difference between successive iterations can be bounded as

$$\begin{aligned} \|\mathbf{U}^{[k]} - \mathbf{U}^{[k-1]}\|_{\Omega^S} &\le \|\mathbf{U}^{[k]} - \mathbf{u}\|_{\Omega^S} + \|\mathbf{U}^{[k-1]} - \mathbf{u}\|_{\Omega^S} \\ &\le C_0 2^{-k+1} + C_1(N^{-1}\ln N)^2. \end{aligned} \tag{8}$$

Table 1 Error estimates and computed rates of convergence

	$N=64$	$N=128$	$N=256$	$N=512$	$N=1{,}024$	$N=2{,}048$	
$\varepsilon_1=10^0$	6.16e−04	5.80e−05	1.89e−05	8.67e−06	7.81e−07	2.71e−07	
	3.41	1.62	1.12	3.47	1.53		$\rho^N_{\varepsilon_1}$
$\varepsilon_1=10^{-1}$	8.44e−04	2.42e−04	5.34e−05	1.50e−05	3.35e−06	9.41e−07	
	1.80	2.18	1.83	2.17	1.83		$\rho^N_{\varepsilon_1}$
$\varepsilon_1=10^{-2}$	1.30e−02	3.32e−03	8.39e−04	2.10e−04	5.26e−05	1.31e−05	
	1.96	1.98	2.00	2.00	2.00		$\rho^N_{\varepsilon_1}$
$\varepsilon_1=10^{-3}$	7.31e−02	3.00e−02	1.05e−02	3.36e−03	1.05e−03	3.18e−04	
	1.28	1.52	1.64	1.68	1.72		$\rho^N_{\varepsilon_1}$
$\vdots$	$\vdots$	$\vdots$	$\vdots$	$\vdots$	$\vdots$	$\vdots$	$\vdots$
$\varepsilon_1=10^{-10}$	7.31e-02	3.00e-02	1.05e-02	3.36e-03	1.05e-03	3.18e-04	
	1.28	1.52	1.64	1.68	1.72		$\rho^N_{\varepsilon_1}$

Table 2 Number of iterations required by the Schwarz algorithm

$N=$	$N=64$	$N=128$	$N=256$	$N=512$	$N=1{,}024$	$N=2{,}048$
$\varepsilon_1=1$	7	9	10	11	13	14
$\varepsilon_1=10^{-1}$	7	8	10	11	12	13
$\varepsilon_1=10^{-2}$	7	8	10	11	12	13
$\varepsilon_1=10^{-3}$	7	8	9	10	11	13
$\vdots$	$\vdots$	$\vdots$	$\vdots$	$\vdots$	$\vdots$	$\vdots$
$\varepsilon_1=10^{-10}$	7	8	9	10	11	12

The algorithm will terminate when $\|\mathbf{U}^{[k]}-\mathbf{U}^{[k-1]}\|_{\Omega^S}\leq\lambda N^{-2}$. Together, these inequalities suggest that, at worst, the number of iterations required is proportional to $2\log_2 N$. In fact, as can be seen from Table 2, in practise $\log_2 N+1$ iterations are required.

Finally, we investigate numerically how the accuracy of the method, and the number of iterations required, depends on the relative magnitude of the perturbation parameters. Therefore we fix $\varepsilon_1=10^{-5}$ and present the estimated errors for $\varepsilon_2\in\{1,10^{-1},10^{-2},\ldots,10^{-10}\}$. These are shown in Table 3, and are computed by comparing $\mathbf{U}^N$ with the piecewise linear interpolant to the solution obtained on a standard Shishkin mesh [2] with 2^{16} mesh points. The rates of convergence, ρ^N_E, are computed as above. Clearly the error estimates are bounded independently of the singular perturbation parameters and are minimised when ε_1 and ε_2 are of the same magnitude.

In Table 4 we show the number of iterations required to obtain the results of Table 3. When the singular perturbation parameters are of the same magnitude the algorithm terminates after only one iteration, as shown in [8].

Table 3 Error estimates and computed rates of convergence for $\varepsilon_1 = 10^{-5}$

	$N = 64$	$N = 128$	$N = 256$	$N = 512$	$N = 1{,}024$	$N = 2{,}048$	
$\varepsilon_2 = 10^0$	5.71e−02	2.28e−02	7.85e−03	2.51e−03	7.78e−04	2.35e−04	
	1.33	1.54	1.65	1.69	1.73		ρ_E^N
$\varepsilon_2 = 10^{-1}$	5.71e−02	2.28e−02	7.85e−03	2.51e−03	7.78e−04	2.35e−04	
	1.33	1.54	1.65	1.69	1.73		ρ_E^N
$\varepsilon_2 = 10^{-2}$	1.21e−01	5.36e−02	2.31e−02	1.00e−02	2.57e−03	9.67e−04	
	1.17	1.22	1.21	1.96	1.41		ρ_E^N
$\varepsilon_2 = 10^{-3}$	1.38e−01	3.75e−02	1.27e−02	4.09e−03	1.29e−03	3.74e−04	
	1.88	1.56	1.64	1.66	1.79		ρ_E^N
$\varepsilon_2 = 10^{-4}$	5.70e−02	2.28e−02	7.85e−03	2.51e−03	7.78e−04	2.35e−04	
	1.33	1.54	1.65	1.69	1.73		ρ_E^N
$\varepsilon_2 = 10^{-5}$	9.83e−03	3.37e−03	1.10e−03	3.49e−04	1.08e−04	3.24e−05	
	1.54	1.61	1.66	1.70	1.73		ρ_E^N
$\varepsilon_2 = 10^{-6}$	1.33e−01	4.97e−02	1.72e−02	5.55e−03	1.72e−03	5.17e−04	
	1.42	1.53	1.63	1.69	1.73		ρ_E^N
⋮	⋮	⋮	⋮	⋮	⋮	⋮	⋮
$\varepsilon_2 = 10^{-10}$	1.33e−01	4.97e−02	1.72e−02	5.55e−03	1.72e−03	5.17e−04	
	1.42	1.53	1.63	1.69	1.73		ρ_E^N

Table 4 Number of iterations required for $\varepsilon_1 = 10^{-5}$

	$N = 64$	$N = 128$	$N = 256$	$N = 512$	$N = 1{,}024$	$N = 2{,}048$
$\varepsilon_2 = 10^0$	7	8	9	10	11	12
$\varepsilon_2 = 10^{-1}$	6	7	8	9	10	11
$\varepsilon_2 = 10^{-2}$	6	7	8	9	10	11
$\varepsilon_2 = 10^{-3}$	5	6	6	7	7	8
$\varepsilon_2 = 10^{-4}$	2	2	2	2	2	2
$\varepsilon_2 = 10^{-5}$	1	1	1	1	1	1
$\varepsilon_2 = 10^{-6}$	2	2	2	2	2	2
$\varepsilon_2 = 10^{-7}$	5	5	5	6	6	7
$\varepsilon_2 = 10^{-8}$	6	7	8	9	10	10
$\varepsilon_2 = 10^{-9}$	6	7	8	9	10	11
$\varepsilon_2 = 10^{-10}$	6	7	8	9	10	11

Most important, however, is that Table 4 demonstrates that Theorem 3.1 is sharp when the magnitudes of ε_1 and ε_2 are different, i.e., one does not obtain the rapid convergence of the iterative algorithm that is observed when $\varepsilon_1 = \varepsilon_2$. The design of a fast converging overlapping Schwarz algorithm for the case $\varepsilon_1 \ll \varepsilon_2$ is the subject of on-going research.

References

1. N. Kopteva, M. Pickett, and H. Purtill, *A robust overlapping Schwarz method for a singularly perturbed semilinear reaction-diffusion problem with multiple solutions*, Preprint, University of Limerick, 2008.
2. T. Linß and N. Madden, *Accurate solution of a system of coupled singularly perturbed reaction-diffusion equations.* Computing **73**(2) (2004), 121–133.
3. T. Linß and N. Madden, *Layer adapted meshes for a system of coupled singularly perturbed reaction-diffusion problems.* IMA J. Numer. Anal. **29**(1) (2009), 109–125.
4. H. MacMullen, J.J.H. Miller, E. O'Riordan, and G.I. Shishkin, *A second-order parameter-uniform overlapping schwarz method for reaction-diffusion problems with boundary layers.* J. Comput. Appl. Math. **130** (2001), 231–244.
5. N. Madden and M. Stynes, *A uniformly convergent numerical method for a coupled system of two singularly perturbed linear-reaction diffusion problems.* IMA J. Numer. Anal. **23** (2003), 627–644.
6. J.J.H. Miller, E. O'Riordan, and G.I. Shishkin, *Fitted numerical methods for singular perturbation problems*, World Scientific, River Edge, NJ, 1996.
7. M. Stephens, Ph.D. thesis, Department of Mathematics, National University of Ireland, Galway (2008).
8. M. Stephens and N. Madden, *A parameter-uniform Schwarz method for a coupled system of reaction diffusion equations.* J. Comput. Appl. Math. (to appear) (2008).

On Numerical Simulation of an Aeroelastic Problem Nearby the Flutter Boundary

P. Sváček

Abstract This paper is devoted to the numerical approximation of fluid-structure interaction problems. To take into account the turbulence effects typically present in the technical applications, the flow is modelled with the aid of the Reynolds Averaged Navier–Stokes (RANS) equations. The numerical approximation of RANS and the turbulence model by the finite element method is described. In order to avoid spurious oscillations stabilization procedures are applied and the robust suitable linearization of the Spalart–Allmaras turbulence model is described. The numerical method developed is applied to a problem of a channel flow over a double circular airfoil profile. The nonlinear post-flutter behaviour is numerically approximated.

1 Introduction

The interaction of fluid flow with an elastic structure is important in many technical disciplines such as aeroelasticity/hydroelasticity; see [1]. Engineering applications frequently use linearized models. Recently, problems of nonlinear aeroelasticity have begun to be important in an increasing number of situations under investigation. In technical applications the use of a suitable model for turbulence effects is necessary. In most cases models based on Reynolds Averaged Navier–Stokes (RANS) equations are used together with the Boussinesq approximation of the Reynolds stresses (see [2]). The turbulence viscosity is then approximated by the solution of a partial differential equation. The approximation of the turbulence model with the aid of the finite element method is complicated owing to the

P. Sváček
Czech Technical University Prague, Faculty of Mechanical Engineering, Karlovo nám. 13, 121 35 Praha 2, Czech Republic, E-mail: Petr.Svacek@fs.cvut.cz

A.F. Hegarty et al. (eds.), *BAIL 2008 - Boundary and Interior Layers.* 257
Lecture Notes in Computational Science and Engineering,
DOI: 10.1007/978-3-642-00605-0,

dominating convection. Moreover, the use of the standard SUPG/GLS method for stabilization of the convective term does not eliminate local undershoots/overshoots. In order to minimize this phenomenon crosswind diffusion must be applied. In the ALE formulation the stabilization terms then needs to be modified with respect to the moving frame. The robustness of this approach is demonstrated by several numerical results for the vibrating double circular airfoil (DCA). The numerical results are compared to the experimental data. Furthermore, the aeroelastic model previously studied in [3] is solved with the aid of the turbulence model for the far field velocity in the post-critical region. Nonlinear effects are shown.

2 Mathematical Model

2.1 Fluid Model

For the mathematical description of turbulent flow we use the Reynolds equations, written in the ALE form

$$\frac{D^{\mathcal{A}}\mathbf{v}}{Dt} - \nabla \cdot \left(\nu_{eff}\left(\nabla \mathbf{v} + (\nabla \mathbf{v})^T\right)\right) + (\bar{\mathbf{w}} \cdot \nabla)\mathbf{v} + \nabla p = 0, \qquad \operatorname{div} \mathbf{v} = 0 \text{ in } \Omega_t, \tag{1}$$

where the vector $\mathbf{v}$ is the mean part of the velocity, p denotes the mean kinematic pressure, $\bar{\mathbf{w}}$ denotes the convection velocity in the ALE frame, i.e., $\bar{\mathbf{w}} = \mathbf{v} - \mathbf{w}_D$ with $\mathbf{w}_D$ the domain velocity, while $\nu_{eff} = (\nu + \nu_T)$ where ν and ν_T denote the kinematic and turbulent viscosity respectively. The *ALE derivative* is denoted by $D^{\mathcal{A}}/Dt$, where $\mathcal{A}_t$ is an ALE mapping from the reference domain Ω_0 onto the computational domain Ω_t at time t. More details about the ALE method can be found in, e.g., [3,4].

The system has boundary conditions prescribed on the mutually disjoint parts of the boundary $\partial\Omega_t$ (see Fig. 1):

$$\text{(a)} \quad \mathbf{v}(x,t) = \mathbf{v}_D(x), \quad x \in \Gamma_D, \qquad \text{(b)} \ \mathbf{v}(x,t) = \mathbf{w}_D(x,t), \quad x \in \Gamma_{Wt},$$
$$\text{(c)} \quad -\nu_{eff}\left(\nabla \mathbf{v} + (\nabla \mathbf{v})^T\right) \cdot \mathbf{n} + (p - p_{ref})\,\mathbf{n} = 0 \text{ on } \Gamma_O. \tag{2}$$

Here Γ_{Wt} denotes the only moving part of the boundary, i.e., the instantaneous position of the airfoil surface at time t. Finally, the system (1) has the initial condition $\mathbf{v}(x,0) = \mathbf{v}_0(x)$ for $x \in \Omega_0$. In practical computations $\mathbf{v}_D, \mathbf{w}_D, \mathbf{v}_0$ are continuous functions. Moreover, the function $\mathbf{v}_0$ should satisfy $\nabla \cdot \mathbf{v}_0 = 0$.

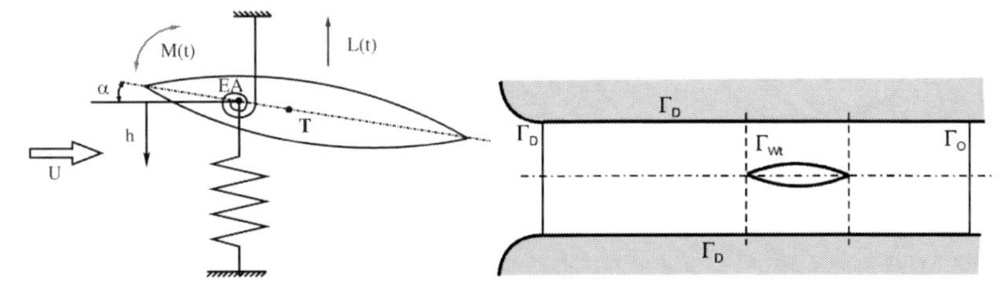

Fig. 1 The elastic support of the airfoil on translational and rotational springs (*left*) and the computational domain for the channel flow over DCA profile (*right*)

2.2 Turbulence Model

In the system of equations (1) the turbulent viscosity ν_T needs further modelling. We consider the Spalart–Allmaras turbulence model [2]. The turbulence viscosity is $\nu_T = \tilde{\nu}\chi^3/(\chi^3 + c_v^3)$, where $\chi = \tilde{\nu}/\nu$ and the additional equation for $\tilde{\nu}$ is

$$\frac{D^{\mathcal{A}}\tilde{\nu}}{Dt} + \bar{\mathbf{w}} \cdot \nabla\tilde{\nu} = \nabla \cdot \left(\frac{\nu + \tilde{\nu}}{\beta}\nabla\tilde{\nu}\right) + \frac{c_{b_2}}{\beta}(\nabla\tilde{\nu})^2 + G(\tilde{\nu}) - Y(\tilde{\nu}), \qquad (3)$$

where $G(\tilde{\nu}) = c_{b_1}\widetilde{S}\,\tilde{\nu}$, $\widetilde{S} = \left(S + \frac{\tilde{\nu}}{\kappa^2 y^2} f_{v2}\right)$, $f_{v_2} = 1 - \frac{\chi}{1+\chi f_{v_1}}$, $Y(\tilde{\nu}) = c_{w_1}\frac{\tilde{\nu}^2}{y^2}\left(\frac{1+c_{w_3}^6}{1+c_{w_3}^6/g^6}\right)^{\frac{1}{6}}$, $g = r + c_{w_2}(r^6 - r)$, $r = \frac{\tilde{\nu}}{\widetilde{S}\kappa^2 y^2}$, $S = \sqrt{2\sum_{i,j}\omega_{ij}^2}$, y denotes the distance from the wall and $\omega_{ij} = \frac{1}{2}(\partial_j v_i - \partial_i v_j)$. The following constants are used: $c_{b_1} = 0.1355$, $c_{b_2} = 0.622$, $\beta = \frac{2}{3}$, $c_v = 7.1$, $c_{w_2} = 0.3$, $c_{w_3} = 2.0$, $\kappa = 0.41$, $c_{w_1} = c_{b_1}/\kappa^2 + (1 + c_{b_2})/\beta$.

2.3 Structure Model

The structure is considered to be a flexibly supported airfoil that can be vertically displaced and rotated. Figure 1 shows the elastic support of the airfoil on translational and rotational springs and its placement in the channel. The governing nonlinear equations are written in the form (see [1, 3])

$$\begin{aligned} m\ddot{h} + S_\alpha\ddot{\alpha}\cos\alpha - S_\alpha\dot{\alpha}^2\sin\alpha + k_{hh}h &= -L(t), \\ S_\alpha\ddot{h}\cos\alpha + I_\alpha\ddot{\alpha} + k_{\alpha\alpha}\alpha &= M(t). \end{aligned} \qquad (4)$$

where m is the mass of the airfoil, S_α is the static moment of the airfoil around the elastic axis, I_α is the inertia moment of the airfoil around the elastic axis, and k_{hh} and $k_{\alpha\alpha}$ are the bending stiffness and torsional stiffness, respectively. The pressure and viscous forces acting on the vibrating airfoil immersed in fluid result in the lift force $L(t)$ and the torsional moment $M(t)$ defined by

$$L = -l \int_{\Gamma_{Wt}} \sum_{j=1}^{2} \tau_{2j} n_j \, dS, \qquad M = l \int_{\Gamma_{Wt}} \sum_{i,j=1}^{2} \tau_{ij} n_j r_i^{\mathrm{ort}} \, dS, \tag{5}$$

where

$$\tau_{ij} = \rho \left[-p\delta_{ij} + \nu \left(\frac{\partial v_i}{\partial x_j} + \frac{\partial v_j}{\partial x_i} \right) \right], \; r_1^{\mathrm{ort}} = -(x_2 - x_{\mathrm{EA2}}), \; r_2^{\mathrm{ort}} = x_1 - x_{\mathrm{EA1}}.$$

Here τ_{ij} denotes the components of the fluid stress tensor, δ_{ij} is the Kronecker symbol, $\mathbf{n} = (n_1, n_2)$ is the unit outer normal to the domain occupied by surrounding fluid Ω_t on instantaneous airfoil boundary Γ_{Wt} (i.e., pointing into the airfoil) and $x_{\mathrm{EA}} = (x_{\mathrm{EA1}}, x_{\mathrm{EA2}})$ is the position of the elastic axis (lying in the interior of the airfoil). Relations (5) and (6) define the coupling of the fluid dynamical model with the structural model.

3 Numerical Approximation

3.1 Fluid Model

To discretize the flow model we consider an equidistant partition of the time interval $[0, T]$ and approximate the solution $\mathbf{v}(\cdot, t_n)$ and $p(\cdot, t_n)$ (defined in Ω_{t_n}) at time t_n by $\mathbf{v}^n$ and p^n, respectively. A second-order two-step implicit scheme is used for the time discretization. For each fixed time $t = t_{n+1}$ define the spaces $\mathcal{W} = \mathbf{H}^1(\Omega_{n+1})$, $\mathcal{Q} = L^2(\Omega_{n+1})$ and $\mathcal{X} = \left\{ \mathbf{z} \in \mathcal{W} : \mathbf{z} = 0 \text{ on } \Gamma_D \cup \Gamma_{W t_{n+1}} \right\}$.

The ALE velocity $\mathbf{w}_D(t_{n+1})$ is approximated by $\mathbf{w}_D{}^{n+1}$ and the notation $\widehat{\mathbf{v}}^i = \mathbf{v}^i \circ \mathcal{A}_{t_i} \circ \mathcal{A}_{t_{n+1}}^{-1}$ is employed, where $\circ$ denotes the composition of functions. Then on each time level t_{n+1}, the ALE time discretized flow model is: Find $\mathbf{v} = \mathbf{v}^{n+1}$ and $p = p^{n+1}$ defined in Ω_{n+1} such that

$$\frac{3\mathbf{v} - 4\widehat{\mathbf{v}}^n + \widehat{\mathbf{v}}^{n-1}}{2\tau} + \bar{\mathbf{w}}^{n+1} \cdot \nabla \mathbf{v} - \nabla \cdot \left(\nu_{\mathit{eff}} \left(\nabla \mathbf{v} + (\nabla \mathbf{v})^T \right) \right) + \nabla p = 0,$$
$$\operatorname{div} \mathbf{v} = 0, \tag{6}$$

hold in Ω_{n+1} and $\mathbf{v}$ satisfies the boundary conditions (2a–b). Here $\bar{\mathbf{w}}^{n+1}$ denotes the convection velocity in the ALE frame at time t_{n+1}, i.e., $\bar{\mathbf{w}}^{n+1} = \mathbf{v} - \mathbf{w}_D{}^{n+1}$. The problem (6) is weakly formulated: Find $U = (\mathbf{v}, p)$ satisfying

$$a(U; U, V) = f(V) \quad \text{for all } \; V = (\mathbf{z}, q) \in \mathcal{X} \times \mathcal{Q}, \tag{7}$$

with **v** satisfying the conditions (2a–b). The forms used in (7) are defined by

$$
\begin{aligned}
a(U^*;U,V) &= \left(\frac{3\mathbf{v}}{2\tau},\mathbf{z}\right)_{\Omega_{n+1}} + \left(\bar{\mathbf{w}}^{n+1}\cdot\nabla\mathbf{v},\mathbf{z}\right)_{\Omega_{n+1}} \\
&+ \left(\nu_{eff}\left(\nabla\mathbf{v}+(\nabla\mathbf{v})^T\right),\nabla\mathbf{z}\right)_{\Omega_{n+1}} - (p,\nabla\cdot\mathbf{z})_{\Omega_{n+1}} \\
&+ (\nabla\cdot\mathbf{v},q)_{\Omega_{n+1}} \\
f(V) &= \int_{\Omega_{n+1}} \frac{4\widehat{\mathbf{v}}^n - \widehat{\mathbf{v}}^{n-1}}{2\tau}\cdot\mathbf{z}\,\mathrm{d}x - \int_{\Gamma_O} p_{\mathrm{ref}}\mathbf{z}\cdot\mathbf{n}\,\mathrm{dS},
\end{aligned}
$$

where $U = (\mathbf{v}, p)$, $V = (\mathbf{z}, q)$, $U^* = (\mathbf{v}^*, p)$, $\bar{\mathbf{w}}^{n+1} = \mathbf{v}^* - {\mathbf{w}_D}^{n+1}$.

3.1.1 Spatial Discretization

To apply the Galerkin FEM, we approximate the spaces $\mathcal{W}$, $\mathcal{X}$, $\mathcal{Q}$ of the weak formulation by finite dimensional subspaces $\mathcal{W}_\Delta$, $\mathcal{X}_\Delta$, $\mathcal{Q}_\Delta$, $\Delta \in (0,\Delta_0)$, $\Delta_0 > 0$, $\mathcal{X}_\Delta = \{\mathbf{v}_\Delta \in \mathcal{W}_\Delta; \mathbf{v}_\Delta|_{\Gamma_D\cap\Gamma_{Wt}} = 0\}$, where the spaces W_Δ, $\mathcal{X}_\Delta$ and $\mathcal{Q}_\Delta$ are formed by piecewise linear functions defined over an admissible triangulation $\mathcal{T}_\Delta$; see [5]. To overcome the instability caused by the incompatibility of the pressure and velocity pairs, the pressure stabilizing/Petrov–Galerkin method [6] is applied together with the Galerkin Least Squares (GLS) method [7].

First, the local element residual terms $\mathcal{R}_K^a$, $\mathcal{R}_K^f$ are defined on $K \in \mathcal{T}_\Delta$ by

$$
\mathcal{R}_K^a(\bar{\mathbf{w}}^{n+1};\mathbf{v},p) = \frac{3}{2\Delta t}\mathbf{v} - \nabla\cdot\left(\nu_{eff}\left(\nabla\mathbf{v}+(\nabla\mathbf{v})^T\right)\right) + \left(\bar{\mathbf{w}}^{n+1}\cdot\nabla\right)\mathbf{v} + \nabla p \quad (8)
$$

and

$$
\mathcal{R}_K^f(\hat{\mathbf{v}}^n,\hat{\mathbf{v}}^{n-1}) = \frac{4}{2\Delta t}\hat{\mathbf{v}}^n - \frac{1}{2\Delta t}\hat{\mathbf{v}}^{n-1}. \quad (9)
$$

The GLS stabilizing terms are taken with respect to the transport velocity $\bar{\mathbf{w}}^{n+1}$, i.e.,

$$
\begin{aligned}
\mathcal{L}(U_\Delta^*;U_\Delta,V_\Delta) &= \sum_{K\in\mathcal{T}_\Delta} \delta_K\left(\mathcal{R}_K^a(\bar{\mathbf{w}}^{n+1};\mathbf{v},p),\left(\bar{\mathbf{w}}^{n+1}\cdot\nabla\right)\mathbf{z}+\nabla q\right)_K, \\
\mathcal{F}(V_\Delta) &= \sum_{K\in\mathcal{T}_\Delta} \delta_K\left(\mathcal{R}_K^f(\hat{\mathbf{v}}^n,\hat{\mathbf{v}}^{n-1}),\left(\bar{\mathbf{w}}^{n+1}\cdot\nabla\right)\mathbf{z}+\nabla q\right)_K, \quad (10)
\end{aligned}
$$

with $\delta_K = h_K^2/\tau_K$, and grad-div stabilization is used:

$$
\mathcal{P}_\Delta(U,V) = \sum_{K\in\mathcal{T}_\Delta} \tau_K(\nabla\cdot\mathbf{v},\nabla\cdot\mathbf{z})_K \quad (11)
$$

where $\tau_K = \nu_K \left(1 + Re^{loc} + \frac{h_K^2}{\nu_K \Delta t}\right)$, $Re^{loc} = \frac{h|\bar{\mathbf{w}}^{n+1}|_{0,K}}{2\nu_K}$, $\nu_K = |\nu + \nu_T|_{0,K}$ is the maximal element viscosity, $|\cdot|_{0,K}$ is the $L^2(K)$ norm and h_K the local element size.

The *GLS stabilized discrete problem* is then: Find $U_\Delta = (\mathbf{v}, p) \in \mathcal{W}_\Delta \times \mathcal{Q}_\Delta$ such that $\mathbf{v}$ satisfies approximately the Dirichlet boundary conditions (2a–b) and the equation

$$a(U_\Delta; U_\Delta, V_\Delta) + \mathcal{L}(U_\Delta; U_\Delta, V_\Delta) + \mathcal{P}_\Delta(U_\Delta, V_\Delta) = f(V_\Delta) + \mathcal{F}(V_\Delta), \tag{12}$$

holds for all $V_\Delta = (\mathbf{z}, q) \in \mathcal{X}_\Delta \times \mathcal{Q}_\Delta$.

3.2 Numerical Discretization of the Turbulence Model

The equation (3) is discretized in time at each fixed time level $t = t_{n+1}$ with the aid of a second-order backward difference formula, i.e.,

$$\frac{D^{\mathcal{A}}\tilde{\nu}}{Dt} \approx \frac{3\tilde{\nu}^{n+1} - 4\widehat{\tilde{\nu}^n} + \widehat{\tilde{\nu}^{n-1}}}{2\Delta t}$$

in Ω_{n+1}, where $\widehat{\tilde{\nu}^i} = \tilde{\nu}^i \circ \mathcal{A}_{t_i} \circ \mathcal{A}^{-1}_{t_{n+1}}$. The numerical solution of the time-discretized Spalart–Allmaras turbulence problem is obtained from the finite element method applied to the linearized problem. Here we use

$$s(\tilde{\nu}^{n+1})^2 \approx 2s\,\tilde{\nu}^n\,\tilde{\nu}^{n+1} - s\,(\tilde{\nu}^n)^2, \quad (\nabla\tilde{\nu}^{n+1})^2 \approx \nabla\tilde{\nu}^n \cdot \nabla\tilde{\nu}^{n+1}.$$

The linearized problem is then weakly formulated. Furthermore, a finite element subspace $\mathcal{V}_\Delta \subset \mathcal{V}$ of the piecewise linear continuous functions over $\mathcal{T}_\Delta$ is constructed and the SUPG stabilization is applied. The SUPG terms and parameters are modified with respect to the ALE moving frame, which means that the stabilization terms and parameters need to be taken with respect to the modified transport velocity $\bar{\mathbf{w}}^{n+1} = \mathbf{v}^{n+1} - \mathbf{w}_D{}^{n+1}$. Furthermore the transport velocity is affected by the linearization procedure. The SUPG-stabilized linearized problem reads: Find $\tilde{\nu}_\Delta \in \mathcal{V}_\Delta$ such that $B(\tilde{\nu}_\Delta, \varphi) = L(\varphi)$ for all $\varphi \in \mathcal{V}_\Delta$, where

$$B(\tilde{\nu}_\Delta, \varphi) = \left(\frac{3\tilde{\nu}^{n+1}}{2\Delta t} + \mathbf{b}\cdot\nabla\tilde{\nu}^{n+1} + 2s\,\tilde{\nu}^n\,\tilde{\nu}^{n+1}, \varphi\right)_{\Omega_{n+1}} + \left(\varepsilon\,\nabla\tilde{\nu}^{n+1}, \nabla\varphi\right)_{\Omega_{n+1}}$$
$$+ \sum_{K\in\mathcal{T}_\Delta} \alpha_K\Big(\big(\frac{3}{2\Delta t} + 2s\,\tilde{\nu}^n_\Delta\big)\tilde{\nu} + \mathbf{b}\cdot\nabla\tilde{\nu}_\Delta + \nabla\cdot(\varepsilon\,\nabla\tilde{\nu}_\Delta), (\mathbf{b}\cdot\nabla)\varphi\Big)_K,$$

$$L(\varphi) = \left(s\,\tilde{\nu}^n\,\tilde{\nu}^n + \frac{4\widehat{\tilde{\nu}^n} - \widehat{\tilde{\nu}^{n-1}}}{2\Delta t} + c_{b_1}\tilde{S}\,\tilde{\nu}^n, \varphi\right)_{\Omega_{n+1}}$$
$$+ \sum_{K\in\mathcal{T}_\Delta} \alpha_K\left(s\,\tilde{\nu}^n_\Delta\,\tilde{\nu}^n_\Delta + \frac{4\widehat{\tilde{\nu}^n_\Delta} - \widehat{\tilde{\nu}^{n-1}_\Delta}}{2\Delta t} + c_{b_1}\tilde{S}, (\mathbf{b}\cdot\nabla)\varphi)\right)_K;$$

here the transport velocity $\mathbf{b} := \mathbf{v}^{n+1} - \mathbf{w}_D{}^{n+1} - \frac{c_{b2}}{\beta}\nabla\tilde{\nu}_\Delta^n$ is the fluid velocity $\mathbf{v}^{n+1}$ modified by the velocity coming from the mesh motion and the linearization of the quadratic term $(\nabla\tilde{\nu})^2$. We choose the parameter $\alpha_K = \left(\frac{4|\varepsilon|_{0,K}}{h_K^2} + \frac{2|\mathbf{b}|_{0,K}}{h_K} + |s|_{0,K}\right)^{-1}$. The SUPG/GLS stabilization does not exclude local oscillations near sharp layers. Such oscillations are suppressed by the application of an additional *crosswind* diffusion; cf. [8].

4 Numerical Results

4.1 Computational Meshes

To obtain physically admissible solutions one should use meshes that are refined in the boundary layer and wake regions. The computational meshes used in this paper are generated with the aid of our code, which refines the boundary layer region *a priori* and uses anisotropic mesh refinement [9] elsewhere. Different meshes are used for the DCA profile (where the leading edge is sharp and the so-called H-mesh can be used) and the NACA 0012 profile (where the so-called C-mesh needs to be employed). In the latter case we follow the procedure described in [10]. The H-mesh details generated around the DCA profile are shown in Fig. 2. The distance of the first point "perpendicular" to the airfoil boundary is chosen to satisfy $Y^+ \approx 1$, where $Y^+ = \frac{u_\tau Y}{\nu}$ and Y is the distance to the wall and u_τ is the friction velocity on the wall.

4.1.1 Flow Over Vibrating DCA Profile

First, the numerical method was applied to a problem of channel flow over a moving DCA profile with prescribed vibrations: see Fig. 1. The harmonic vibrations of the airfoil are prescribed with frequency of $f_H = 20.4$ Hz around the elastic axis positioned at one third of the airfoil. The position of the airfoil is then given by its angle of rotation $\alpha(t)$ and vertical displacement $h(t)$; in this case we considered $\alpha(t) = -1.5° - 4.5° \sin(2\pi f_H t)$ and $h(t) = -1.5 - 4.5\sin(2\pi f_H t)$ mm. The far

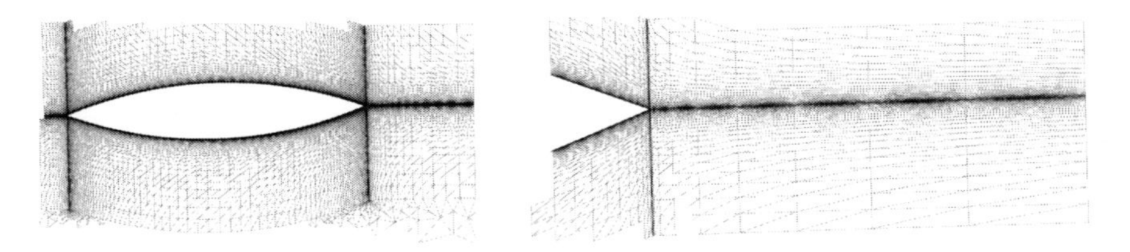

Fig. 2 Computational mesh around the DCA profile

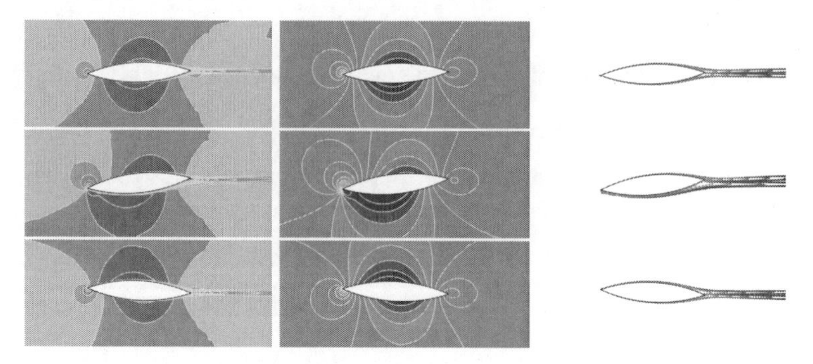

Fig. 3 The fluid velocity isolines (*left*), pressure isolines (*middle*) and the turbulent viscosity (*right*) distributions around vibrating DCA airfoil

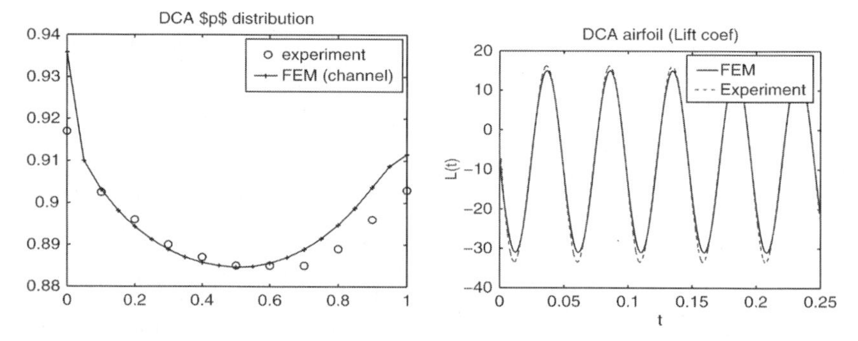

Fig. 4 Comparison of the pressure distribution over the DCA airfoil (*left*) and the lift coefficient (*right*)

field velocity $U_\infty = 120\,\mathrm{m\,s}^{-1}$ and the reference length $c = 0.12\,\mathrm{m}$ were used. This model description corresponds to the experimental measurement of nonlinear aeroelastic instability – Limit Cycle Oscillations (LCO). Numerical results for this model problem are shown in Figs. 3 and 4. The numerical results were computed on a triangular mesh with 19,205 nodes and 38,012 elements (76,820 unknowns). The results were compared to the experimental data: Fig. 4 on the left shows the distribution of the mean pressure coefficient $c_p = \frac{p-p_0}{\frac{1}{2}\rho U^2}$ on the surface of the airfoil from experimental measurement and from the numerical simulation. These data are in very good agreement. Figure 4 on the right then shows the dependence of the lift coefficient on the time from experimental data and from numerical experiment. Figure 3 then shows the instantaneous distribution of velocity magnitude, pressure and turbulent viscosity around the vibrating DCA airfoil at time instants corresponding to the positions $\alpha = 0°, 3°, -3°$.

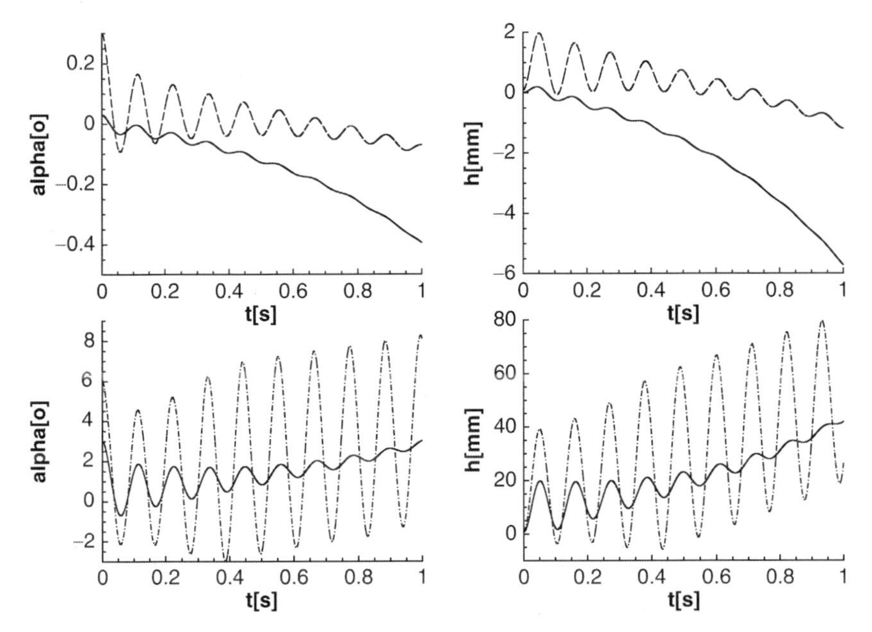

Fig. 5 The aeroelastic response $\alpha(t)$ (*left*) and $h(t)$ (*right*) for the post-flutter velocity $U_\infty = 40\mathrm{m\,s^{-1}}$ and initial conditions $h(0) = 0$ and $\alpha(0) = \alpha_0$: $\alpha_0 = 0.03°$ (*up, solid line*) $\alpha_0 = 0.3°$ (*up, dashed line*), and $3°$ (*down, solid*), $6°$ (*down, dashdot*)

4.1.2 Flow Induced Airfoil Vibrations

This section presents the results of the numerical simulation of flow-induced vibrations obtained for the NACA 0012 airfoil. The following quantities are considered: $m = 8.6622 \times 10^{-2}$ kg, $S_\alpha = 7.79673 \times 10^{-4}$ kg m, $I_\alpha = 4.87291 \times 10^{-4}$ kg m^2, $k_{hh} = 105.109$ N m^{-1}, $k_{\alpha\alpha} = 3.695582$ N m rad^{-1}, $l = 0.05$ m, $c = 0.3$ m, $\rho = 1.225$ kg m^{-3}, $\nu = 1.5 \cdot 10^{-5}$ m s^{-2}. Linear theory predicts the critical velocity $U_{crit} = 37.7\,\mathrm{m\,s^{-1}}$; see [3]. Numerical results were computed for different far field velocities U_∞ and were in agreement with NASTRAN computations [3, 11]. Here, the numerical approximation of a problem after the loss of stability is computed on an anisotropically refined mesh with 19,430 nodes and 38,428 elements. Figure 5 shows the aeroelastic response of the system for the far field velocity $U_\infty = 40\,\mathrm{m\,s^{-1}}$, where different initial conditions $\alpha(0)$ were applied; one sees typical divergence instability behaviour for low values of the initial condition $\alpha(0)$ but for the value $\alpha(0) = 6°$ a combination of divergence and flutter-type behaviour can be observed.

Acknowledgement This research was supported under the Research Plan MSM 6840770003 of the Ministry of Education of the Czech Republic.

References

1. E. H. Dowell. *A Modern Course in Aeroelasticity*. Kluwer, Dodrecht, 1995.
2. D. C. Wilcox. *Turbulence Modeling for CFD*. DCW Industries, 1993.
3. M. Feistauer, J. Horáček, and P. Sváček. Numerical simulation of flow induced airfoil vibrations with large amplitudes. *J. Fluids Struct.*, 23(3):391–411, 2007.
4. T. Nomura and T. J. R. Hughes. An arbitrary Lagrangian-Eulerian finite element method for interaction of fluid and a rigid body. *Comp. Meth. Appl. Mech. Eng.*, 95:115–138, 1992.
5. P. G. Ciarlet. *The FEM for Elliptic Problems*. North-Holland, Amsterdam, 1979.
6. T. J. R. Hughes, L. P. Franca, and M. Balestra. A new finite element formulation for computational fluid dynamics: V. circumventing the Babuška-Breezzi condition: a stable Petrov-Galerkin formulation of the Stokes problem accomodating equal order interpolation. *Comp. Meth. Appl. Mech. Eng.*, 59:85–89, 1986.
7. T. Gelhard, G. Lube, M. A. Olshanskii, and J.-H. Starcke. Stabilized finite element schemes with LBB-stable elements for incompressible flows. *J. Comp. Appl. Math.*, 177:243–267, 2005.
8. R. Codina. A discontinuity capturing crosswind-dissipation for FE solution of the convection diffusion equation. *Comp. Meth. Appl. Mech. Eng.*, 110:325–342, 1993.
9. V. Dolejší. Anisotropic mesh adaptation technique for viscous flow simulation. *East-West J. Num. Math.*, 9:1–24, 2001.
10. A. Shmilovich and D. A. Caughey. Grid generation for wing-tail-fuselage configurations. *J. Aircraft*, 6(22):467–472, 1985.
11. Sváček, P. , M. Feistauer, and J. Horáček. Numerical modelling of flow induced airfoil vibrations problem. In *Int. Conf. on Innovation and Integration in Aerospace Sci*, page 10 pp. Belfast: CEIAT, Queen's University Belfast, 2005.

A Parameter–Uniform Finite Difference Method for a Singularly Perturbed Initial Value Problem: A Special Case

S. Valarmathi and J.J.H. Miller

Abstract A system of singularly perturbed ordinary differential equations of first order with given initial conditions is considered. The leading term of each equation is multiplied by a small positive parameter. These parameters are assumed to be distinct. The components of the solution exhibit overlapping layers. A Shishkin piecewise–uniform mesh is constructed, which is used, in conjunction with a classical finite difference discretisation, to form a new numerical method for solving this problem. It is proved, in a special case, that the numerical approximations obtained from this method are essentially first order convergent uniformly in all of the parameters. Numerical results are presented in support of the theory.

1 Introduction

We consider the initial value problem for the singularly perturbed system of linear first order differential equations

$$E\mathbf{u}'(t) + A(t)\mathbf{u}(t) = \mathbf{f}(t), \quad t \in (0, T], \quad \mathbf{u}(0) \text{ given.} \tag{1}$$

Here $\mathbf{u}$ is a column n-vector, E and $A(t)$ are $n \times n$ matrices, $E = diag(\varepsilon)$, $\boldsymbol{\varepsilon} = (\varepsilon_1, \ldots, \varepsilon_n)$ with ε_i distinct and $0 < \varepsilon_i \leq 1$ for all $i = 1 \ldots n$. For convenience we assume the ordering

$$\varepsilon_1 < \cdots < \varepsilon_n.$$

J.J.H. Miller (✉)
Institute for Numerical Computation and Analysis, Dublin, Ireland, E-mail: jm@incaireland.org

A.F. Hegarty et al. (eds.), *BAIL 2008 - Boundary and Interior Layers.*
Lecture Notes in Computational Science and Engineering,
DOI: 10.1007/978-3-642-00605-0,

Cases with some of the parameters coincident are not considered here. We write the problem in the operator form

$$\mathbf{Lu} = \mathbf{f}, \quad \mathbf{u}(0) \text{ given,}$$

where the operator $\mathbf{L}$ is defined by

$$\mathbf{L} = ED + A(t) \text{ and } D = \frac{d}{dt}.$$

We assume that, for all $t \in [0, T]$, the components $a_{ij}(t)$ of $A(t)$ satisfy the inequalities

$$a_{ii}(t) > \sum_{\substack{j \neq i \\ j=1}}^{n} |a_{ij}(t)| \text{ for } i = 1, \ldots, n, \text{ and } a_{ij}(t) \leq 0 \text{ when } i \neq j.$$

We take α to be any number such that

$$0 < \alpha < \min_{\substack{t \in (0,1] \\ i=1, \ldots, n}} \left(\sum_{j=1}^{n} a_{ij}(t) \right).$$

We also assume that $T \geq 2 \max_i(\varepsilon_i)/\alpha$, which ensures that we are solving over a domain that includes all of the layers. For this it suffices to take $T \geq 2/\alpha$. We introduce the norms $\| \mathbf{V} \| = \max_{1 \leq k \leq n} |V_k|$ for any n-vector $\mathbf{V}$, $\| y \| = \sup_{0 \leq t \leq T} |y(t)|$ for any scalar-valued function y and $\| \mathbf{y} \| = \max_{1 \leq k \leq n} \| y_k \|$ for any vector-valued function $\mathbf{y}$. Throughout the paper C denotes a generic positive constant, which is independent of t and of all singular perturbation and discretisation parameters.

The initial value problems considered here arise in many areas of applied mathematics; see for example [1]. Parameter uniform numerical methods for simpler problems of this kind, when all the singular perturbation parameters are equal, were considered in [2]. For a reaction-diffusion boundary value problem in the case $n = 2$ a parameter uniform numerical method was constructed in [3] and in the case of general n in [4]. A general introduction to parameter uniform numerical methods is given in [5] and [6].

2 Analytical Results

The operator $\mathbf{L}$ satisfies the following maximum principle

Lemma 1. *Let the above assumptions on the matrix $A(t)$ hold. Let $\psi(t)$ be any function in the domain of $\mathbf{L}$ such that $\psi(0) \geq 0$. Then $\mathbf{L}\psi(t) \geq 0$ for all $t \in (0, T]$ implies that $\psi(t) \geq 0$ for all $t \in [0, T]$.*

We remark that the maximum principle is not necessary for the results that follow, but it is a convenient tool in the proof of the following stability result.

Lemma 2. *Let the above assumptions on the matrix $A(t)$ hold. If $\psi(t)$ is any function in the domain of* **L**, *then*

$$\| \psi(t) \| \leq \max\left\{ \| \psi(0) \|, \frac{1}{\alpha} \| \mathbf{L}\psi \| \right\}, \qquad t \in [0,T].$$

The Shishkin decomposition of the solution $\mathbf{u}$ of (1) is given by $\mathbf{u} = \mathbf{v} + \mathbf{w}$ where $\mathbf{v}$ is the solution of $\mathbf{L}\mathbf{v} = \mathbf{f}$ on $(0,T]$ with $\mathbf{v}(0) = A^{-1}(0)\mathbf{f}(0)$ and $\mathbf{w}$ is the solution of $\mathbf{L}\mathbf{w} = \mathbf{0}$ on $(0,T]$ with $\mathbf{w}(0) = \mathbf{u}(0) - \mathbf{v}(0)$. Here $\mathbf{v}, \mathbf{w}$ are, respectively, the smooth and singular components of $\mathbf{u}$.
Bounds on the smooth component and its derivatives are contained in

Lemma 3. *There exists a constant C, such that for each $i = 1, \ldots, n$, $\| v_i^{(k)} \| \leq C$ for $k = 0, 1$ and $\| \varepsilon_i v_i'' \| \leq C$.*

We define the layer functions $B_i, i = 1, \ldots, n$, associated with the solution $\mathbf{u}$ by

$$B_i(t) = e^{-\alpha t/\varepsilon_i}, \; t \in [0,\infty).$$

Some elementary properties of the layer functions are given in

Lemma 4. *Let $1 \leq i < j \leq n$ and $0 \leq s < t < \infty$. Then*

$$B_i(t) < B_j(t), \; \textit{for all} \;\; t > 0,$$

$$B_i(s) > B_i(t), \; \textit{for all} \;\; 0 \leq s < t \leq T,$$

$$B_i(0) = 1 \;\; \textit{and} \;\; 0 < B_i(t) < 1 \;\; \textit{for all} \;\; t > 0.$$

Bounds on the singular component and its derivatives are contained in

Lemma 5. *There exists a constant C such that, for each $t \in [0,T]$ and $i = 1, 2, 3$*

$$|w_i(t)| \leq CB_3(t), \;\; |w_i'(t)| \leq C\left[\varepsilon_i^{-1}B_i(t) + \cdots + \varepsilon_3^{-1}B_3(t)\right],$$
$$\left|\varepsilon_i w_i''(t)\right| \leq C\left[\varepsilon_1^{-1}B_1(t) + \varepsilon_2^{-1}B_2(t) + \varepsilon_3^{-1}B_3(t)\right].$$

Proof. The bounds on the w_i are obtained by applying Lemma 1 to the functions $\psi^{\pm} = CB_3\mathbf{e} \pm \mathbf{w}$. The bound on w_3' follows from the third equation of the system satisfied by $\mathbf{w}$. The first two equations of this system form an inhomogeneous system for the components w_1, w_2. The required bounds on w_1', w_2' are obtained by a Shishkin decomposition of w_1, w_2 followed by the application of Lemmas 2.3 and 2.4 in [7]. Finally, the bounds on the w_i'' follow immediately from the system satisfied by $\mathbf{w}$. □

For each $i \neq j$ we now define the point $t_{i,j}$ by

$$\frac{B_i(t_{i,j})}{\varepsilon_i} = \frac{B_j(t_{i,j})}{\varepsilon_j}.$$

It is easy to see that this point exists and is unique for each i and j, since for $i < j$ we have $\varepsilon_i < \varepsilon_j$ and the ratio of the two sides of this equation, namely

$$\frac{B_i(t)}{\varepsilon_i}\frac{\varepsilon_j}{B_j(t)} = \frac{\varepsilon_j}{\varepsilon_i}\exp\left(-\alpha t\left(\frac{1}{\varepsilon_i} - \frac{1}{\varepsilon_j}\right)\right),$$

is monotonically decreasing from $\frac{\varepsilon_j}{\varepsilon_i} > 1$ to 0 as t increases from 0 to ∞.
Also, the following inequalities hold, for all i, j with $1 \leq i < j \leq n$

$$\varepsilon_i^{-1} B_i(t) > \varepsilon_j^{-1} B_j(t) \qquad \text{on } [0, t_{i,j}), \tag{2}$$

$$\varepsilon_i^{-1} B_i(t) < \varepsilon_j^{-1} B_j(t) \qquad \text{on } (t_{i,j}, \infty) \tag{3}$$

and if $\varepsilon_i \leq \varepsilon_j/2$ then $t_{i,j} \in (0, T]$.

Lemma 6. *The points $t_{i,j}$ satisfy the following inequalities*

$$t_{i,j} < t_{i+1,j}, \text{ if } i+1 < j$$

and

$$t_{i,j} < t_{i,j+1}, \text{ if } i < j$$

Proof. It is not hard to see that the point $t_{i,j}$ is given by

$$t_{i,j} = \frac{\ln\left(\frac{1}{\varepsilon_i}\right) - \ln\left(\frac{1}{\varepsilon_j}\right)}{\alpha\left(\frac{1}{\varepsilon_i} - \frac{1}{\varepsilon_j}\right)}.$$

We can write $\varepsilon_k = \exp(-p_k)$ for some $p_k > 0$ for all k. Then

$$t_{i,j} = \frac{p_i - p_j}{\alpha(\exp p_i - \exp p_j)}.$$

The inequality $t_{i,j} < t_{i+1,j}$ is equivalent to

$$\frac{p_i - p_j}{\exp p_i - \exp p_j} < \frac{p_{i+1} - p_j}{\exp p_{i+1} - \exp p_j},$$

which can be written in the form

$$(p_{i+1} - p_j)\exp(p_i - p_j) + (p_i - p_{i+1}) - (p_i - p_j)\exp(p_{i+1} - p_j) > 0.$$

Writing $a = p_i - p_j$ and $b = p_{i+1} - p_j$ we have $a > b > 0$ and $a - b = p_i - p_{i+1}$. Moreover, the previous inequality is then equivalent to

$$\frac{\exp a - 1}{a} > \frac{\exp b - 1}{b},$$

which is true because $a > b$.
The second part of the lemma is proved by a similar argument. □

3 The Discrete Problem

We construct a piecewise uniform mesh with N mesh-intervals and mesh-points $\{t_i\}_{i=0}^N$ by dividing the interval $[0, T]$ into $n + 1$ sub-intervals as follows

$$[0, T] = [0, \sigma_1] \cup (\sigma_1, \sigma_2] \cup \dots (\sigma_{n-1}, \sigma_n] \cup (\sigma_n, T].$$

On the sub-interval $[0, \sigma_1]$ a uniform mesh with $\frac{N}{2^n}$ mesh-intervals is placed, similarly on $(\sigma_i, \sigma_{i+1}]$, $1 \le i \le n-1$, a uniform mesh of $\frac{N}{2^{n-i+1}}$ mesh-intervals and on $(\sigma_n, T]$ a uniform mesh of $\frac{N}{2}$ mesh-intervals. The n transition points between the uniform meshes are defined by

$$\sigma_i = \min\left\{\frac{\sigma_{i+1}}{2}, \frac{\varepsilon_i}{\alpha}\ln N\right\}$$

for $i = 1, \dots, n-1$ and

$$\sigma_n = \min\left\{\frac{T}{2}, \frac{\varepsilon_n}{\alpha}\ln N\right\}.$$

Clearly

$$0 < \sigma_1 < \dots < \sigma_n \le \frac{T}{2}.$$

This construction leads to a class of 2^n possible Shishkin piecewise uniform meshes $M_{\mathbf{b}}$, where $\mathbf{b}$ denotes an n–vector with $b_i = 0$ if $\sigma_i = \frac{\sigma_{i+1}}{2}$ and $b_i = 1$ otherwise. Writing $\delta_j = t_j - t_{j-1}$ we remark that, on any such mesh, we have

$$\delta_j \le CN^{-1}, \quad 1 \le j \le N$$

and

$$\sigma_i \le C\varepsilon_i \ln N, \quad 1 \le i \le n.$$

On these meshes we now consider the discrete solutions defined by the backward Euler finite difference scheme

$$ED^-\mathbf{U} + A(t)\mathbf{U} = \mathbf{f}, \qquad \mathbf{U}(0) = \mathbf{u}(0),$$

or in operator form

$$\mathbf{L}^N\mathbf{U} = \mathbf{f}, \qquad \mathbf{U}(0) = \mathbf{u}(0),$$

where

$$\mathbf{L}^N = ED^- + A(t)$$

and D^- is the backward difference operator

$$D^-\mathbf{U}(t_j) = \frac{\mathbf{U}(t_j) - \mathbf{U}(t_{j-1})}{\delta_j}.$$

We have the following discrete maximum principle analogous to the continuous case

Lemma 7. *Let the above assumptions on the matrix $A(t)$ hold. Then, for any mesh function $\boldsymbol{\Psi}$, the inequalities $\boldsymbol{\Psi}(0) \geq \mathbf{0}$ and $\mathbf{L}^N\boldsymbol{\Psi}(t_j) \geq \mathbf{0}$ for $1 \leq j \leq N$, imply that $\boldsymbol{\Psi}(t_j) \geq \mathbf{0}$ for $0 \leq j \leq N$.*

An immediate consequence of this is the following discrete stability result.

Lemma 8. *Let the above assumptions on the matrix $A(t)$ hold. Then, for any mesh function $\boldsymbol{\Psi}$,*

$$\| \boldsymbol{\Psi}(t_j) \| \leq \max \left\{ \| \boldsymbol{\Psi}(0) \|, \frac{1}{\alpha} \| \mathbf{L}^N \boldsymbol{\Psi} \| \right\}, 0 \leq j \leq N.$$

4 The Local Truncation Error

From Lemma 8, we see that in order to bound the error $\| \mathbf{U} - \mathbf{u} \|$ it suffices to bound $\mathbf{L}^N(\mathbf{U} - \mathbf{u})$. But this expression satisfies

$$\mathbf{L}^N(\mathbf{U} - \mathbf{u}) = \mathbf{L}^N(\mathbf{U}) - \mathbf{L}^N(\mathbf{u}) = \mathbf{f} - \mathbf{L}^N(\mathbf{u}) = \mathbf{L}(\mathbf{u}) - \mathbf{L}^N(\mathbf{u})$$
$$= (\mathbf{L} - \mathbf{L}^N)\mathbf{u} = -E(D^- - D)\mathbf{u},$$

which is the local truncation of the first derivative. We have

$$E(D^- - D)\mathbf{u} = E(D^- - D)\mathbf{v} + E(D^- - D)\mathbf{w}$$

and so, by the triangle inequality,

$$\| \mathbf{L}^N(\mathbf{U} - \mathbf{u}) \| \leq \| E(D^- - D)\mathbf{v} \| + \| E(D^- - D)\mathbf{w} \| .$$

Thus, we can treat the smooth and singular components of the local truncation error separately We note first that, for any smooth function ψ, we have the following two distinct estimates of the local truncation error of the first derivative

$$|(D^- - D)\psi(t_j)| \leq \max_{s \in [t_{j-1}, t_j]} |\psi''(s)| \frac{\delta_j}{2} \tag{4}$$

and

$$|(D^- - D)\psi(t_j)| \leq 2 \max_{s \in [t_{j-1}, t_j]} |\psi'(s)|. \tag{5}$$

5 Error Estimate: The Special Case n = 3

Here we establish the error estimate when $n = 3$. The same approach suffices for $n = 1$ and $n = 2$ and is similar to that used in [3] for the reaction-diffusion problem. For general n additional techniques are required and will be the topic of future work. The technique applied in [4] for the general reaction-diffusion problem uses discrete Green's functions instead of the Shishkin decompositons used in this paper. We estimate the smooth component of the local truncation error in the following lemma.

Lemma 9. *For each* $i = 1, \ldots, n$ *and* $j = 1, \ldots, N$ *we have*

$$|\varepsilon_i (D^- - D) v_i(t_j)| \le C N^{-1}.$$

For the singular component we obtain a similar estimate, but we must distinguish between the different types of mesh. We need the following preliminary lemmas.

Lemma 10. *On each mesh* $M_{\mathbf{b}}$*, for* $i = 1, 2, 3$ *we have the estimate*

$$|\varepsilon_i (D^- - D) w_i(t_j)| \le C \frac{\delta_j}{\varepsilon_1}.$$

In what follows we make use of second degree polynomials of the form

$$p_{i;\theta} = \sum_{k=0}^{2} \frac{(t - t_\theta)^k}{k!} w_i^{(k)}(t_\theta)$$

where θ denotes a pair of integers separated by a comma.

Lemma 11. *On each mesh of the form* $M_{1b_2b_3}$*, for* $i = 1, 2, 3$ *there exists a decomposition*

$$w_i = w_{i,1} + w_{i,2},$$

for which we have the estimates

$$|\varepsilon_i w'_{i,1}(t)| \le C B_1(t),$$

$$|\varepsilon_i w''_{i,1}(t)| \le C \frac{B_1(t)}{\varepsilon_1}, \qquad |\varepsilon_i w''_{i,2}(t)| \le C \left(\frac{B_2(t)}{\varepsilon_2} + \frac{B_3(t)}{\varepsilon_3} \right).$$

Furthermore

$$|\varepsilon_i (D^- - D) w_i(t_j)| \le C \left(B_1(t_{j-1}) + \frac{\delta_j}{\varepsilon_2} \right).$$

Proof. Since $b_1 = 1$ we have $\varepsilon_1 \le \varepsilon_2/2$, so $t_{1,2} \in (0, T]$ and we can define the components of the decomposition by

$$w_{i,2} = \begin{cases} p_{i;1,2} & \text{on } [0, t_{1,2}) \\ w_i & \text{otherwise} \end{cases}$$

$$w_{i,1} = w_i - w_{i,2} \quad \text{on } [0, \mathrm{T}]$$

The proof is completed using the ideas in Lemma 2.6 in [7]. □

Lemma 12. *On each mesh of the form $M_{b_1 1 b_3}$, for $i = 1, 2, 3$ there exists a decomposition*

$$w_i = w_{i,1} + w_{i,2} + w_{i,3},$$

for which we have the estimates

$$|\varepsilon_i w'_{i,j}(t)| \le CB_j(t) \text{ for } j = 1, 2,$$

$$|\varepsilon_i w''_{i,j}(t)| \le C\frac{B_j(t)}{\varepsilon_j} \text{ for } j = 1, 2, 3.$$

Furthermore

$$|\varepsilon_i(D^- - D)w_i(t_j)| \le C\left(B_2(t_{j-1}) + \frac{\delta_j}{\varepsilon_3}\right).$$

Proof. Since $b_2 = 1$ we have $\varepsilon_2 \le \varepsilon_3/2$, so $t_{2,3} \in (0, T]$ and we can define the components of the decomposition by

$$w_{i,3} = \begin{cases} p_{i;2,3} & \text{on } [0, t_{2,3}) \\ w_i & \text{otherwise} \end{cases}$$

$$w_{i,2} = \begin{cases} p_{i;1,2} & \text{on } [0, t_{1,2}) \\ w_i - w_{i,3} & \text{otherwise} \end{cases}$$

$$w_{i,1} = w_i - w_{i,2} - w_{i,3} \text{ on } [0, T]$$

The proof is completed by a simple generalisation of the proof of the previous lemma. □

Lemma 13. *On each mesh $M_{\mathbf{b}}$, for $i = 1, 2, 3$ we have the estimate*

$$|\varepsilon_i(D^- - D)w_i(t_j)| \le CB_3(t_{j-1}).$$

Using the above preliminary lemmas on appropriate subintervals we obtain the desired estimate of the singular component of the local truncation error in the following.

Lemma 14. *For $i = 1, 2, 3$ and $j = 1, \ldots, N$, we have the estimate*

$$|\varepsilon_i(D^- - D)w_i(t_j)| \le CN^{-1}\ln N.$$

Proof. On any subinterval $[0, t]$ we have $\frac{\delta_j}{\varepsilon_1} \le CN^{-1}\frac{t}{\varepsilon_1}$. It follows at once from Lemma 10 that the desired estimate holds on the mesh M_{000} in $[0, T]$ because $\frac{T}{\varepsilon_1} \le C \ln N$; on the mesh M_{001} in $[0, \sigma_3]$ because $\sigma_1 = \frac{\sigma_3}{4}$ hence $\frac{\sigma_3}{\varepsilon_1} \le C \ln N$; on the meshes M_{010}, M_{011} in $[0, \sigma_2]$ because $\sigma_1 = \frac{\sigma_2}{2}$ hence $\frac{\sigma_2}{\varepsilon_1} \le C \ln N$; on any mesh in $[0, \sigma_1]$ because $\frac{\sigma_1}{\varepsilon_1} \le C \ln N$.

On any mesh of the form $M_{1b_2b_3}$ we have $\sigma_1 = \frac{\varepsilon_1}{\alpha} \ln N$ and so in any subinterval of the form $(\sigma_1, t]$ we have $B_1(t_{j-1}) \le B_1(\sigma_1) = N^{-1}$ and $\frac{\delta_j}{\varepsilon_2} \le CN^{-1}\frac{t-\sigma_1}{\varepsilon_2}$. It follows at once from Lemma 11 that the desired estimate holds on the mesh M_{100}

in the subinterval $(\sigma_1, T]$, on the mesh M_{101} in the subinterval $(\sigma_1, \sigma_3]$ and on the meshes M_{110}, M_{111} in the subinterval $(\sigma_1, \sigma_2]$.

Similarly, on any mesh of the form $M_{b_1 1 b_3}$ we have $\sigma_2 = \frac{\varepsilon_2}{\alpha} \ln N$ and so in any subinterval of the form $(\sigma_2, t]$ we have $B_2(t_{j-1}) \leq B_2(\sigma_2) = N^{-1}$ and $\frac{\delta_j}{\varepsilon_3} \leq CN^{-1} \frac{t-\sigma_2}{\varepsilon_3}$. It follows at once from Lemma 12 that the desired estimate holds on the meshes M_{010}, M_{110} in the subinterval $(\sigma_2, T]$ and on the meshes M_{011}, M_{111} in the subinterval $(\sigma_2, \sigma_3]$.

On any mesh of the form $M_{b_1 b_2 1}$ we have $\sigma_3 = \frac{\varepsilon_3}{\alpha} \ln N$ and so in the subinterval $(\sigma_3, T]$ we have $B_3(t_{j-1}) \leq B_3(\sigma_3) = N^{-1}$. It follows at once from Lemma 13 that the desired estimate holds. □

Let $\mathbf{u}$ denote the exact solution of (1) and $\mathbf{U}$ the discrete solution. Then, using Lemmas 9 and 14, we have the following ε-uniform error estimate

Theorem 1. *There exists a constant C such that*

$$\| \mathbf{U} - \mathbf{u} \| \leq CN^{-1} \ln N$$

for all $N > 1$

Table 1 Values of D_ε^N, D^N, p^N, p^*, and $C_{p^*}^N$ for various ε_1 and N with fixed $\varepsilon_2 = 2^{-6}$, $\varepsilon_3 = 2^{-4}$

	Number of mesh points N								
ε_1	128	256	512	1024	2048	4096	8192	16384	32768
2^{-7}	0.135-2	0.832-3	0.485-3	0.276-3	0.154-3	0.853-4	0.466-4	0.253-4	0.136-4
2^{-11}	0.195-2	0.118-2	0.688-3	0.391-3	0.215-3	0.117-3	0.625-4	0.332-4	0.175-4
2^{-15}	0.230-2	0.136-2	0.808-3	0.469-3	0.262-3	0.145-3	0.792-4	0.430-4	0.232-4
2^{-19}	0.232-2	0.138-2	0.810-3	0.476-3	0.266-3	0.147-3	0.805-4	0.436-4	0.235-4
2^{-23}	0.232-2	0.138-2	0.810-3	0.477-3	0.266-3	0.147-3	0.806-4	0.437-4	0.236-4
2^{-27}	0.232-2	0.138-2	0.810-3	0.477-3	0.266-3	0.147-3	0.806-4	0.437-4	0.236-4
2^{-31}	0.232-2	0.138-2	0.810-3	0.477-3	0.266-3	0.147-3	0.806-4	0.437-4	0.236-4
2^{-35}	0.232-2	0.138-2	0.810-3	0.477-3	0.266-3	0.147-3	0.806-4	0.437-4	0.236-4
2^{-39}	0.232-2	0.138-2	0.810-3	0.477-3	0.266-3	0.147-3	0.806-4	0.437-4	0.236-4
2^{-43}	0.232-2	0.138-2	0.810-3	0.477-3	0.266-3	0.147-3	0.806-4	0.437-4	0.236-4
D^N	0.232-2	0.138-2	0.810-3	0.477-3	0.266-3	0.147-3	0.806-4	0.437-4	0.236-4
p^N	0.753+0	0.767+0	0.765+0	0.842+0	0.855+0	0.867+0	0.882+0	0.891+0	
$C_{0.753}^N$	0.221+0	0.221+0	0.219+0	0.217+0	0.204+0	0.190+0	0.176+0	0.161+0	0.146+0

Computed order of ε_1–uniform convergence = 0.753

Computed ε_1–uniform error constant = 0.221

6 Numerical Results

The above numerical method is applied to the following singularly perturbed initial value problem

$$\varepsilon_1 u_1{}'(t) + 4u_1(t) - u_2(t) - u_3(t) = t \qquad (6)$$

$$\varepsilon_2 u_2{}'(t) - u_1(t) + 4u_2(t) - u_3(t) = 1 \qquad (7)$$

$$\varepsilon_3 u_3{}'(t) - u_1(t) - u_2(t) + 4u_3(t) = 1 + t^2 \qquad (8)$$

for $t \in (0,1]$ and $\mathbf{u}(0) = 0$. For various values of ε_1, fixed values $\varepsilon_2 = 2^{-6}$, $\varepsilon_3 = 2^{-4}$ and $N = 2^r, r = 7, \ \dots \ 15$, the computed order of ε_1–uniform convergence and the computed ε_1–uniform error constant are found using the general methodology from [5], [6]. The results, presented in Table 1 below, exhibit the behaviour expected from an ε_1–uniform method.

Similar numerical experiments illustrate separate ε_2– and ε_3– uniform behaviour.

Acknowledgement The first author acknowledges the support of the UGC, New Delhi, India under the Minor Research Project-X Plan period. Both authors acknowledge partial conference travel support from INCA, Dublin.

References

1. A. C. Athanasios, Approximation of Large-Scale Dynamical Systems. SIAM, Philadelphia (2005).
2. S. Hemavathi, S. Valarmathi, *A parameter uniform numerical method for a system of singularly perturbed ordinary differential equations.* Proceedings of the International Conference on Boundary and Interior Layers, BAIL 2006, Goettingen (2006).
3. N. Madden, M. Stynes, *A uniformly convergent numerical method for a coupled system of two singularly perturbed reaction-diffusion problems.* IMA J. Num. Anal., 23, 627–644 (2003).
4. T. Linss, N. Madden, *Layer-adapted meshes for a linear system of coupled singularly perturbed reaction-diffusion problems.* IMA J. Num. Anal., 29, 109–125 (2009).
5. P. A. Farrell, A. Hegarty, J. J. H. Miller, E. O'Riordan, G. I. Shishkin, *Robust Computational Techniques for Boundary Layers*, Applied Mathematics and Mathematical Computation. (Eds. R. J. Knops and K. W. Morton), Chapman & Hall/CRC, Boca Raton (2000).
6. J. J. H. Miller, E. O'Riordan, G. I. Shishkin, *Fitted Numerical Methods for Singular Perturbation Problems*, World Scientific, Singapore (1996).
7. P. Maragatha Meenakshi, T. Bhuvaneswari, S. Valarmathi, J. J. H. Miller, *Parameter-uniform finite difference method for a singularly perturbed linear dynamical system* Report Series, Mathematics Department, Trinity College Dublin TCDMATH 07-11 (2007).

Boundary Shock Problems and Singularly Perturbed Riccati Equations

Relja Vulanović

Abstract A quasilinear singularly perturbed boundary-value problem is considered under conditions which guarantee that the solution has a boundary shock. The problem is initially transformed to a Riccati initial-value problem which is then solved numerically using the backward Euler scheme on a Shishkin-type mesh. For this method, a robust error estimate is proved and illustrated by numerical experiments.

1 Introduction

Consider the problem of finding a $C^2[0,1]$-function $u = u(x)$ such that

$$Tu := -\varepsilon u'' - (u^2)' + k(x,u) = 0, \quad x \in X = [0,1], \quad u(0) = 0, \quad u(1) = B, \tag{1}$$

where $' = \mathrm{d}/\mathrm{d}x$, $0 < \varepsilon << 1$, B is a positive constant, and k is a function of the form $k(x,u) = 2uc(x,u)$. It is assumed that c is a sufficiently smooth function on $X \times U$, where $U = [0, B]$, and that

$$c(x,u) \geq 0, \quad c_u(x,u) \geq 0, \quad x \in X, \quad u \in U. \tag{2}$$

Section 2 shows that this problem has a unique solution which, under additional assumptions, exhibits a boundary layer at $x = 0$.

Problems similar to (1) are considered in [Vul90] and [ZI90]. In [Vul90], a somewhat more general problem is solved numerically by applying a layer-resolving transformation which renders the derivatives of the transformed solution bounded uniformly in ε. The transformed problem is then solved using finite-difference schemes on a uniform mesh. The layer-resolving transformation corresponds to

R. Vulanović
Department of Mathematical Sciences, Kent State University Stark Campus,
6000 Frank Ave NW, North Canton, OH 44720, USA, E-mail: rvulanov@kent.edu

A.F. Hegarty et al. (eds.), *BAIL 2008 - Boundary and Interior Layers.*
Lecture Notes in Computational Science and Engineering,
DOI: 10.1007/978-3-642-00605-0,

mesh-generating functions used to create special meshes, dense in the boundary layer, for discretizing the problem (1) directly, cf. [Vul07]. Numerical results obtained by this method show pointwise ε-uniform convergence of first order. However, only L^1 first-order ε-uniform convergence is proved in the paper. The same is proved in [ZI90], but for an exponentially-fitted equidistant finite-difference scheme.

A different method is considered in the present paper and a robust error estimate in the maximum norm is derived. This is achieved by applying the approach from [LS89], in which the differential equation in (1) is integrated from x to 1 and the resulting integral is approximated using the solution of the corresponding reduced problem. The same method is used in [Vul91] for a quasilinear problem without turning points and in [Lin91] for a quasilinear turning-point problem. After the described transformation, the problem (1) becomes an initial-value problem for a Riccati equation. Parameter-uniform numerical methods for such problems are readily available, see [OR87] and [OR05]. An equidistant exponentially-fitted scheme is analyzed in [OR87], whereas the method in [OR05] uses the simple backward scheme on a Shishkin-type mesh. I present here numerical results for the latter, finding this approach more useful because it has mesh points inside the layer and because it guarantees global uniform convergence and not simply nodal uniform convergence.

The error of the approximate solution obtained this way can be estimated at each point of interval X as $M[\varepsilon + N^{-1}(\ln N)^2]$, where N is the number of mesh steps and M is used throughout the paper to denote a generic positive constant independent of both ε and N.

The problem (1) can be referred to as a *boundary shock problem* in contrast to the interior shock problems for which the boundary condition at $x = 0$ looks like $u(0) = A < 0$, see [KC80] and [Lor84] for instance. The difficulty in trying to obtain ε-uniform pointwise accuracy for interior shock problems lies in the fact that the interior shock of the numerical solution is shifted from the original location. The method of the present paper can be applied to interior shock problems only if the position of the shock is known; then the interior shock problem can be broken down to two problems of type (1).

The rest of the paper is organized as follows. The problem (1) and its reduced solution are analyzed in Sect. 2. The transformation to the Riccati equation is presented in Sect. 3. Finally, the numerical method is given in Sect. 4 together with numerical results which confirm the theoretical ones.

2 The Boundary Shock Problem

Since

$$Tu = 0 = T0 \text{ on } X \text{ and } u(t) \geq 0 \text{ for } t = 0, 1,$$

0 is a lower solution of problem (1). Also, the first inequality in (2) implies that B is an upper solution of (1):

$$TB \geq 0 = Tu \ \text{ on } X \text{ and } \ B \geq u(t) \ \text{ for } t = 0, 1.$$

Then Nagumo's result [Nag37] (see also [CH84, pp. 6–7]) guarantees that problem (1) has a solution $u \in C^2(X)$ satisfying $u(x) \in U = [0, B]$ for $x \in X$. Condition (2) implies that $k_u(x, u) \geq 0$ for $x \in X$ and $u \in U$, and therefore inverse-monotonicity arguments [Lor82] give that the solution u is unique.

Let

$$c^* \geq c(x, u) \geq c_* \geq 0, \ x \in X, \ u \in U.$$

Like in [Vul90], another assumption is

$$B > c^* + \sqrt{c^*(c^* - c_*)}. \tag{3}$$

The conditions (2) and (3) are assumed throughout the paper.

The reduced problem corresponding to (1) is the terminal-value problem

$$-u_0' + c(x, u_0) = 0, \ x \in X, \ u_0(1) = B. \tag{4}$$

The upper and lower solutions of (4) are respectively $c_*(x-1)+B$ and $c^*(x-1)+B$. Because of (3), $B > c^*$, which implies that both the upper and lower solutions have values in U when $x \in X$. Therefore, the reduced problem (4) has a solution u_0 which satisfies

$$c^*(x - 1) + B \leq u_0(x) \leq c_*(x - 1) + B, \ x \in X. \tag{5}$$

This solution is unique because of the second inequality in (2). Since $u_0'(x) \geq 0$ for $x \in X$, it follows that

$$u_0(x) \geq u_0(0) \geq \alpha := B - c^* > 0, \ x \in X, \tag{6}$$

and this is why u has a boundary layer at $x = 0$.

The condition (3) may seem technical, but in the constant-coefficient case $c = c^* = c_*$, it reduces to $B > c$, which is essential for the existence of a layer at $x = 0$. In this case, the reduced solution is $u_0(x) = B + c(x - 1)$ and $B > c$ is equivalent to $u_0(0) > 0$. If $B \leq c$, there is no layer. When $B = c$, $u \equiv u_0$ and when $B < c$, the so-called *interior crossing phenomenon* occurs, cf. [CH84, Sect. 4.3 and p. 138].

Estimates of the derivatives of u are proved in [Vul90]. In particular, it holds that

$$0 \leq u'(x) \leq M \left(1 + \varepsilon^{-1} e^{-mx/\varepsilon}\right), \ x \in X, \tag{7}$$

where m is some positive constant independent of ε. Here, like in [Vul90], it is sufficient to know that the constant m exists. However, it is also interesting to see how m relates to B, c^*, and c_*. A closer inspection of the proof of (7) in [Vul90] gives that m can be determined as

$$m = 2\beta e^{-2B} \text{ with } \beta = B^2 - 2c^* B + c^* c_*. \tag{8}$$

Note that $\beta > 0$ because of (3).

Here, the proof of the following estimate is given,

$$|u(x) - u_0(x)| \leq M \left(\varepsilon + e^{-\alpha x/\varepsilon} \right), \ x \in X, \tag{9}$$

with α defined in (6).

Theorem 1. *The solutions u and u_0 of problems (1) and (4), respectively, satisfy (9).*

Proof. For an arbitrary $C^2(X)$-function v, define the linear operator

$$Lv := -\varepsilon v'' - p(x)v' + q(x)v$$

with $p(x) = u(x) + u_0(x)$ and

$$q(x) = \int_0^1 k_u(x, u_0(x) + s[u(x) - u_0(x)]) \, \mathrm{d}s - u'(x) - u_0'(x).$$

Because of (6) and $u(x) \geq 0$, it follows that $p(x) \geq \alpha > 0$, $x \in X$. The inequality $q(x) + p'(x) \geq 0$ also holds true for $x \in X$. Then the conditions for case II of [Lor82] are fulfilled. This implies that

$$|v(x)| \leq M \left[|v(0)| e^{-\alpha x/\varepsilon} + \int_0^1 |Lv(t)| \, \mathrm{d}t \right], \ x \in X,$$

provided $v(1) = 0$. Inequality (9) now follows if v is replaced with $u - u_0$. This is because $L(u - u_0) = Tu - Tu_0 = \varepsilon u_0''$. □

3 The Riccati Equation

In this section, the problem (1) is transformed to a singularly perturbed Riccati initial-value problem. Integrate from x to 1 the differential equation in (1) to get the following problem:

$$\varepsilon u' + u^2 = f(x) := \varepsilon u'(1) + B^2 - \int_x^1 k(t, u(t)) \, \mathrm{d}t, \ x \in X, \ u(0) = 0. \tag{10}$$

This Riccati problem is then approximated by

$$\varepsilon y' + y^2 = g(x) := B^2 - \int_x^1 k(t, u_0(t)) \, \mathrm{d}t, \ x \in X, \ y(0) = 0. \tag{11}$$

The main result of this section is

$$|u(x) - y(x)| \leq M\varepsilon,\ x \in X. \tag{12}$$

The proof of (12) requires several lemmas.

Lemma 1. *The function g defined in (11) satisfies*

$$g(x) \geq \beta > 0,\ x \in X,$$

where β is defined in (8).

Proof. It holds true that

$$k(x, u_0(x)) \leq 2c^* u_0(x) \leq 2c^*[c_*(x-1) + B],\ x \in X,$$

where the second inequality in (5) is used. Therefore, when $x \in X$, it follows that

$$\begin{aligned} g(x) &\geq B^2 - 2c^* \int_x^1 [c_*(t-1) + B]\mathrm{d}t \\ &\geq B^2 - c^*[c_* t^2 + 2(B - c_*)t]_0^1 = \beta. \quad \square \end{aligned}$$

Lemma 2. *The Riccati problem (11) has a $C^1(X)$-solution y which satisfies*

$$z(x) := \gamma\left(1 - e^{-\gamma x/\varepsilon}\right) \leq y(x) \leq B,\ x \in X,$$

where $\gamma = \sqrt{\beta}$. This solution y is unique.

Proof. According to [O'Reg97, p. 19], as cited in [OR05], it should be proved that B and $z(x)$ are respectively upper and lower solutions of (11) (note that $z(x) \leq B$ for $x \in X$ because of (3)). The upper solution is easy to verify. As for z, it holds that

$$\varepsilon z' + z^2 = \gamma^2\left(1 - e^{-\gamma x/\varepsilon} + e^{-2\gamma x/\varepsilon}\right) \leq \gamma^2 \leq g(x),\ x \in X.$$

Like for problem (1), the uniqueness of the solution is a consequence of inverse monotonicity. $\square$

Lemma 3. *Functions f and g defined in (10) and (11) satisfy*

$$|f(x) - g(x)| \leq M_*\varepsilon,\ x \in X,$$

where M_* is some positive constant independent of ε.

Proof. This follows because of (7) and (9) $\square$.

Theorem 2. *The solutions u and y of the Riccati problems (10) and (11), respectively, satisfy (12).*

Proof. Define the linear operator

$$\Lambda v := \varepsilon v' + [u(x) + y(x)]v,$$

so that

$$\Lambda[u(x) - y(x)] = f(x) - g(x).$$

Let I_ℓ denote the operator $I_\ell v = v(\ell)$ for any $\ell \in X$. Since $u(x) + y(x) \geq 0$ on X, operator (Λ, I_ℓ) is inverse monotone on any interval $[\ell, r]$, $r \leq 1$.

Consider now $x \in [0, \varepsilon]$. Because of Lemma 3,

$$\Lambda M_* x = \varepsilon M_* + [u(x) + y(x)]M_* x \geq \varepsilon M_* \geq \pm[f(x) - g(x)].$$

Then inverse monotonicity of (Λ, I_0) on $[0, \varepsilon]$ implies that

$$|u(x) - y(x)| \leq M_* x \leq M_* \varepsilon, \; x \in [0, \varepsilon].$$

It remains to prove

$$|u(x) - y(x)| \leq M\varepsilon, \quad x \in [\varepsilon, 1]. \tag{13}$$

To this end, let M^* be a sufficiently large constant independent of ε and define

$$w(x) = M^* \varepsilon \left(1 - e^{-2Bx/\varepsilon}\right).$$

It follows that

$$\Lambda w(x) = M^* \varepsilon \left[[2B - u(x) - y(x)]e^{-2Bx/\varepsilon} + u(x) + y(x) \right] \geq M^* \varepsilon y(x).$$

However, because of Lemma 2, $y(x) \geq z(x)$, which on interval $[\varepsilon, 1]$ gives

$$\Lambda w(x) \geq M^* \varepsilon \gamma \left(1 - e^{-\gamma}\right).$$

Then by choosing

$$M^* = M_* \max \left\{ [\gamma (1 - e^{-\gamma})]^{-1}, \left(1 - e^{-2B}\right)^{-1} \right\},$$

we get

$$\Lambda w(x) \geq \pm[f(x) - g(x)],$$

using Lemma 3, and also

$$w(\varepsilon) \geq \pm[u(\varepsilon) - y(\varepsilon)].$$

Then (13) follows from inverse monotonicity of (Λ, I_ε) on $[\varepsilon, 1]$. □

4 The Numerical Method

Because of Lemma 2, the result from [OR05] applies immediately to the problem (11). The numerical method used in [OR05] is now described. A piecewise equidistant Shishkin-type mesh is used to discretize (11). Interval X is divided into N subintervals with J equidistant subintervals in the fine part of the mesh covering the boundary layer. It is assumed that $Q := J/N$ is a fixed constant ($Q = 1/2$, used in [OR05], is a frequent choice). The transition point between the fine and coarse parts of the mesh is

$$\tau = \min\left\{Q, \frac{\varepsilon}{\gamma}\ln N\right\},$$

where γ is the constant from Lemma 2. Therefore, the mesh points are defined by

$$x_i = \frac{\tau}{J}i, \quad i = 0, 1, \dots J,$$

and

$$x_i = \tau + \frac{1-\tau}{N-J}(i-J), \quad i = J+1, J+2, \dots, N.$$

Let $h_i = x_i - x_{i-1}, i = 1, 2, \dots, N$. The problem (11) is discretized on this mesh using the backward difference scheme,

$$\varepsilon\frac{Y_i - Y_{i-1}}{h_i} + Y_i^2 = g(x_i), \quad i = 1, 2, \dots, N, \quad Y_0 = 0.$$

This discrete problem can be solved directly,

$$Y_i = \frac{-\varepsilon + \sqrt{\varepsilon^2 + 4h_i(\varepsilon Y_{i-1} + h_i g(x_i))}}{2h_i}, \quad i = 1, 2, \dots, N, \quad Y_0 = 0. \tag{14}$$

Theorem 3. *Let N be sufficiently large but independent of ε, let u be the solution of the continuous problem (1) and let $\bar{Y}$ be the piecewise linear interpolant of the numerical solution given in (14). Then the following error estimate holds true:*

$$|u(x) - \bar{Y}(x)| \le M[\varepsilon + N^{-1}(\ln N)^2], \ x \in X.$$

Proof. Theorem 8 in [OR05] proves that

$$|y(x) - \bar{Y}(x)| \le MN^{-1}(\ln N)^2, \ x \in X, \tag{15}$$

where y is the solution of (11). The assertion then follows from Theorem 2. □

The result of Theorem 3 is now illustrated by some numerical experiments. For the test problem

$$-\varepsilon u'' - (u^2)' + u = 0 \text{ on } X, \quad u(0) = 0, \quad u(1) = 1, \tag{16}$$

an asymptotic solution can be given in the form

$$u_A(x) = u_0(x) - \frac{e^{-x/\varepsilon}}{1 + e^{-x/\varepsilon}},$$

where $u_0(x) = \frac{1}{2}(x+1)$ is the solution of the reduced problem corresponding to (16). For u_A and the solution u of problem (16), it holds true that

$$|u(x) - u_A(x)| \le M\varepsilon, \quad x \in X,$$

see [ZI90], or, more generally, [VB73] and [KC80]. Therefore, when $\varepsilon << N^{-1}$, the numerical solution (14) can be compared to u_A. The function g, given in (11) and used in (14), can be evaluated exactly (otherwise, a quadrature formula may be used): $g(x) = \frac{1}{4}(x+1)^2$ and $\gamma = \frac{1}{2}$.

Let

$$E^N = \max_{0 \le i \le N} |u_A(x_i) - Y_i^N|,$$

where the superscript indicates that N mesh steps are used. The numerical order of convergence is estimated by

$$p^N = \log_2 \frac{E^N}{E^{2N}}.$$

The results are presented in Table 1 for two different values of Q. The density of the mesh in the layer is greater if Q is greater. This is why the results for $Q = \frac{3}{4}$ are somewhat better.

The three considered values of ε are all very small and produce identical errors. It can be expected that (15) is the dominant term in the error estimate of Theorem 3. In fact, the reported numerical orders of convergence are better and correspond more closely to $MN^{-1} \ln N$. This is shown in Table 2. If it is assumed that the error is of the form

$$E^N \approx MN^{-1}(\ln N)^s,$$

for some positive constant s, then s can be found from

$$s \approx s^N := \frac{\ln(2E^{2N}) - \ln E^N}{\ln(\ln 2N) - \ln(\ln N)}.$$

Table 1 Results for $\varepsilon = 10^{-6}, 10^{-9}, 10^{-12}$

	$Q = 1/2$		$Q = 3/4$	
N	E^N	p^N	E^N	p^N
16	2.89E−2	.63	1.98E−2	.64
32	1.87E−2	.71	1.26E−2	.71
64	1.15E−2	.75	7.74E−3	.76
128	6.80E−3	.79	4.57E−3	.80
256	3.93E−3	.82	2.63E−3	.82
512	2.22E−3	—	1.49E−3	—

Table 2 Values of s^N for $\varepsilon = 10^{-6}, 10^{-9}, 10^{-12}$

N	s^N for $Q = 1/2$	s^N for $Q = 3/4$
16	1.16	1.08
32	1.14	1.13
64	1.09	1.07
128	1.08	1.05
256	1.04	1.06

As reported in Table 2, the values of s^N are well below 2 in this numerical example. They indicate that $s \approx 1$.

References

[CH84] Chang, K.W., Howes, F.A.: Nonlinear Singular Perturbation Phenomena. Springer, New York (1984)

[KC80] Kevorkian, J., Cole, J.D.: Perturbation Methods in Applied Mathematics. Springer, New York (1980)

[Lin91] Lin P.: A numerical method for quasilinear singular perturbation problems with turning points. Computing, **46**, 155–164 (1991)

[LS89] Lin P., Su Y.: Numerical solution of quasilinear singularly perturbed ordinary differential equation without turning points. Appl. Math. Mech., **10**, 1005–1010 (1989)

[Lor82] Lorenz, J.: Stability and monotonicity properties of stiff quasilinear boundary problems. Zb. Rad. Prirod. -Mat. Fak. Univ. u Novom Sadu, **12**, 151–175 (1982)

[Lor84] Lorenz, J.: Analysis of difference schemes for a stationary shock problem. SIAM J. Numer. Anal., **21**, 1038–1053 (1984)

[Nag37] Nagumo, M.: Über die Differentialgleichung $y'' = f(x, y, y')$. Proc. Phys. Math. Soc. Jpn., **19**, 861–866 (1937)

[O'Reg97] O'Regan, D.: Existence Theory for Nonlinear Ordinary Differential Equations. Kluwer Academic, Dordrecht (1997)

[OR87] O Reilly, M.J.: A uniform scheme for the singularly perturbed Riccati equation. Numer. Math., **50**, 483–501 (1987)

[OR05] O Reilly, M.J., O'Riordan, E.: A Shishkin mesh for a singularly perturbed Riccati equation. J. Comput. Appl. Math., **182**, 372–387 (2005)

[VB73] Vasil'eva, A.B., Butuzov, V.F.: Asymptoticheskie razlozheniya resheniy singulyarno vozmushchennykh uravneniy. Nauka, Moskva (1973)

[Vul90] Vulanović, R.: Continuous and numerical analysis of a boundary shock problem. Bull. Austral. Math. Soc., **41**, 75–86 (1990)

[Vul91] Vulanović R.: A second order numerical method for nonlinear singular perturbation problems without turning points. Zh. Vychisl. Mat. i Mat. Fiz., **31**, 522–532 (1991)

[Vul07] Vulanović, R.: The layer-resolving transformation and mesh generation for quasilinear singular perturbation problems. J. Comput. Appl. Math., **203**, 177–189 (2007)

[ZI90] Zadorin, A.I., Ignat'ev, V.N.: Raznostnaya skhema dlya nelineynogo singulyarno vozmushchennogo uravneniya vtorogo poryadka. Zh. Vychisl. Mat. i Mat. Fiz., **30**, 1425–1430 (1990)

Electrochemical Pickling: Asymptotics and Numerics

M. Vynnycky and N. Ipek

Abstract Electrochemical pickling in the manufacture of stainless steel strips is characterized by simultaneous multi-ionic transport, driven by diffusion, migration and convection, heterogeneous electrochemical and homogeneous chemical reactions. In this contribution, we summarize recent numerical and asymptotic results in the development of a 8-ion model for the process. In addition, a preliminary asymptotic analysis for the inclusion of homogeneous chemical reactions in the model, which had been omitted hitherto in analytical work for simplicity, is carried out and is found to agree qualitatively with earlier numerics.

1 Introduction

Electrochemical pickling is an important example of an industrial process that involves electrochemical cells in which an electrolyte and an electric current are used to drive reactions so as to yield desired products; furthermore, it exemplifies a complex system in which ionic transport, by diffusion, migration and convection, heterogeneous electrochemical reactions and homogeneous chemical reactions occur simultaneously. In the process, a steel strip having an undesired surface oxide layer is passed between pairs of anodic and cathodic electrodes in an electrochemically neutral electrolyte, usually sodium sulphate (Na_2SO_4); a schematic for the process can be found in [ICV07]. When a current is passed through the cell, the 'pickling' reaction, i.e. the removal of the oxide layer, thought to consist predominantly of chromium oxide (Cr_2O_3), occurs according to [Bra80]

$$Cr_2O_3 + 4H_2O \rightarrow Cr_2O_7^{2-} + 8H^+ + 6e^-, \tag{1}$$

M. Vynnycky (✉)
MACSI, Department of Mathematics and Statistics, University of Limerick, Limerick, Ireland,
E-mail: michael.vynnycky@ul.ie

A.F. Hegarty et al. (eds.), *BAIL 2008 - Boundary and Interior Layers.*
Lecture Notes in Computational Science and Engineering,
DOI: 10.1007/978-3-642-00605-0,

as do other electrochemical reactions that result in the evolution of oxygen and hydrogen,

$$2H_2O \rightarrow O_2 + 4H^+ + 4e^-, \quad 2H_2O + 2e^- \rightarrow H_2 + 2OH^-, \tag{2}$$

respectively. In addition to these heterogeneous reactions, the following homogeneous chemical reactions are thought to be important:

1. Water protolysis:

$$H_2O \rightleftharpoons H^+ + OH^- \tag{3}$$

2. The dissociation of sulphuric acid, according to

$$H_2SO_4 \rightleftharpoons H^+ + HSO_4^-, \quad HSO_4^- \rightleftharpoons H^+ + SO_4^{2-} \tag{4}$$

3. Chromium buffering, described by

$$CrO_4^{2-} + H^+ \rightleftharpoons HCrO_4^-, \quad 2HCrO_4^- \rightleftharpoons Cr_2O_7^{2-} + H_2O \tag{5}$$

Further information on many aspects of the pickling of austenitic stainless steels can be found in a recent survey by Li and Celis [LC03] and the thesis by Ipek [Ipe06].

In the modelling of this electrochemical system, a convenient first approximation is to assume that the electrolyte solution is dilute; in this case (see e.g. [NTA04]), the molar flux, $\mathbf{N}_i$, of the ionic species i can be expressed via the Nernst–Planck equation as

$$\mathbf{N}_i = c_i \mathbf{u} - \frac{z_i F c_i D_i}{RT} \nabla \Phi^{(e)} - D_i \nabla c_i, \tag{6}$$

where $\mathbf{u}$ is the hydrodynamic velocity of the electrolyte, c_i is the concentration of species i, $\Phi^{(e)}$ is the electric potential, D_i is the diffusion coefficient for species i in the solvent and z_i the charge number for ionic species i. The quantities $F(=96485\,\mathrm{C\,mol^{-1}})$, $R(=8.314\ \mathrm{J\ mol\ s^{-1}})$ and T are the Faraday constant, the universal gas constant and the absolute temperature, respectively. In steady state, the differential material balance for species i is given by

$$\nabla \cdot \mathbf{N}_i = R_i, \quad i = 1, .., N, \tag{7}$$

where N is the number of ionic species present and R_i describes the homogeneous chemical reactions. In addition, the solution is assumed to be electrically neutral, which is expressed by

$$\sum_{i=1}^{N} z_i c_i = 0. \tag{8}$$

If $\mathbf{u}$ is assumed to be known, equations (7) and (8) then provide a consistent description of transport processes in the dilute electrolyte, since there are $N + 1$ equations for $N + 1$ unknowns. An important quantity which can be calculated from the flux of charged species *a posteriori* is the current density, $\mathbf{i}$; this is given by Faraday's law as

$$\mathbf{i} = F\sum_{i=1}^{N} z_i \mathbf{N}_i. \tag{9}$$

Boundary conditions are then necessary to close the problem. Such systems usually consist of three types of boundary: an inlet through which an electrolyte is pumped; electrodes at which electrochemical reactions occur, and an outlet through which products and unused electrolyte can exit. At an inlet, it is reasonable to prescribe the incoming composition, so we take

$$c_i = c_i^{eq}, \quad i = 1, .., N, \tag{10}$$

where c_i^{eq} denote the concentrations' equilibrium values at the inlet. The consideration of heterogeneous electrochemical reactions at reacting surfaces will lead to relations of the form

$$\mathbf{N}_i \cdot \mathbf{n} = f_i\left(c_{1,..}, c_N, \Phi^{(e)}, \Phi_s\right), \quad i = 1, .., N, \tag{11}$$

where Φ_s is the electric potential of the reacting surface and $(f_i)_{i=1,..,N}$ are generally taken to be Butler–Volmer (or Tafel laws); an insulated surface can be thought of as a special case of this, where $f_i = 0$ for all $i = 1, .., N$. At an outlet, the molar flux is usually dominated by convection, so that

$$\left(\frac{z_i F c_i D_i}{RT}\nabla\Phi^{(e)} + D_i \nabla c_i\right) \cdot \mathbf{n} = 0, \quad i = 1, .., N. \tag{12}$$

At each boundary, the electroneutrality condition is also required, giving a total of $N + 1$ boundary conditions. In (11) and (12), $\mathbf{n}$ denotes the unit normal vector at the domain boundary.

In the rest of this contribution, we review specific numerical and asymptotic developments in our earlier modelling of electrochemical pickling, as well as including new analytical considerations for the inclusion of hitherto-omitted homogeneous chemical reactions.

2 Summary of Earlier Work

The original formulation of a model for the process [Ipe06] was for $N = 8$, although subsequent numerical solutions, implemented using the finite-element solver Comsol Multiphysics for the canonical geometry shown in Fig. 1, were for at most $N = 6$. In [ICV07], three variants were explored:

- Reduced model (1), where $N = 5$, $i = H^+, OH^-, SO_4^{2-}, Cr_2O_7^{2-}, Na^+$, with $R_i = 0$
- Reduced model (2), where $N = 5$, $i = H^+, OH^-, SO_4^{2-}, Cr_2O_7^{2-}, Na^+$, with $R_{H^+}, R_{OH^-} \neq 0$, $R_i = 0$ otherwise

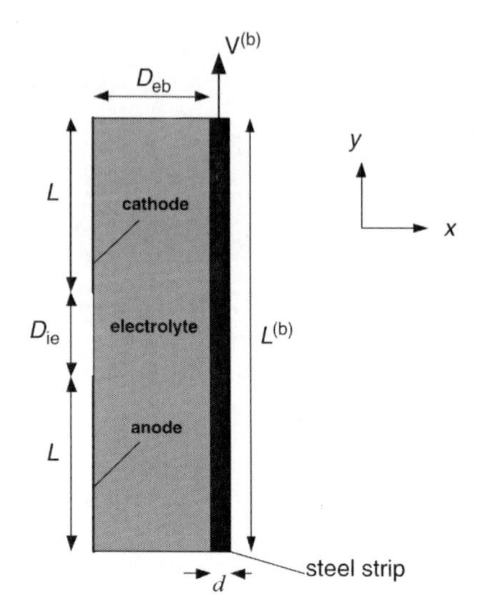

Fig. 1 A cross section of the model geometry for the vertical pickling process, showing the anode, the cathode and the steel strip in the channel. The electrodes to the left are separated by an electrically insulated, impermeable boundary of vertical extent D_{ie}

- A so-called full model, where $N = 6$, $i = H^+, OH^-, SO_4^{2-}, Cr_2O_7^{2-}, Na^+, HSO_4^-$, with $R_{H^+}, R_{OH^-}, R_{SO_4^{2-}}, R_{HSO_4^-} \neq 0$, $R_i = 0$ otherwise

in fact, the case originally formulated in [Ipe06] with

$$N = 8, \quad i = H^+, OH^-, SO_4^{2-}, Cr_2O_7^{2-}, Na^+, HSO_4^-, HCrO_4^-, CrO_4^{2-} \; R_i \neq 0,$$

was never actually solved numerically; the basic logic in this development was that it was more important to include electrochemical reactions than bulk reactions. However, there are potentially other ions present also [LC03], e.g. Fe^{3+}, Ni^{2+}, which may in future require further expansion of the model.

The solely numerical approach did, however, have drawbacks: computation times were lengthy already for $N = 5$ and it was cumbersome to attempt the parameter studies for the process that one would have desired; it was not possible to understand the physical and mathematical reasons for the simplicity of the profile obtained for $\mathbf{i}$ at the steel strip, which was essentially piecewise constant along the surface of the strip. Subsequently, an asymptotic approach was adopted for reduced model (1) in [VI08]. Nondimensionalization of the governing equations led to five dimensionless parameters: $\delta, \varepsilon, \widetilde{Pe}, \Pi, \epsilon$. The first two are geometrical: δ is the aspect width:height ratio of the electrolyte region $\left(D_{eb}/L^{(b)}\right)$, whereas ε is the width:length ratio of the section of steel strip $\left(d/L^{(b)}\right)$. $\widetilde{Pe}$ $\left(:= V^{(b)}L^{(b)}\delta^2/D_{Na^+}\right)$ is the reduced Peclet number, whereas $\Pi = FU/RT$, with U denoting the potential difference between the anode and cathode electrodes. ϵ $\left(\sim 10^{-7}\right)$ is the ratio of the bulk concentration of the ions involved in electrochemical reactions and those which are not; for reduced model (1), these were $(H^+, OH^-, Cr_2O_7^{2-})$ and $\left(Na^+, SO_4^{2-}\right)$, respectively. For pickling, $\delta, \varepsilon, \epsilon \ll 1$, whereas $\widetilde{Pe}, \Pi \gg 1$. In particular, the fact that $\epsilon \ll 1$ normally gives rise to supporting electrolyte theory: it can be shown [Lev42, NTA04] that, at leading order in ϵ, $\Phi^{(e)}$ is constant, and all dependent variables can be expressed as regular perturbation expansions in ϵ. For a recent application of this, see [BV08].

However, Vynnycky and Ipek [VI08] demonstrate that, whilst electrochemical pickling occurs in the presence of a supporting electrolyte, the classical theory cannot hold; this is already evident from [ICV07], where the electric field in the electrolyte is not constant. A re-working of the theory indicates that whilst $\left(H^+, OH^-, Cr_2O_7^{2-}\right)$ and $\left(Na^+, SO_4^{2-}\right)$ can be thought of, respectively, as minority and majority ions in the bulk, as is done in the classical theory, a minority ion can become a majority ion near an electrode at which it is produced; furthermore, as it is advected downstream with the flow, it can even be a majority ion elsewhere. In mathematical terms, whereas the asymptotic expansions for Na^+, SO_4^{2-} and $\Phi^{(e)}$ are regular, singular perturbation expansions are necessary for H^+, OH^- and $Cr_2O_7^{2-}$. An additional quirk of the pickling model is that $(f_i)_{i=1,..,N}$ in equation (11) does not depend on any of the ionic concentrations; consequently, for this particular case, the potential problem for the electric fields in the electrolyte and in the strip decouples completely from the boundary-layer equations valid adjacent to the vertical boundaries in the geometry. Work on the numerical solution of this system of equations, which is analogous to the potential flow/momentum boundary layer system in fluid mechanics, is currently ongoing. Thus, although the model has turned out to have a remarkably simple asymptotic structure due to the form of $(f_i)_{i=1,..,N}$, it is still the case that, even if $(f_i)_{i=1,..,N}$ were concentration-dependent, this approach would lead to considerably shorter computing times and smaller memory requirements than the numerical solution of the originally specified system.

The next consideration is whether the inclusion of reaction terms affects the asymptotic structure given in [VI08]. On the one hand, whilst electrochemical problems involving convection, diffusion and migration give rise to boundary layers whose thicknesses can be easily classified, i.e. $\widetilde{Pe}^{-\frac{1}{3}}$ for a stationary electrode or $\widetilde{Pe}^{-\frac{1}{2}}$ for a moving one, it is clear that the inclusion of reaction terms necessitates a case-by-case approach; it is notable that whilst authors commonly refer to a thin reaction layer adjacent to an electrode, a qualitative estimate for its thickness is never given [NBCL07, ICV07, YYW91].

3 Inclusion of Homogeneous Chemical Reactions

To guide us in how to proceed, we show in Fig. 2 the profiles for C_{H^+} $\left(:= c_{H^+}/c_{Na^+}^{eq}\right)$ and C_{OH^-} $\left(:= c_{OH^-}/c_{Na^+}^{eq}\right)$ at $Y = 0.25$, as computed for reduced model (2) in [ICV07]. First, we nondimensionalize the x- and y-coordinates shown in Fig. 1 through

$$X = x/D_{eb}, \quad Y = y/L^{(b)}.$$

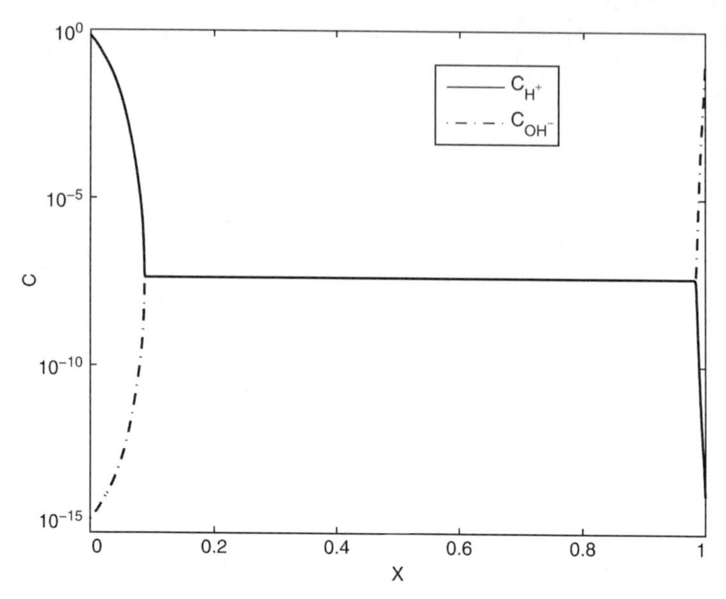

Fig. 2 $C_{\mathrm{H}+}$ and $C_{\mathrm{OH}-}$ at $Y = 0.25$ (reduced model 2)

Near $X = 0$, we see that whilst $C_{\mathrm{H}+}$ is $O(1)$, as in reduced model (1), $C_{\mathrm{OH}-}$ is plausibly $O(\epsilon^2)$; now, we try to reconcile this with the asymptotics. The material balance equations for H^+ and OH^- ions are, in dimensionless form and removing the second derivatives in Y,

$$X\frac{\partial C_i}{\partial Y} = \frac{\mathcal{D}_i}{\widetilde{Pe}}\left(z_i\frac{\partial}{\partial X}\left(C_i\frac{\partial\tilde{\phi}}{\partial X}\right) + \frac{\partial^2 C_i}{\partial X^2}\right) + \tilde{R}_i, \tag{13}$$

where $\tilde{R}_{\mathrm{H}+} = \tilde{R}_{\mathrm{OH}-} = L^{(b)}\left(k^f_{\mathrm{H_2O}}c_{\mathrm{H_2O}} - k^b_{\mathrm{H_2O}}\left(c^{eq}_{\mathrm{Na}+}\right)^2 C_{\mathrm{H}+}C_{\mathrm{OH}-}\right)/c^{eq}_{\mathrm{Na}+}V^{(b)}$, and $\tilde{\phi}$ is related to $\Phi^{(e)}$ by $\Phi^{(e)} = \Phi^{(e)}(0, Y) + \Pi^{-1}\tilde{\phi}$. If there is now to be a balance at leading order between reaction and diffusion terms, then we should have, for H^+,

$$\frac{1}{\widetilde{Pe}\,[X_{\mathrm{H}+}]^2} \sim \frac{L^{(b)}k^f_{\mathrm{H_2O}}c_{\mathrm{H_2O}}}{c^{eq}_{\mathrm{Na}+}V^{(b)}},$$

where $[X_{\mathrm{H}+}]$ denotes the thickness of the proposed reaction layer for H^+, and where we have assumed that $C_{\mathrm{H}+}\sim O(1)$. Hence,

$$[X_{\mathrm{H}+}] \sim \frac{1}{D_{eb}}\left(\frac{D_{\mathrm{Na}+}c^{eq}_{\mathrm{Na}+}}{k^f_{\mathrm{H_2O}}c_{\mathrm{H_2O}}}\right)^{\frac{1}{2}} \sim 0.04,$$

which is numerically of the same order of magnitude as the thickness of a boundary layer based on a diffusive-convective balance, i.e. $\widetilde{Pe}^{-\frac{1}{3}}$. For OH^-, on the other hand, we have

$$[X_{OH^-}] \sim \frac{\epsilon}{D_{eb}} \left(\frac{D_{Na^+} c_{Na^+}^{eq}}{k_{H_2O}^f c_{H_2O}} \right)^{\frac{1}{2}},$$

where $[X_{OH^-}]$ denotes the thickness of the proposed reaction layer for OH^-, and where we have assumed that $C_{OH^-} \sim O(\epsilon^2)$. Clearly, $[X_{OH^-}] \ll [X_{H^+}]$, which means that we would need $\tilde{R}_{OH^-} = 0$ when $X \sim [X_{H^+}]$. Hence, $\tilde{R}_{H^+} = 0$, which then gives that

$$C_{OH^-} C_{H^+} \equiv k_{H_2O}^f c_{H_2O} / k_{H_2O}^b \left(c_{Na^+}^{eq} \right)^2 \tag{14}$$

when $X \gg [X_{OH^-}]$; there will then need to be an additional boundary layer for C_{OH^-}, of thickness $[X_{OH^-}]$, in which C_{OH^-} will have to be solved for using the material balance equation, but this will not be of importance for the leading order behaviour of the cell. Nonetheless, this analysis is consistent with the numerical results.

Near $X = 1$, the roles of H^+ and OH^- are reversed, with now

$$[X_{H^+}] \sim \frac{\epsilon}{D_{eb}} \left(\frac{D_{Na^+} c_{Na^+}^{eq}}{k_{H_2O}^f c_{H_2O}} \right)^{\frac{1}{2}}, \quad [X_{OH^-}] \sim \frac{1}{D_{eb}} \left(\frac{D_{Na^+} c_{Na^+}^{eq}}{k_{H_2O}^f c_{H_2O}} \right)^{\frac{1}{2}},$$

which will once again lead to (14). A slight difference in the analysis, although ultimately of no consequence for the asymptotic structure, is that since $[X_{OH^-}] \gg \widetilde{Pe}^{-\frac{1}{2}}$, the actual leading order balance is convective-diffusive. In summary, the inclusion of a homogeneous reaction in the model has affected the results in a somewhat surprising way: in both layers considered, the reaction terms have vanished at leading order in ϵ, although their inclusion has ensured that the concentration of minority ion that is not being produced in the electrochemical reaction at the adjacent electrode has a much smaller magnitude than in the corresponding case, i.e. reduced model (1), when the reaction terms are excluded. Note also that this analysis in no way affects the conclusions in our earlier work on reduced model (1) concerning $\Phi^{(e)}$; we would therefore expect the current density obtained from reduced model (2) to be the same as that for reduced model (1), and this was indeed shown to be the case in [ICV07].

4 Conclusions

In this paper, we have summarized recent numerical and analytical developments in our modelling of electrochemical pickling. Whilst our earlier work had begun to reconcile earlier numerical and asymptotic trends for a reduced model which did not include homogeneous chemical reaction terms, here we considered preliminary steps in including them. The analysis here was for the lower part of the cell (see Fig. 1); however, it will be radically different in the upper part, since the ions produced in the lower part will be advected there. A further open issue is how the analysis changes when homogeneous chemical reactions involving ions that do not participate in the electrochemical reactions are included in the model, i.e. reduced model (3) and the full model; the numerical results from [ICV07] suggest that the leading order current density is unaffected, although it is not clear at present how this can be shown asymptotically.

More generally, this contribution has shown a practical example of how the underlying structure of an apparently complex electrochemical system with several transport and reaction mechanisms can be unravelled using a combination of asymptotics and numerics.

Acknowledgements The first author wishes to acknowledge the support of the Mathematics Applications Consortium for Science and Industry (www.macsi.ul.ie) funded by the Science Foundation Ireland Mathematics Initiative Grant 06/MI/005.

References

[Bra80] Braun, E.: How to improve pickling of stainless steel strip. Iron Steel Eng., **57**, 79–81 (1980)

[BV08] Bark, F.H., Vynnycky, M.: A note on electrolysis with forced convection at large Peclet number in a channel and an excess of supporting electrolyte. Russ. J. Electrochem., **44**, 470–478 (2008)

[ICV07] Ipek, N., Cornell, A., Vynnycky, M.: A mathematical model for the electrochemical pickling of steel. J. Electrochem. Soc., **154**, P108–P119 (2007)

[Ipe06] Ipek, N.: Mathematical modelling and experimental studies of the electrolytic pickling of stainless steel. Ph.D. thesis, Royal Institute of Technology, Stockholm (2006)

[LC03] Li, L.F., Celis, J.P.: Pickling of austenitic stainless steels (a review). Can. Metall. Q., **42**, 365–376 (2003)

[Lev42] Levich, V.: The theory of concentration overpotential. Acta Physicochimica U.R.S.S., **17**, 257–307 (1942)

[NBCL07] Nylen, L., Behm, M., Cornell, A., Lindbergh, G.: Investigation of the oxygen evolving electrode in pH-neutral electrolytes. Modelling and experiments of the RDE cell. Electrochimica Acta, **52**, 4513–4524 (2007)

[NTA04] Newman, J.S., Thomas-Alyea, K.E.: Electrochemical Systems. 3 ed., Wiley, New Jersey (2004)

[VI08] Vynnycky, M., Ipek, N.: Asymptotic analysis of a model for the electrochemical pickling of steel (2008), submitted to SIAM Journal on Applied Maths

[YYW91] Yin, K.M., Yeu, T., White, R.E.: A mathematical model of electrochemical reactions coupled with homogeneous chemical reactions. J. Electrochem. Soc., **138**, 1051–1054 (1991)

Energy Norm A-Posteriori Error Estimates for a Discontinuous Galerkin Scheme Applied to Elliptic Problems with an Interface

Paolo Zunino

Abstract It is well known that the solution of second order elliptic problems with interfaces may feature internal layers and/or singularities. We present an adaptive discontinuous Galerkin (DG) method to suitably approximate such problems. First, we introduce the weighted interior penalty method, which generalizes the classical interior penalty DG schemes by replacing the arithmetic means with suitably weighted averages where the weights depend on the coefficients of the problem. Then, we discuss the construction of a family of residual based local error indicators for the energy norm, applied to advection–diffusion-reaction equations featuring a diffusivity parameter that may be discontinuous along an interface. In particular, we demonstrate how the weights can incorporate into the scheme some a-priori knowledge of the exact solution that improves the efficacy of the estimator and of the corresponding adapted mesh. The theoretical results are confirmed by means of numerical experiments.

1 Introduction and Problem Setting

We aim to approximate u, solution of the following boundary value problem,

$$-\epsilon\Delta u + \beta \cdot \nabla u + \mu u = f \text{ in } \Omega \subset \mathbb{R}^2, \quad u = 0 \text{ on } \partial\Omega, \tag{1}$$

where Ω is a convex polygonal domain, $\mu \in L^\infty(\Omega)$ is a positive function and $\beta \in [W^{1,\infty}(\Omega)]^2$ is a vector function such that $\nabla \cdot \beta = 0$. Let Γ be a single planar interface subdividing Ω in two subregions Ω_i, $i = 1, 2$. By consequence, each subregion still is a convex polygon. For simplicity, the coefficient ϵ is defined on each subregion by a positive, possibly small, constant. Given $V := H_0^1(\Omega)$, the weak

P. Zunino
MOX - Department of Mathematics - Politecnico di Milano, Italy, E-mail: paolo.zunino@polimi.it

A.F. Hegarty et al. (eds.), *BAIL 2008 - Boundary and Interior Layers.*
Lecture Notes in Computational Science and Engineering,
DOI: 10.1007/978-3-642-00605-0,

formulation of problem (1) corresponds to find $u \in V$ such that

$$a(u,v) := \int_{\Omega} \Big(\epsilon \nabla u \cdot \nabla v - \beta u \cdot \nabla v + \mu u v\Big) = F(v) := \int_{\Omega} f v, \ \forall v \in V. \quad (2)$$

A transmission problem for Poisson equation has already been addressed by means of Nitsche type mortaring in [1] and [2] encompassing more general domains than in the present case. Here, following [3–5], we introduce a discontinuous Galerkin method that automatically accounts for the presence of an interface, provided that it is conforming with the computational mesh. A similar technique, has also been recently adopted in [6] for the discretization of symmetric Friedrichs systems. In this setting, we develop an a-posteriori local error indicator for the energy norm. In particular, we focus on the derivation of an estimator that is robust with respect to the jump of the coefficient ϵ at the interface. The seek of robust a-posteriori error estimators for singularly perturbed problems is an active field of research. In the framework of conforming finite element methods, we mention the seminal work by Verfürth [7] . For the specific case of discontinuous coefficients, we refer to [8] for Crouzeix–Raviart elements and to [9] for fully discontinuous elements. A vivid literature also concerns finite difference methods. We refer to [10] for a recent contribution.

2 Numerical Approximation

For the numerical approximation of problem (2) we consider a shape regular family of triangulations, T_h, of Ω that are conforming with the interface Γ. Let e be an edge of the element $K \in T_h$, which is a triangle in Ω. Let h_e be the size of an edge and h_K be the one of an element. We denote with F_h the collection of all edges of T_h, with F_h^i and $F_h^{\partial\Omega}$ the collections of all the internal edges and of all the boundary edges respectively. For any interior edge of the mesh we denote with n_e its unit normal vector, and with n the unit normal vector with respect to $\partial\Omega$. Then, we introduce a totally discontinuous approximation space, $V_h^p := \{v_h \in L^2(\Omega);\ v_h|_K \in \mathbb{P}^p,\ \forall K \in T_h\}$, with $p > 0$. For any function v that is discontinuous on the inter-element interface e, we define $v(x)|_e^{\pm} := \lim_{\delta \to 0^+} v(x \pm \delta n_e)$ for a.e. $x \in e$ and we will use the abridged notation $v^{\pm}$. The jump over edges is defined as $[\![v]\!]_e := v^- - v^+$, while we denote with $\{v\}$ the arithmetic mean of v^- and v^+. We also introduce the weighted averages, $\{v\}_w := w_e^- v^- + w_e^+ v^+$, $\{v\}^w := w_e^+ v^- + w_e^- v^+$, for all $e \in F_h^i$, where the weights are positive and necessarily satisfy $w_e^- + w_e^+ = 1$. Setting $v|_e^+ = 0,\ w_e^- = 1,\ w_e^+ = 0$ and $n_e = n$ for all $e \in F_h^{\partial\Omega}$, we define jumps and averages also on $F_h^{\partial\Omega}$. As a result of that $[\![v]\!]_e = v^-,\ \{v\}_w = v^-$, and $\{v\}^w = 0$ on $\partial\Omega$. The idea of exploiting a tilted average instead of the standard arithmetic mean is not completely new. Indeed, it has already been proposed by Heinrich for mortar methods, see for instance [2] and references therein. Here, we aim to apply a weighing technique to obtain a robust scheme for problems featuring a discontinuous and locally vanishing diffusivity. To this purpose, it is convenient to choose the weights depending on the coefficients of the problem, as in [3–5]. First, we introduce the *heterogeneity factor*, which quantifies the variation of ϵ on each inter-element interface, $\lambda(x)|_{\partial K} : \partial K \setminus \partial\Omega \to (-1,1)$ such that $\lambda(x)|_{\partial K} := [\![\epsilon(x)]\!]_{\partial K}/2\{\epsilon(x)\}$. Second, to construct tilted weights starting from the heterogeneity factor, we define a suitable weighing function ϕ. Observing that $\lambda \in (-1,1)$, we propose $\phi(t) := \frac{1}{2}(1 + \mathrm{sign}(t)|t|^{\alpha})$, where $\alpha \in \mathbb{R}^+$ plays the role of *tilting factor* and we define $w_e^{\pm} := \phi(\pm\lambda)$. Then, we introduce the following bilinear form,

$$
\begin{aligned}
a_h^{\alpha}(u_h, v_h) :=& \int_{T_h} \Big[(\epsilon \nabla u_h - \beta u_h) \cdot \nabla v_h + \mu u_h v_h \Big] \\
&+ \int_{F_h} \Big[\{\beta u_h\}_w \cdot n_e [\![v_h]\!] - \{\epsilon \nabla u_h\}_w \cdot n_e [\![v_h]\!] - \{\epsilon \nabla v_h\}_w \cdot n_e [\![u_h]\!] \\
&+ \big(\tfrac{1}{2}|\beta \cdot n_e| - \tfrac{1}{2}\beta \cdot n_e (w_e^- - w_e^+) + \gamma \{\epsilon\}_w h_e^{-1}\big) [\![u_h]\!][\![v_h]\!] \Big],
\end{aligned}
$$

where γ is a positive constant and where we have applied the abridged notation $\int_{T_h} := \sum_{K \in T_h} \int_K$ etc. The weighted interior penalty method reads as follows: find $u_h \in V_h^p$ such that,

$$
a_h^{\alpha}(u_h, v_h) = F(v_h), \ \forall v_h \in V_h^p. \tag{3}
$$

We observe that (3) represents a family of numerical methods, depending on the parameter α. The weighing function $\phi(t)$ has been explicitly designed to make sure that for small values of α, the tilting effect is very pronounced, while for $\alpha \to \infty$ the method coincides with the standard symmetric interior penalty method, based on arithmetic averages. Another significant value is $\alpha = 1$. In this case, $\{\epsilon\}_w$ coincides with the harmonic average and it corresponds to the stiffness of two sequential springs of modulus ϵ^- and ϵ^+. This seems to be a more natural choice than the standard average. For further details on the weighing technique, we refer to [5]. Owing to the restrictions on the shape of Ω and Γ, we assert that $u \in W := V \cap H^2(T_h)$. By consequence, the analysis of the consistency of (3) with respect to (2) is straightforward, as well as the proof of the well posedness of the scheme, which relies on the positivity of the bilinear form $a_h^{\alpha}(\cdot,\cdot)$, provided that γ is large enough, see [5]. The choice of γ is independent on the diameter of the elements $K \in T_h$, but may depend on their stretching and on the finite element polynomial order, p. For linear elements on a shape regular triangulation, an estimate of the optimal value of γ is provided in [11], and accordingly we set $\gamma = 2$ for the tests of Sect. 4.

3 Residual Based A-Posteriori Error Analysis

We develop a residual based a-posteriori error estimate for the energy norm, aiming to extend the technique proposed by Karakashian and Pascal for the Poisson problem, see [12], to advection-diffusion-reaction equations. A similar study is also pursued in [9], with a different approach. Given $\|v\|_{0,T_h}^2 := \int_{T_h} v^2$, we define the energy norm associated with (3), where the subscripts h,α remind the influence of the mesh size and the tilting factor,

$$|||v|||_{h,\alpha}^2 := \|\epsilon^{\frac{1}{2}}\nabla v\|_{0,T_h}^2 + \|\mu^{\frac{1}{2}} v\|_{0,T_h}^2 + \|(\tfrac{1}{2}|\beta\cdot n_e| + \gamma\{\epsilon\}_w h_e^{-1})^{\frac{1}{2}}[\![v]\!]\|_{0,F_h}^2.$$

Let us denote with π_h^0 the L^2 projection operator from $W(h) := W \oplus V_h^p$ into V_h^0, that is the space of piecewise constant functions on T_h. For any $v \in H^1(K)$ it satisfies the following error estimates,

$$h_K^{-1}\|v - \pi_h^0 v\|_{0,K} \le C_K\|\nabla v\|_{0,K}, \quad h_e^{-\frac{1}{2}}\|v - \pi_h^0 v\|_{0,e} \le C_e\|\nabla v\|_{0,K},$$

where C_K and C_e are positive constants only dependent on the shape regularity of the computational mesh. For simplicity, we apply the following notation, $C_{T_h} := \max_{K\in T_h} C_K$, $C_{F_h} := \max_{e\in F_h} C_e$. Furthermore, let $I_{h,0}$ be a H_0^1-conformal (Oswald type) quasi-interpolation operator from V_h^p to $H_0^1 \cap V_h^p$. From now on, let $e := u - u_h$ be the error of the method, where u is the solution of (2) and $u_h \in V_h^p$ the one of (3). A representation of the error suitable to our purpose is addressed in the following lemma.

Lemma 1. *Given $\zeta := e - \pi_h^0 e$ and $\xi := u_h - I_{h,0}u_h$ we have, $|||e|||_{h,\alpha}^2 = r_h^\alpha(u_h,\zeta) + s_h^\alpha(\xi,e)$ where,*

$$\begin{aligned} r_h^\alpha(u_h,\zeta) :=& -\int_{T_h}(-\epsilon\Delta u_h + \beta\cdot\nabla u_h + \mu u_h - f)\zeta \\ &-\int_{F_h}\Big[[\![\epsilon\nabla u_h - \beta u_h]\!]\cdot n_e\{\zeta\}^w \\ &-(\tfrac{1}{2}|\beta\cdot n_e| - \tfrac{1}{2}\beta\cdot n_e(w_e^- - w_e^+) + \gamma h_e^{-1}\{\epsilon\}_w)[\![u_h]\!][\![\zeta]\!]\Big], \\ s_h^\alpha(\xi,e) :=& -\int_{T_h}\Big[(\epsilon\nabla\xi + \beta\xi)\cdot\nabla e + \mu\xi e\Big] \\ &+\int_{F_h}\Big[\{\beta\xi\}^w\cdot n_e[\![e]\!] + \{\epsilon\nabla\xi\}_w\cdot n_e[\![e]\!] \\ &-(\tfrac{1}{2}|\beta\cdot n_e| - \tfrac{1}{2}\beta\cdot n_e(w_e^- - w_e^+) + \gamma\{\epsilon\}_w h_e^{-1})[\![\xi]\!][\![e]\!]\Big]. \end{aligned}$$

Clearly, $r_h^\alpha(u_h,\zeta)$ accounts for the residuals of the problem, while $s_h^\alpha(\xi,e)$ depends on the nonconformity of the approximation, quantified by ξ.

Proof. First, because the numerical scheme is consistent, namely $a_h^\alpha(e,v_h) = 0$ for all $v_h \in V_h^p$, and because the advective part of the bilinear form $a_h^\alpha(u_h,v_h)$ is skew-symmetric, we assert that

$$|||e|||_{h,\alpha}^2 = a_h^\alpha(e,e) + \int_{F_h} 2\{\epsilon\nabla e\}_w\cdot n_e[\![e]\!] = a_h^\alpha(e,\zeta-\xi) + \int_{F_h} 2\{\epsilon\nabla e\}_w\cdot n_e[\![e]\!].$$

Owing to the identities $\nabla\zeta = \nabla e$ on any $K \in T_h$, $[\![\xi]\!] = [\![u_h]\!] = -[\![e]\!]$ on any $e \in F_h$, the symmetry terms of $a_h^\alpha(e,\zeta)$ can be combined with the consistency terms of $a_h^\alpha(e,\xi)$ as follows,

$$-\int_{F_h} \{\epsilon\nabla\zeta\}_w \cdot n_e [\![e]\!] + \int_{F_h} \{\epsilon\nabla e\}_w \cdot n_e [\![\xi]\!] = -2\int_{F_h} \{\epsilon\nabla e\}_w \cdot n_e [\![e]\!].$$

Let $a_h^{\alpha,i}(u_h, v_h)$ be the bilinear form $a_h^\alpha(u_h, v_h)$ deprived of the term $\{\epsilon\nabla v_h\}_w \cdot n_e [\![u_h]\!]$, which corresponds to the so called *incomplete* version of the interior penalty method. Combining the previous equalities we obtain,

$$|||e|||_{h,\alpha}^2 = a_h^{\alpha,i}(e,\zeta) - \int_{T_h}\Big[\big(\epsilon\nabla e - \beta e\big)\cdot\nabla\xi + \mu e\xi\Big] - \int_{F_h}\Big[\{\beta e\}_w \cdot n_e [\![\xi]\!]$$
$$- \{\epsilon\nabla\xi\}_w \cdot n_e [\![e]\!] + \big(\tfrac{1}{2}|\beta\cdot n_e| - \tfrac{1}{2}\beta\cdot n_e(w_e^- - w_e^+) + \gamma\{\epsilon\}_w h_e^{-1}\big)[\![e]\!][\![\xi]\!]\Big].$$

Finally, the result follows from the previous identity after integration by parts of $\int_{T_h}\beta e\cdot\nabla\xi$ and $\int_{T_h}\big(\epsilon\nabla e - \beta e\big)\cdot\nabla\zeta$ over each element $K \in T_h$. □

We notice that the right hand side of the error representation formula of lemma 1 is not directly computable, because both $r_h^\alpha(u_h,\zeta)$ and $s_h^\alpha(\xi,e)$ depend on the error, through ζ and e. Conversely, the quantity ξ is computable on a post-processing phase, after the solution of the discrete problem. Then, to obtain a computable upper bound for the energy norm, we derive suitable estimates for $r_h^\alpha(u_h,\zeta)$ and $s_h^\alpha(\xi,e)$, aiming to separate the contribution of u_h and ξ from ζ and e, respectively. Since we address problems with discontinuous coefficients, we will pay attention on how to distribute the error generated at the interface on the neighboring elements. As shown in [5], suitably exploiting the weighted interior penalty technique it is possible to obtain a tilted distribution. By this way, we develop a family of estimators depending on the tilting factor, α, which feature different behavior with respect to the heterogeneity of ϵ. To set up such estimators, without lack of generality we adopt a local reference system for any element $K \in T_h$ and we assume that n_e coincides with the outer unit normal vector with respect to K. Let K^+ be the elements of T_h that share an edge with K. Accordingly, we denote with ϵ^- and ϵ^+ the inner and outer values of ϵ with respect to K, with the simplified notation $\epsilon^- = \epsilon$ when clear from the context. Furthermore, we define a local Péclét number relative to each edge of the computational mesh, $\mathrm{Pe}_e := \|\beta\cdot n_e\|_{\infty,e} h_e/(2\{\epsilon\}_w)$. Then, we introduce the following upper bounds for $r_h^\alpha(u_h,\zeta)$ and $s_h^\alpha(\xi,e)$, where we highlight the influence of the heterogeneity of ϵ^- and ϵ^+ on the multiplicative constants of the residuals.

Lemma 2. *For any $\delta > 0$ we have,*

$$r_h^\alpha(u_h,\zeta) \le \delta\big(C_{T_h}^2 + 9C_{F_h}^2\big)\sum_{K\in T_h}\|\epsilon^{\frac{1}{2}}\nabla e\|_{0,K}^2 + \frac{1}{4\delta}\sum_{K\in T_h}\big(\eta_K^{r,\alpha}(u_h)\big)^2, \tag{4}$$

$$\big(\eta_K^{r,\alpha}(u_h)\big)^2 := \frac{h_K^2}{\epsilon}\| -\epsilon\Delta u_h + \beta\cdot\nabla u_h + \mu u_h - f\|_{0,K}^2$$

$$+\sum_{e\in\partial K}\frac{\{\epsilon\}_w}{\epsilon^-}\Big[\gamma^2\|\{\epsilon\}_w^{\frac{1}{2}}h_e^{-\frac{1}{2}}[\![u_h]\!]\|_{0,e}^2$$

$$+\mathrm{Pe}_e\|(\tfrac{1}{2}|\beta\cdot n_e|-\tfrac{1}{2}\beta\cdot n_e(w_e^- - w_e^+))^{\frac{1}{2}}[\![u_h]\!]\|_{0,e}^2\Big]$$

$$+\sum_{e\in\partial K\setminus\partial\Omega}\frac{\{\epsilon\}_w}{\epsilon^-}(w_e^-)^2\Big(\frac{h_e}{\{\epsilon\}_w}\Big)\Big(\frac{\epsilon^-}{\epsilon^+}\Big)\|[\![\epsilon\nabla u_h-\beta u_h]\!]\cdot n_e\|_{0,e}^2.$$

Proof. First, we rearrange the terms of $r_h^\alpha(u_h,\zeta)$ in order to highlight the contribution on each element of the terms living on the mesh skeleton, F_h^i. The discussion presented in [5] shows that, among all the possible splitting strategies, the most effective is the following,

$$r_h^\alpha(u_h,\zeta)=-\sum_{K\in T_h}\Big[\int_K(-\epsilon\Delta u_h+\beta\cdot\nabla u_h+\mu u_h-f)\zeta$$

$$-\sum_{e\in\partial K}\int_e(\tfrac{1}{2}|\beta\cdot n_e|-\tfrac{1}{2}\beta\cdot n_e(w_e^- - w_e^+)+\gamma h_e^{-1}\{\epsilon\}_w)[\![u_h]\!]\zeta^-$$

$$+\sum_{e\in\partial K\setminus\partial\Omega}\int_e[\![\epsilon\nabla u_h-\beta u_h]\!]\cdot n_e w_e^-\zeta^+\Big].$$

Second, we provide an upper bound for each row on the right hand side,

$$\int_K(-\epsilon\Delta u_h+\beta\cdot\nabla u_h+\mu u_h-f)\zeta$$

$$\leq\delta C_K^2\|\epsilon^{\frac{1}{2}}\nabla e\|_{0,K}^2+\frac{1}{4\delta}\frac{h_K^2}{\epsilon}\|-\epsilon\Delta u_h+\beta\cdot\nabla u_h+\mu u_h-f\|_{0,K}^2,$$

$$\sum_{e\in\partial K\setminus\partial\Omega}\int_e[\![\epsilon\nabla u_h-\beta u_h]\!]\cdot n_e w_e^-\zeta^+\leq 3\delta C_e^2\|(\epsilon^+)^{\frac{1}{2}}\nabla e\|_{0,K^+}^2$$

$$+\frac{1}{4\delta}\sum_{e\in\partial K\setminus\partial\Omega}\frac{(w_e^-)^2h_e}{\epsilon^+}\|[\![\epsilon\nabla u_h-\beta u_h]\!]\cdot n_e\|_{0,e}^2,$$

$$\sum_{e\in\partial K}\int_e(\tfrac{1}{2}|\beta\cdot n_e|-\tfrac{1}{2}\beta\cdot n_e(w_e^- - w_e^+))[\![u_h]\!]\zeta^-\leq 3\delta C_e^2\|\epsilon^{\frac{1}{2}}\nabla e\|_{0,K}^2$$

$$+\frac{1}{4\delta}\sum_{e\in\partial K}\frac{\{\epsilon\}_w}{\epsilon^-}\mathrm{Pe}_e\|(\tfrac{1}{2}|\beta\cdot n_e|-\tfrac{1}{2}\beta\cdot n_e(w_e^- - w_e^+))^{\frac{1}{2}}[\![u_h]\!]\|_{0,e}^2,$$

$$\sum_{e\in\partial K}\int_e\gamma h_e^{-1}\{\epsilon\}_w[\![u_h]\!]\zeta^-\leq 3\delta C_e^2\|\epsilon^{\frac{1}{2}}\nabla e\|_{0,K}^2$$

$$+\frac{1}{4\delta}\sum_{e\in\partial K}\gamma^2\frac{\{\epsilon\}_w}{\epsilon^-}\|\{\epsilon\}_w^{\frac{1}{2}}h_e^{-\frac{1}{2}}[\![u_h]\!]\|_{0,e}^2.\qquad\square$$

Lemma 3. *For any $\delta > 0$ we have,*

$$s_h^\alpha(\xi,e) \le \delta \sum_{K\in T_h} \Big[2\|\epsilon^{\frac{1}{2}}\nabla e\|_{0,K}^2 + \|\mu^{\frac{1}{2}} e\|_{0,K}^2 + \sum_{e\in\partial K} 3\|\{\epsilon\}_w^{\frac{1}{2}} h_e^{-\frac{1}{2}} [\![e]\!]\|_{0,e}^2 \Big] + \frac{1}{4\delta} \sum_{K\in T_h} \big(\eta_K^{s,\alpha}(u_h)\big)^2, \tag{5}$$

$$\begin{aligned}\big(\eta_K^{s,\alpha}(u_h)\big)^2 := {} & \|\epsilon^{\frac{1}{2}}\nabla\xi\|_{0,K}^2 + \|\mu^{\frac{1}{2}}\xi\|_{0,K}^2 + \||\beta|\epsilon^{-\frac{1}{2}}\xi\|_{0,K}^2 \\ & + \sum_{e\in\partial K} \Big[\Big(\frac{w_e^-\epsilon^-}{\{\epsilon\}_w}\Big)^2 \|\{\epsilon\}_w^{\frac{1}{2}} h_e^{\frac{1}{2}} \nabla\xi\cdot n_e\|_{0,e}^2 + \gamma^2 \|\{\epsilon\}_w^{\frac{1}{2}} h_e^{-\frac{1}{2}}\xi\|_{0,e}^2 \\ & + \mathrm{Pe}_e \|\big(\tfrac{1}{2}|\beta\cdot n_e| - \tfrac{1}{2}\beta\cdot n_e\big)^{\frac{1}{2}}\xi\|_{0,e}^2 \Big].\end{aligned}$$

Proof. First, we split the error components on the edges over the neighboring elements,

$$\begin{aligned} s_h^\alpha(\xi,e) = {} & - \sum_{K\in T_h} \Big[\int_K \Big((\epsilon\nabla\xi + \beta\xi)\cdot\nabla e + \mu\xi e \Big) \\ & + \sum_{e\in\partial K} \int_e \Big(\big(\tfrac{1}{2}|\beta\cdot n_e| - \tfrac{1}{2}\beta\cdot n_e\big)[\![e]\!]\xi^- + \gamma\{\epsilon\}_w h_e^{-1}[\![e]\!]\xi^- - w_e^-\epsilon^-\nabla\xi^-\cdot n_e[\![e]\!] \Big) \Big], \end{aligned}$$

where we have exploited the identity,

$$\int_{F_h} \{\beta\xi\}^w \cdot n_e [\![e]\!] + \tfrac{1}{2}(w_e^- - w_e^+)[\![\xi]\!][\![e]\!] = \int_{F_h} \{\beta\xi\}\cdot n_e [\![e]\!].$$

We notice that the last two rows in the definition of $s_h^\alpha(\xi,e)$ are already fully computable, because $[\![u_h]\!] = -[\![e]\!]$ and $\xi = u_h - I_{h,0}u_h$, but we will anyway provide an upper bound for them, in order to obtain a more usual expression for the estimator. Conversely, in the first row we have to split the contribution of ξ and e by means of the following estimate,

$$\begin{aligned}\int_K (\epsilon\nabla\xi + \beta\xi)\cdot\nabla e + \mu\xi e \le {} & \delta\Big(2\|\epsilon^{\frac{1}{2}}\nabla e\|_{0,K}^2 + \|\mu^{\frac{1}{2}}e\|_{0,K}^2\Big) \\ & + \frac{1}{4\delta}\Big(\|\epsilon^{\frac{1}{2}}\nabla\xi\|_{0,K}^2 + \|\mu^{\frac{1}{2}}\xi\|_{0,K}^2 + \||\beta|\epsilon^{-\frac{1}{2}}\xi\|_{0,K}^2\Big).\end{aligned}$$

For the remaining terms, we propose the following upper bound, that together with the previous estimate gives (5),

$$\begin{aligned} & \sum_{e\in\partial K} \int_e \Big(\big(\tfrac{1}{2}|\beta\cdot n_e| - \tfrac{1}{2}\beta\cdot n_e\big)[\![e]\!]\xi^- + \gamma\{\epsilon\}_w h_e^{-1}[\![e]\!]\xi^- - w_e^-\epsilon^-\nabla\xi^-\cdot n_e[\![e]\!] \Big) \\ & \le \frac{1}{4\delta} \sum_{e\in\partial K} \Big[\Big(\frac{w_e^-\epsilon^-}{\{\epsilon\}_w}\Big)^2 \|\{\epsilon\}_w^{\frac{1}{2}} h_e^{\frac{1}{2}}\nabla\xi\cdot n_e\|_{0,e}^2 + \gamma^2\|\{\epsilon\}_w^{\frac{1}{2}} h_e^{-\frac{1}{2}}\xi\|_{0,e}^2 \\ & + \mathrm{Pe}_e\|\big(\tfrac{1}{2}|\beta\cdot n_e| - \tfrac{1}{2}\beta\cdot n_e\big)^{\frac{1}{2}}\xi\|_{0,e}^2 \Big] + 3\delta \sum_{e\in\partial K} \|\{\epsilon\}_w^{\frac{1}{2}} h_e^{-\frac{1}{2}}[\![e]\!]\|_{0,e}^2. \end{aligned}$$

□

Combining the error representation formula of lemma 1, with (4) and (5) of lemmas 2 and 3 respectively, we immediately obtain the following result.

Theorem 1. *Provided that* $\delta(C^2_{T_h} + 9C^2_{F_h} + 3) < 1$ *there exists a positive constant* C *independent of* h *and of the coefficients* ϵ, β, μ, *such that* $|||e|||_{h,\alpha} \leq C\sqrt{\sum_{T_h} \eta^\alpha_K(u_h)^2}$, *where the local error estimators* $\eta^\alpha_K(u_h)$ *are defined as* $\eta^\alpha_K(u_h) := \sqrt{\eta^{r,\alpha}_K(u_h)^2 + \eta^{s,\alpha}_K(u_h)^2}$.

We finally define $I_{h,0}$, a H^1_0-conformal quasi-interpolation operator from V^p_h to $H^1_0 \cap V^p_h$. Several options are discussed in [8] and references therein. The simplest one is the so called Oswald operator, which involves the arithmetic average of the multiple values of $u_h \in V^p_h$ at each node. However, this definition is not effective when the considered node belongs to the interface of discontinuity of ϵ. In this case, a weighted average seems to be more suited. First, for any node x_i belonging to $F^{\partial\Omega}_h$, we immediately set $I_{h,0}u_h(x_i) := 0$. Let now x_i be any node on the interior of the skeleton, namely $x_i \in F^i_h$. We denote with F_i the collection of faces that share x_i, more precisely $F_i := \{e_j \in F^i_h : x_i \in \bar{e}_j\}$ being $|F_i|$ its cardinal. Then, we average the multiple values of u_h on each interface, namely we consider $u_j := \{u_h(x_i)\}^w|_{e_j}$. Let λ_j be the heterogeneity factor associated to $e_j \in F_i$. The nodal values of the H^1_0-conformal quasi-interpolator $I_{h,0}u_h$ are then defined by means of the following multiple weighted average, $I_{h,0}u_h(x_i) := \lim_{\delta_j \to |\lambda_j|^+} \sum_{j=1}^{|F_i|} \delta_j u_j / \sum_{j=1}^{|F_i|} \delta_j$.

4 Numerical Results and Conclusions

We consider a one dimensional problem where we split the domain $\Omega = (0,1)$ into two subregions, $\Omega_1 = (0, \frac{1}{2})$, $\Omega_2 = (\frac{1}{2}, 1)$. The diffusivity $\epsilon(x)$ is a discontinuous function across the interface $x = \frac{1}{2}$, precisely $\epsilon_1 = 2^{-i}$ with $i = 1, 5, 10$ in Ω_1 and $\epsilon_2 = 1$ in Ω_2. In the case $\beta = 1$, $\mu = 0$, $f = 0$ with the boundary conditions $u_1(x=0) = 1$, $u_2(x=1) = 0$, the exact solution of the problem can be easily computed. We refer to [3] for an explicit formula of $u(x)$.

We preform numerical experiments exploiting linear finite elements, $p = 1$, and we compare in table 1 the true error and its estimator for different values of the tilting factor. We notice that the tilted weights strongly influence the estimator and allow us to tune its sensitivity with respect to the jump of ϵ across the interface. Let $\sqrt{\sum_K (\eta_K)^2} / |||e|||_{h,\alpha}$ be the *effectivity* of the estimator, denoted with *eff.* in table 1. In the case of standard interior penalties, we notice that the indicator *eff.* considerably increases when ϵ becomes heterogeneous, namely when ϵ_1 decreases. Both terms $\eta^{r,\alpha}_K$, $\eta^{s,\alpha}_K$ behave similarly, but the former dominates because it is not robust with respect to the ratio ϵ^+/ϵ^-. As shown in table 1, the magnitude of $\eta^{r,\alpha}_K$ is sensibly reduced with the introduction of the tilted weights, $\alpha < \infty$, improving the effectivity of the estimator. This behavior can be related to the expression of

Table 1 Comparison of the error and its estimator for different values of the tilting factor α. In particular, $\alpha \to \infty$ corresponds to standard interior penalties and the tilting effect increases for smaller values of α

α	i	$\epsilon_1 = 2^{-i}$	$\sqrt{\sum_K (\eta_K^{r,\alpha})^2}$	$\sqrt{\sum_K (\eta_K^{s,\alpha})^2}$	$\sqrt{\sum_K (\eta_K^{\alpha})^2}$	$\|\|\|e\|\|\|_{h,\alpha}$	eff.
	1	5.0e−01	3.886e−02	9.697e−03	4.005e−02	3.223e−03	12.4
∞	5	3.1e−02	4.252e−01	5.689e−02	4.290e−01	2.583e−02	16.6
	10	9.8e−04	6.703e+00	1.421e+00	6.852e+00	2.689e−01	25.5
	1	5.0e−01	3.892e−02	9.907e−03	4.016e−02	3.322e−03	12.1
1	5	3.1e−02	3.892e−01	1.331e−01	4.113e−01	4.879e−02	8.4
	10	9.8e−04	7.834e−01	1.582e+00	1.765e+00	2.804e−01	6.3
	1	5.0e−01	4.097e−02	1.601e−02	4.399e−02	4.805e−03	9.2
10^{-2}	5	3.1e−02	3.517e−01	2.792e−01	4.490e−01	9.040e−02	5.0
	10	9.8e−04	3.251e−01	1.661e+00	1.693e+00	2.911e−01	5.8

$\eta_K^{r,\alpha}$, where the multiplicative coefficients of the residuals on F_h^i are robust with respect to ϵ^+/ϵ^- for any value of α when $\epsilon^- \to 0$, except for the limit case $\alpha \to \infty$. Conversely, because of the term $\|\||\beta|\epsilon^{-\frac{1}{2}}\xi\|_{0,K}$, the contribution of $\eta_K^{s,\alpha}$ is not sensitive with respect to the tilted weights. This suggests that the advantage of the tilted weights is maximum for a given range of ϵ_1, as confirmed by the last three rows of table 1. Although this brief discussion is not exhaustive, table 1 confirms that the weighted interior penalties turn out to be more effective than the standard scheme in all cases. For additional experiments on the influence of the tilting factor for a model problem in two space dimensions, we refer to [5].

References

1. R. Becker, P. Hansbo, and R. Stenberg. A finite element method for domain decomposition with non-matching grids. *M2AN Math. Model. Numer. Anal.*, 37(2):209–225, 2003.
2. B. Heinrich and S. Nicaise. The Nitsche mortar finite-element method for transmission problems with singularities. *IMA J. Numer. Anal.*, 23(2):331–358, 2003.
3. E. Burman and P. Zunino. A domain decomposition method based on weighted interior penalties for advection-diffusion-reaction problems. *SIAM J. Numer. Anal.*, 44(4):1612–1638, 2006.
4. A. Ern, A. Stephansen, and P. Zunino. A discontinuous Galerkin method with weighted averages for advection-diffusion equations with locally small and anisotropic diffusivity. *IMA J. Numer. Anal.*, 29(2):235–256, 2009.
5. P. Zunino. Discontinuous Galerkin methods based on weighted interior penalties for second order PDEs with non-smooth coefficients. *J. Sci. Comput.*, 38(1):99–126, 2009.
6. D.A. Di Pietro, A. Ern, and J.-L. Guermond. Discontinuous Galerkin methods for anisotropic semidefinite diffusion with advection. *SIAM J. Numer. Anal.*, 46(2):805–831, 2008.
7. R. Verfürth. Robust a posteriori error estimates for stationary convection-diffusion equations. *SIAM J. Numer. Anal.*, 43(4):1766–1782, 2005.
8. M. Ainsworth. Robust a posteriori error estimation for nonconforming finite element approximation. *SIAM J. Numer. Anal.*, 42(6):2320–2341, 2005.

9. A. Ern and A.F. Stephansen. A posteriori energy-norm error estimates for advection-diffusion equations approximated by weighted interior penalty methods. *J. Comput. Math.*, 26(4):488–510, 2008.
10. N. Kopteva. Maximum norm a posteriori error estimate for a 2D singularly perturbed semilinear reaction-diffusion problem. *SIAM J. Numer. Anal.*, 46(3):1602–1618, 2008.
11. P. Hansbo and M.G. Larson. Discontinuous Galerkin methods for incompressible and nearly incompressible elasticity by Nitsche's method. *Comput. Methods Appl. Mech. Engrg.*, 191 (17–18):1895–1908, 2002.
12. O.A. Karakashian and F. Pascal. A posteriori error estimates for a discontinuous Galerkin approximation of second-order elliptic problems. *SIAM J. Numer. Anal.*, 41(6):2374–2399, 2003.

Editorial Policy

1. Volumes in the following three categories will be published in LNCSE:

i) Research monographs
ii) Lecture and seminar notes
iii) Conference proceedings

Those considering a book which might be suitable for the series are strongly advised to contact the publisher or the series editors at an early stage.

2. Categories i) and ii). These categories will be emphasized by Lecture Notes in Computational Science and Engineering. **Submissions by interdisciplinary teams of authors are encouraged.** The goal is to report new developments – quickly, informally, and in a way that will make them accessible to non-specialists. In the evaluation of submissions timeliness of the work is an important criterion. Texts should be well-rounded, well-written and reasonably self-contained. In most cases the work will contain results of others as well as those of the author(s). In each case the author(s) should provide sufficient motivation, examples, and applications. In this respect, Ph.D. theses will usually be deemed unsuitable for the Lecture Notes series. Proposals for volumes in these categories should be submitted either to one of the series editors or to Springer-Verlag, Heidelberg, and will be refereed. A provisional judgment on the acceptability of a project can be based on partial information about the work: a detailed outline describing the contents of each chapter, the estimated length, a bibliography, and one or two sample chapters – or a first draft. A final decision whether to accept will rest on an evaluation of the completed work which should include

- at least 100 pages of text;
- a table of contents;
- an informative introduction perhaps with some historical remarks which should be accessible to readers unfamiliar with the topic treated;
- a subject index.

3. Category iii). Conference proceedings will be considered for publication provided that they are both of exceptional interest and devoted to a single topic. One (or more) expert participants will act as the scientific editor(s) of the volume. They select the papers which are suitable for inclusion and have them individually refereed as for a journal. Papers not closely related to the central topic are to be excluded. Organizers should contact Lecture Notes in Computational Science and Engineering at the planning stage.

In exceptional cases some other multi-author-volumes may be considered in this category.

4. Format. Only works in English are considered. They should be submitted in camera-ready form according to Springer-Verlag's specifications.
Electronic material can be included if appropriate. Please contact the publisher.
Technical instructions and/or LaTeX macros are available via http://www.springer.com/authors/book+authors?SGWID=0-154102-12-417900-0. The macros can also be sent on request.

General Remarks

Lecture Notes are printed by photo-offset from the master-copy delivered in camera-ready form by the authors. For this purpose Springer-Verlag provides technical instructions for the preparation of manuscripts. See also *Editorial Policy.*

Careful preparation of manuscripts will help keep production time short and ensure a satisfactory appearance of the finished book.

The following terms and conditions hold:

Categories i), ii), and iii):
Authors receive 50 free copies of their book. No royalty is paid. Commitment to publish is made by letter of intent rather than by signing a formal contract. Springer- Verlag secures the copyright for each volume.

For conference proceedings, editors receive a total of 50 free copies of their volume for distribution to the contributing authors.

All categories:
Authors are entitled to purchase further copies of their book and other Springer mathematics books for their personal use, at a discount of 33.3% directly from Springer-Verlag.

Addresses:

Timothy J. Barth
NASA Ames Research Center
NAS Division
Moffett Field, CA 94035, USA
e-mail: barth@nas.nasa.gov

Michael Griebel
Institut für Numerische Simulation
der Universität Bonn
Wegelerstr. 6
53115 Bonn, Germany
e-mail: griebel@ins.uni-bonn.de

David E. Keyes
Department of Applied Physics
and Applied Mathematics
Columbia University
200 S. W. Mudd Building
500 W. 120th Street
New York, NY 10027, USA
e-mail: david.keyes@columbia.edu

Risto M. Nieminen
Laboratory of Physics
Helsinki University of Technology
02150 Espoo, Finland
e-mail: rni@fyslab.hut.fi

Dirk Roose
Department of Computer Science
Katholieke Universiteit Leuven
Celestijnenlaan 200A
3001 Leuven-Heverlee, Belgium
e-mail: dirk.roose@cs.kuleuven.ac.be

Tamar Schlick
Department of Chemistry
Courant Institute of Mathematical
Sciences
New York University
and Howard Hughes Medical Institute
251 Mercer Street
New York, NY 10012, USA
e-mail: schlick@nyu.edu

Mathematics Editor at Springer:
Martin Peters
Springer-Verlag
Mathematics Editorial IV
Tiergartenstrasse 17
D-69121 Heidelberg, Germany
Tel.: *49 (6221) 487-8185
Fax: *49 (6221) 487-8355
e-mail: martin.peters@springer.com

Lecture Notes in Computational Science and Engineering

1. D. Funaro, *Spectral Elements for Transport-Dominated Equations.*
2. H. P. Langtangen, *Computational Partial Differential Equations.* Numerical Methods and Diffpack Programming.
3. W. Hackbusch, G. Wittum (eds.), *Multigrid Methods V.*
4. P. Deuflhard, J. Hermans, B. Leimkuhler, A. E. Mark, S. Reich, R. D. Skeel (eds.), *Computational Molecular Dynamics: Challenges, Methods, Ideas.*
5. D. Kröner, M. Ohlberger, C. Rohde (eds.), *An Introduction to Recent Developments in Theory and Numerics for Conservation Laws.*
6. S. Turek, *Efficient Solvers for Incompressible Flow Problems.* An Algorithmic and Computational Approach.
7. R. von Schwerin, ***M**ulti **B**ody **S**ystem **SIM**ulation.* Numerical Methods, Algorithms, and Software.
8. H.-J. Bungartz, F. Durst, C. Zenger (eds.), *High Performance Scientific and Engineering Computing.*
9. T. J. Barth, H. Deconinck (eds.), *High-Order Methods for Computational Physics.*
10. H. P. Langtangen, A. M. Bruaset, E. Quak (eds.), *Advances in Software Tools for Scientific Computing.*
11. B. Cockburn, G. E. Karniadakis, C.-W. Shu (eds.), *Discontinuous Galerkin Methods.* Theory, Computation and Applications.
12. U. van Rienen, *Numerical Methods in Computational Electrodynamics.* Linear Systems in Practical Applications.
13. B. Engquist, L. Johnsson, M. Hammill, F. Short (eds.), *Simulation and Visualization on the Grid.*
14. E. Dick, K. Riemslagh, J. Vierendeels (eds.), *Multigrid Methods VI.*
15. A. Frommer, T. Lippert, B. Medeke, K. Schilling (eds.), *Numerical Challenges in Lattice Quantum Chromodynamics.*
16. J. Lang, *Adaptive Multilevel Solution of Nonlinear Parabolic PDE Systems.* Theory, Algorithm, and Applications.
17. B. I. Wohlmuth, *Discretization Methods and Iterative Solvers Based on Domain Decomposition.*
18. U. van Rienen, M. Günther, D. Hecht (eds.), *Scientific Computing in Electrical Engineering.*
19. I. Babuška, P. G. Ciarlet, T. Miyoshi (eds.), *Mathematical Modeling and Numerical Simulation in Continuum Mechanics.*
20. T. J. Barth, T. Chan, R. Haimes (eds.), *Multiscale and Multiresolution Methods.* Theory and Applications.
21. M. Breuer, F. Durst, C. Zenger (eds.), *High Performance Scientific and Engineering Computing.*
22. K. Urban, *Wavelets in Numerical Simulation.* Problem Adapted Construction and Applications.
23. L. F. Pavarino, A. Toselli (eds.), *Recent Developments in Domain Decomposition Methods.*

24. T. Schlick, H. H. Gan (eds.), *Computational Methods for Macromolecules: Challenges and Applications.*

25. T. J. Barth, H. Deconinck (eds.), *Error Estimation and Adaptive Discretization Methods in Computational Fluid Dynamics.*

26. M. Griebel, M. A. Schweitzer (eds.), *Meshfree Methods for Partial Differential Equations.*

27. S. Müller, *Adaptive Multiscale Schemes for Conservation Laws.*

28. C. Carstensen, S. Funken, W. Hackbusch, R. H. W. Hoppe, P. Monk (eds.), *Computational Electromagnetics.*

29. M. A. Schweitzer, *A Parallel Multilevel Partition of Unity Method for Elliptic Partial Differential Equations.*

30. T. Biegler, O. Ghattas, M. Heinkenschloss, B. van Bloemen Waanders (eds.), *Large-Scale PDE-Constrained Optimization.*

31. M. Ainsworth, P. Davies, D. Duncan, P. Martin, B. Rynne (eds.), *Topics in Computational Wave Propagation.* Direct and Inverse Problems.

32. H. Emmerich, B. Nestler, M. Schreckenberg (eds.), *Interface and Transport Dynamics.* Computational Modelling.

33. H. P. Langtangen, A. Tveito (eds.), *Advanced Topics in Computational Partial Differential Equations.* Numerical Methods and Diffpack Programming.

34. V. John, *Large Eddy Simulation of Turbulent Incompressible Flows.* Analytical and Numerical Results for a Class of LES Models.

35. E. Bänsch (ed.), *Challenges in Scientific Computing - CISC 2002.*

36. B. N. Khoromskij, G. Wittum, *Numerical Solution of Elliptic Differential Equations by Reduction to the Interface.*

37. A. Iske, *Multiresolution Methods in Scattered Data Modelling.*

38. S.-I. Niculescu, K. Gu (eds.), *Advances in Time-Delay Systems.*

39. S. Attinger, P. Koumoutsakos (eds.), *Multiscale Modelling and Simulation.*

40. R. Kornhuber, R. Hoppe, J. Périaux, O. Pironneau, O. Wildlund, J. Xu (eds.), *Domain Decomposition Methods in Science and Engineering.*

41. T. Plewa, T. Linde, V.G. Weirs (eds.), *Adaptive Mesh Refinement – Theory and Applications.*

42. A. Schmidt, K.G. Siebert, *Design of Adaptive Finite Element Software.* The Finite Element Toolbox ALBERTA.

43. M. Griebel, M.A. Schweitzer (eds.), *Meshfree Methods for Partial Differential Equations II.*

44. B. Engquist, P. Lötstedt, O. Runborg (eds.), *Multiscale Methods in Science and Engineering.*

45. P. Benner, V. Mehrmann, D.C. Sorensen (eds.), *Dimension Reduction of Large-Scale Systems.*

46. D. Kressner, *Numerical Methods for General and Structured Eigenvalue Problems.*

47. A. Boriçi, A. Frommer, B. Joó, A. Kennedy, B. Pendleton (eds.), *QCD and Numerical Analysis III.*

48. F. Graziani (ed.), *Computational Methods in Transport.*

49. B. Leimkuhler, C. Chipot, R. Elber, A. Laaksonen, A. Mark, T. Schlick, C. Schütte, R. Skeel (eds.), *New Algorithms for Macromolecular Simulation.*

50. M. Bücker, G. Corliss, P. Hovland, U. Naumann, B. Norris (eds.), *Automatic Differentiation: Applications, Theory, and Implementations.*

51. A.M. Bruaset, A. Tveito (eds.), *Numerical Solution of Partial Differential Equations on Parallel Computers.*

52. K.H. Hoffmann, A. Meyer (eds.), *Parallel Algorithms and Cluster Computing.*

53. H.-J. Bungartz, M. Schäfer (eds.), *Fluid-Structure Interaction.*

54. J. Behrens, *Adaptive Atmospheric Modeling.*

55. O. Widlund, D. Keyes (eds.), *Domain Decomposition Methods in Science and Engineering XVI.*

56. S. Kassinos, C. Langer, G. Iaccarino, P. Moin (eds.), *Complex Effects in Large Eddy Simulations.*

57. M. Griebel, M.A Schweitzer (eds.), *Meshfree Methods for Partial Differential Equations III.*

58. A.N. Gorban, B. Kégl, D.C. Wunsch, A. Zinovyev (eds.), *Principal Manifolds for Data Visualization and Dimension Reduction.*

59. H. Ammari (ed.), *Modeling and Computations in Electromagnetics: A Volume Dedicated to Jean-Claude Nédélec.*

60. U. Langer, M. Discacciati, D. Keyes, O. Widlund, W. Zulehner (eds.), *Domain Decomposition Methods in Science and Engineering XVII.*

61. T. Mathew, *Domain Decomposition Methods for the Numerical Solution of Partial Differential Equations.*

62. F. Graziani (ed.), *Computational Methods in Transport: Verification and Validation.*

63. M. Bebendorf, *Hierarchical Matrices.* A Means to Efficiently Solve Elliptic Boundary Value Problems.

64. C.H. Bischof, H.M. Bücker, P. Hovland, U. Naumann, J. Utke (eds.), *Advances in Automatic Differentiation.*

65. M. Griebel, M.A. Schweitzer (eds.), *Meshfree Methods for Partial Differential Equations IV.*

66. B. Engquist, P. Lötstedt, O. Runborg (eds.), *Multiscale Modeling and Simulation in Science.*

67. I.H. Tuncer, Ü. Gülcat, D.R. Emerson, K. Matsuno (eds.), *Parallel Computational Fluid Dynamics.*

68. S. Yip, T. Diaz de la Rubia (eds.), *Scientific Modeling and Simulations.*

69. A. Hegarty, N. Kopteva, E. O'Riordan, M. Stynes (eds.), *BAIL* 2008 – *Boundary and Interior Layers.*

For further information on these books please have a look at our mathematics catalogue at the following URL: `www.springer.com/series/3527`

Monographs in Computational Science and Engineering

1. J. Sundnes, G.T. Lines, X. Cai, B.F. Nielsen, K.-A. Mardal, A. Tveito, *Computing the Electrical Activity in the Heart.*

For further information on this book, please have a look at our mathematics catalogue at the following URL: `www.springer.com/series/7417`

Texts in Computational Science and Engineering

1. H. P. Langtangen, *Computational Partial Differential Equations.* Numerical Methods and Diffpack Programming. 2nd Edition
2. A. Quarteroni, F. Saleri, *Scientific Computing with MATLAB and Octave.* 2nd Edition
3. H. P. Langtangen, *Python Scripting for Computational Science.* 3rd Edition
4. H. Gardner, G. Manduchi, *Design Patterns for e-Science.*
5. M. Griebel, S. Knapek, G. Zumbusch, *Numerical Simulation in Molecular Dynamics.*

For further information on these books please have a look at our mathematics catalogue at the following URL: `www.springer.com/series/5151`